# Lecture Notes in Bioinformatics　3745

Edited by S. Istrail, P. Pevzner, and M. Waterman

Subseries of Lecture Notes in Computer Science

T0216442

José Luis Oliveira
Víctor Maojo
Fernando Martin-Sanchez
António Sousa Pereira (Eds.)

# Biological and Medical Data Analysis

6th International Symposium, ISBMDA 2005
Aveiro, Portugal, November 10-11, 2005
Proceedings

 Springer

Series Editors

Sorin Istrail, Brown University, Providence, RI, USA
Pavel Pevzner, University of California, San Diego, CA, USA
Michael Waterman, University of Southern California, Los Angeles, CA, USA

Volume Editors

José Luis Oliveira
António Sousa Pereira
University of Aveiro
Department of Electronics and Telecommunications (DET/IEETA)
Campus Santiago, 3810 193 Aveiro, Portugal
E-mail: {jlo,asp}@det.ua.pt

Víctor Maojo
Polytechnical University of Madrid
School of Computer Science, Artificial Intelligence Lab
Boadilla del Monte, 28660 Madrid, Spain
E-mail: vmaojo@infomed.dia.fi.upm.es

Fernando Martin-Sanchez
Institute of Health Carlos III
Department of Medical Bioinformatics
Ctra. Majadahonda a Pozuelo, km. 2, 28220 Majadahonda, Madrid, Spain
E-mail: fmartin@isciii.es

Library of Congress Control Number: 2005934196

CR Subject Classification (1998): H.2.8, I.2, H.3, G.3, I.5.1, I.4, J.3, F.1

ISSN       0302-9743
ISBN-10    3-540-29674-3 Springer Berlin Heidelberg New York
ISBN-13    978-3-540-29674-4 Springer Berlin Heidelberg New York

Springer is a part of Springer Science+Business Media

springeronline.com

© Springer-Verlag Berlin Heidelberg 2005
Printed in Germany

Typesetting: Camera-ready by author, data conversion by Olgun Computergrafik
Printed on acid-free paper      SPIN: 11573067       06/3142      5 4 3 2 1 0

# Preface

The sequencing of the genomes of humans and other organisms is inspiring the development of new statistical and bioinformatics tools that we hope can modify the current understanding of human diseases and therapies. As our knowledge about the human genome increases so does our belief that to fully grasp the mechanisms of diseases we need to understand their genetic basis and the proteomics behind them and to integrate the knowledge generated in the laboratory in clinical settings. The new genetic and proteomic data has brought forth the possibility of developing new targets and therapies based on these findings, of implementing newly developed preventive measures, and also of discovering new research approaches to old problems.

To fully enhance our understanding of disease processes, to develop more and better therapies to combat and cure diseases, and to develop strategies to prevent them, there is a need for synergy of the disciplines involved, medicine, molecular biology, biochemistry and computer science, leading to more recent fields such as bioinformatics and biomedical informatics.

The 6th International Symposium on Biological and Medical Data Analysis aimed to become a place where researchers involved in these diverse but increasingly complementary areas could meet to present and discuss their scientific results.

The papers in this volume discuss issues from statistical models to architectures and applications to bioinformatics and biomedicine. They cover both practical experience and novel research ideas and concepts.

We would like to express our gratitude to all the authors for their contributions to preparing and revising the papers as well as the Technical Program Committee who helped put together an excellent program for the conference.

Víctor Maojo

November 2005

José Luís Oliveira

Fernando Martín-Sánchez
António Sousa Pereira

## General Chair

José Luís Oliveira, Univ. Aveiro, Portugal

## Scientific Committee Coordinators

V. Maojo, Univ. Politecnica de Madrid, Spain
F. Martín-Sánchez, Institute of Health Carlos III, Spain
A. Sousa Pereira, Univ. Aveiro, Portugal

## Steering Committee

R. Brause, J.W. Goethe Univ., Germany
D. Polónia, Univ. Aveiro, Portugal
F. Vicente, Institute of Health Carlos III, Spain

## Scientific Committee

A. Babic, Univ. Linkoping, Sweden
R. Baud, Univ. Hospital of Geneva, Switzerland
V. Breton, Univ. Clermont-Ferrand, France
J. Carazo, Univ. Autonoma of Madrid, Spain
A. Carvalho, Univ. São Paulo, Brazil
P. Cinquin, Univ. Grenoble, France
W. Dubitzky, Univ. Ulster, UK
M. Dugas, Univ. Munich, Germany
P. Ghazal, Univ. Edinburgh, UK
R. Guthke, Hans-Knoell Institut, Germany
O. Kohlbacher, Univ. Tübingen, Germany
C. Kulikowski, Rutgers Univ., USA
P. Larranaga, Univ. Basque Country, Spain
N. Maglaveras, Univ. Thessaloniki, Greece
L. Ohno-Machado, Harvard Univ., USA
F. Pinciroli, Politecnico di Milano, Italy
D. Pisanelli, ISTC-CNR, Italy
G. Potamias, ICS-FORTH, Greece
M. Santos, Univ. Aveiro, Portugal
F. Sanz, Univ. Pompeu Fabra, Spain
W. Sauerbrei, Univ. Freiburg, Germany
S. Schulz, Univ. Freiburg, Germany

A. Silva, Univ. Aveiro, Portugal
T. Solomonides, Univ. West of England, UK
B. Zupan, Univ. Ljubljana, Slovenia
J. Zvárová, Univ. Charles, Czech Republic

## Special Reviewers

G. Moura, Univ. Aveiro, Portugal
A. Tomé, Univ. Aveiro, Portugal

# Table of Contents

## Medical Databases and Information Systems

Application of Three-Level Handprinted Documents Recognition
in Medical Information Systems.................................... 1
  *Jerzy Sas and Marek Kurzynski*

Data Management and Visualization Issues
in a Fully Digital Echocardiography Laboratory....................... 13
  *Carlos Costa, José Luís Oliveira, Augusto Silva, Vasco Gama Ribeiro,
  and José Ribeiro*

A Framework Based on Web Services and Grid Technologies
for Medical Image Registration .................................... 22
  *Ignacio Blanquer, Vicente Hernández, Ferran Mas,
  and Damià Segrelles*

Biomedical Image Processing Integration Through INBIOMED:
A Web Services-Based Platform..................................... 34
  *David Pérez del Rey, José Crespo, Alberto Anguita,
  Juan Luis Pérez Ordóñez, Julián Dorado, Gloria Bueno, Vicente Feliú,
  Antonio Estruch, and José Antonio Heredia*

The Ontological Lens:
Zooming in and out from Genomic to Clinical Level ................... 44
  *Domenico M. Pisanelli, Francesco Pinciroli, and Marco Masseroli*

## Data Analysis and Image Processing

Dynamics of Vertebral Column Observed
by Stereovision and Recurrent Neural Network Model ................. 51
  *C. Fernando Mugarra Gonzalez, Stanisław Jankowski, Jacek J. Dusza,
  Vicente Carrilero López, and Javier M. Duart Clemente*

Endocardial Tracking in Contrast Echocardiography Using Optical Flow .. 61
  *Norberto Malpica, Juan F. Garamendi, Manuel Desco,
  and Emanuele Schiavi*

Unfolding of Virtual Endoscopy Using Ray-Template ................... 69
  *Hye-Jin Lee, Sukhyun Lim, and Byeong-Seok Shin*

# Knowledge Discovery and Data Mining

Integration of Genetic and Medical Information
Through a Web Crawler System .................................... 78
    *Gaspar Dias, José Luís Oliveira, Francisco-Javier Vicente,*
    *and Fernando Martín-Sánchez*

Vertical Integration of Bioinformatics Tools
and Information Processing on Analysis Outcome ...................... 89
    *Andigoni Malousi, Vassilis Koutkias, Ioanna Chouvarda,*
    *and Nicos Maglaveras*

A Grid Infrastructure for Text Mining of Full Text Articles
and Creation of a Knowledge Base of Gene Relations .................. 101
    *Jeyakumar Natarajan, Niranjan Mulay, Catherine DeSesa,*
    *Catherine J. Hack, Werner Dubitzky, and Eric G. Bremer*

Prediction of the Performance of Human Liver Cell Bioreactors
by Donor Organ Data ............................................ 109
    *Wolfgang Schmidt-Heck, Katrin Zeilinger, Gesine Pless,*
    *Joerg C. Gerlach, Michael Pfaff, and Reinhard Guthke*

A Bioinformatic Approach to Epigenetic Susceptibility
in Non-disjunctional Diseases ...................................... 120
    *Ismael Ejarque, Guillermo López-Campos, Michel Herranz,*
    *Francisco-Javier Vicente, and Fernando Martín-Sánchez*

Foreseeing Promising Bio-medical Findings for Effective Applications
of Data Mining .................................................. 130
    *Stefano Bonacina, Marco Masseroli, and Francesco Pinciroli*

# Statistical Methods and Tools
# for Biomedical Data Analysis

Hybridizing Sparse Component Analysis with Genetic Algorithms
for Blind Source Separation ...................................... 137
    *Kurt Stadlthanner, Fabian J. Theis, Carlos G. Puntonet,*
    *Juan M. Górriz, Ana Maria Tomé, and Elmar W. Lang*

Hardware Approach to the Artificial Hand Control Algorithm Realization . 149
    *Andrzej R. Wolczowski, Przemyslaw M. Szecówka,*
    *Krzysztof Krysztoforski, and Mateusz Kowalski*

Improving the Therapeutic Performance
of a Medical Bayesian Network Using Noisy Threshold Models .......... 161
    *Stefan Visscher, Peter Lucas, Marc Bonten, and Karin Schurink*

SVM Detection of Premature Ectopic Excitations
Based on Modified PCA .......................................... 173
   *Stanisław Jankowski, Jacek J. Dusza, Mariusz Wierzbowski,*
   *and Artur Oręziak*

## Decision Support Systems

A Text Corpora-Based Estimation of the Familiarity
of Health Terminology ......................................... 184
   *Qing Zeng, Eunjung Kim, Jon Crowell, and Tony Tse*

On Sample Size and Classification Accuracy:
A Performance Comparison ...................................... 193
   *Margarita Sordo and Qing Zeng*

Influenza Forecast: Comparison of Case-Based Reasoning
and Statistical Methods ........................................ 202
   *Tina Waligora and Rainer Schmidt*

Tumor Classification from Gene Expression Data:
A Coding-Based Multiclass Learning Approach ..................... 211
   *Alexander Hüntemann, José C. González, and Elizabeth Tapia*

Boosted Decision Trees for Diagnosis Type of Hypertension ............ 223
   *Michal Wozniak*

Markov Chains Pattern Recognition Approach Applied
to the Medical Diagnosis Tasks .................................. 231
   *Michal Wozniak*

Computer-Aided Sequential Diagnosis Using Fuzzy Relations –
Comparative Analysis of Methods ................................ 242
   *Marek Kurzynski and Andrzej Zolnierek*

## Collaborative Systems in Biomedical Informatics

Service Oriented Architecture for Biomedical Collaborative Research ..... 252
   *José Antonio Heredia, Antonio Estruch, Oscar Coltell,*
   *David Pérez del Rey, Guillermo de la Calle, Juan Pedro Sánchez,*
   *and Ferran Sanz*

Simultaneous Scheduling of Replication and Computation
for Bioinformatic Applications on the Grid ......................... 262
   *Frédéric Desprez, Antoine Vernois, and Christophe Blanchet*

The INFOBIOMED Network of Excellence:
Developments for Facilitating Training and Mobility .................... 274
   *Guillermo de la Calle, Mario Benito, Juan Luis Moreno,*
   *and Eva Molero*

# Bioinformatics: Computational Models

Using Treemaps to Visualize Phylogenetic Trees ...................... 283
  Adam Arvelakis, Martin Reczko, Alexandros Stamatakis,
  Alkiviadis Symeonidis, and Ioannis G. Tollis

An Ontological Approach to Represent Molecular Structure Information .. 294
  Eva Armengol and Enric Plaza

Focal Activity in Simulated LQT2 Models at Rapid Ventricular Pacing:
Analysis of Cardiac Electrical Activity Using Grid-Based Computation ... 305
  Chong Wang, Antje Krause, Chris Nugent, and Werner Dubitzky

# Bioinformatics: Structural Analysis

Extracting Molecular Diversity Between Populations
Through Sequence Alignments ....................................... 317
  Steinar Thorvaldsen, Tor Flå, and Nils P. Willassen

Detection of Hydrophobic Clusters in Molecular Dynamics Protein
Unfolding Simulations Using Association Rules ...................... 329
  Paulo J. Azevedo, Cândida G. Silva, J. Rui Rodrigues,
  Nuno Loureiro-Ferreira, and Rui M.M. Brito

Protein Secondary Structure Classifiers Fusion Using OWA .............. 338
  Majid Kazemian, Behzad Moshiri, Hamid Nikbakht, and Caro Lucas

Efficient Computation of Fitness Function by Pruning
in Hydrophobic-Hydrophilic Model .................................. 346
  Md. Tamjidul Hoque, Madhu Chetty, and Laurence S. Dooley

Evaluation of Fuzzy Measures in Profile Hidden Markov Models
for Protein Sequences .............................................. 355
  Niranjan P. Bidargaddi, Madhu Chetty, and Joarder Kamruzzaman

# Bioinformatics: Microarray Data Analysis

Relevance, Redundancy and Differential Prioritization
in Feature Selection for Multiclass Gene Expression Data ............... 367
  Chia Huey Ooi, Madhu Chetty, and Shyh Wei Teng

Gene Selection and Classification of Human Lymphoma
from Microarray Data ............................................... 379
  Joarder Kamruzzaman, Suryani Lim, Iqbal Gondal, and Rezaul Begg

Microarray Data Analysis and Management in Colorectal Cancer ........ 391
  Oscar García-Hernández, Guillermo López-Campos,
  Juan Pedro Sánchez, Rosa Blanco, Alejandro Romera-Lopez,
  Beatriz Perez-Villamil, and Fernando Martín-Sánchez

**Author Index** ..................................................... 401

# Application of Three-Level Handprinted Documents Recognition in Medical Information Systems

Jerzy Sas[1] and Marek Kurzynski[2]

[1] Wroclaw University of Technology, Institute of Applied Informatics,
Wyb. Wyspianskiego 27, 50-370 Wroclaw, Poland
jerzy.sas@pwr.wroc.pl
[2] Wroclaw University of Technology, Faculty of Electronics,
Chair of Systems and Computer Networks,
Wyb. Wyspianskiego 27, 50-370 Wroclaw, Poland
marek.kurzynski@pwr.wroc.pl

**Abstract.** In this paper the application of novel three-level recognition concept to processing of some structured documents (forms) in medical information systems is presented. The recognition process is decomposed into three levels: character recognition, word recognition and form contents recognition. On the word and form contents level the probabilistic lexicons are available. The decision on the word level is performed using results of character classification based on a character image analysis and probabilistic lexicon treated as a special kind of soft classifier. The novel approach to combining these both classifiers is proposed, where fusion procedure interleaves soft outcomes of both classifiers so as to obtain the best recognition quality. Similar approach is applied on the semantic level with combining soft outcomes of word classifier and probabilistic form lexicon. Proposed algorithms were experimentally applied in medical information system and results of automatic classification of laboratory test order forms obtained on the real data are described.

## 1 Introduction

Automatic analysis of handwritten forms is useful in such applications where direct information insertion into the computer system is not possible or inconvenient. Such situation appears frequently in hospital medical information systems, where physicians or medical staff not always can enter the information directly at the system terminal. Form scanning is considered to be especially useful in laboratory support software, where paper forms are still frequently used as a medium for laboratory test orders representation. Hence, in many commercially available medical laboratory systems a scanning and recognition module is available.

Typical form being considered here has precisely defined structure. It consists of separated data fields, which in turn consist of character fields. In our approach we assume that the whole form contents describes an object from the finite set of items and the ultimate aim of form recognition is selecting of relatively small

J.L. Oliveira et al. (Eds.): ISBMDA 2005, LNBI 3745, pp. 1–12, 2005.
© Springer-Verlag Berlin Heidelberg 2005

subset of objects. Therefore, instead of using the classic pattern recognition approach consisting in indicating a single class, we will apply "soft" recognizer ([3]) which fetches the vector of soft labels of classes, i.e. values of classifying function.

In order to improve the overall form recognition quality, compound recognition methods are applied. Two most widely used categories of compound methods consist in combining classifiers based on different recognition algorithms and different feature sets ([4]). Another approach divides the recognition process into levels in such a way, that the results of classification on lower level are used as features on the upper level ([2]). Two-level approach is typical in handwriting recognition, in which the separate characters are recognized on the lower level and next on the upper level the words are recognized, usually with the use of lexicons.

In this paper, the method which uses both classifier combination and multilevel recognition is described. Probabilistic properties of lexicon and character classifier are typically used to build Hidden Markov Model(HMM) of the language ([11]). We propose another approach to the word recognition, in which probabilistic lexicon is treated as a special kind of classifier based on a word length, and next result of its activity is combined with soft outcomes of character classifier based on recognition of character image. Soft outcomes of a word classifier can be used next as data for semantic level classifier, which - similarly as previously - combined with object lexicon - recognizes the object described by the whole form.

The contents of the work are as follows. Section 2 introduces necessary background. In section 3 the classification methods on successive levels of object recognition problem are presented and concept of fusion strategies of character-based and lexicon-based classifiers are discussed. The proposed algorithms were practically implemented in application for automatic processing of laboratory test order forms in hospital information system. The system architecture and some implementation details are described in section 4. Results of experiments on proposed method efficiency are presented in section 5

## 2    Preliminaries

Let us consider a paper form $F$ designed to be filled by handwritten characters. The form consists of data fields. Each data field contains a sequence of characters of limited length coming from the alphabet $\mathcal{A} = \{c_1, c_2, ..., c_L\}$. We assume that the actual length of filled part of data field can be faultlessly determined. The set $\mathcal{A}$ can be different for each field. Typically we deal with fields that can contain only digits, letters or both of them. For each data field there exists a probabilistic lexicon $\mathcal{L}$. Lexicon contains words that can appear in the data field and their probabilities:

$$\mathcal{L} = \{(W_1, p_1), (W_2, p_2), ..., (W_N, p_N)\}, \tag{1}$$

where $W_j$ is the word consisting of characters from $\mathcal{A}$, $p_j$ is its probability and $N$ is the number of words in the lexicon.

The completely filled form describes an object (e.g. a patient in medical applications) and the data items written in the data fields are its attributes. The form contents, after manual verification is entered to the database, which also contains the information about the objects appearance probability. An example can be a medical information system database, where the forms contain test orders for patients registered in the database. The patients suffering from chronic diseases are more frequently examined, so it is more probable that the form being recognized concerns such a patient. Thus, this data base can be treated as a kind of probabilistic lexicon containing objects recognized in the past and the information about probability of its appearance, viz.

$$\mathcal{L}_{\mathcal{B}} = \{(b_1, \pi_1), (b_2, \pi_2), ..., (b_M, \pi_M)\}. \tag{2}$$

Our aim is to recognize the object $b \in \mathcal{L}_{\mathcal{B}}$ on the base of scanned image of a form $F$ and both lexicons (1), (2). The recognition process can be divided into three levels, naturally corresponding to the three-level form structure:

- character (alphabetical) level – where separate characters are recognized,
- word level – where the contents of data fields is recognized, based on the alphabetical level classification results, their probabilistic properties and probabilistic lexicon (1),
- semantic level – where the relations between fields of the form being processed and lexicon (2) are taken into account to further improve the recognition performance.

In the next section the classification methods used on the successive levels of recognition procedure are described in details.

## 3    Three-Level Form Recognition Method

### 3.1    Character Recognition on the Alphabetical Level

We assume that on character (alphabetical) level classifier $\Psi_C$ is given which gets character image $x$ as its input and assigns it to a class (character label) $c$ from $\mathcal{A}$, i.e., $\Psi_C(x) = c$. Alternatively, we may define the classifier output to be a $L$-dimensional vector with supports for the characters from $\mathcal{A}$ ([4]), i.e.

$$\Psi_C(x) = [d_1(x), d_2(x), ..., d_L(x)]^T. \tag{3}$$

Without loss of generality we can restrict $d_i(x)$ within the interval $[0, 1]$ and additionally $\sum_i d_i(x) = 1$. Thus, $d_i(x)$ is the degree of support given by classifier $\Psi_C$ to the hypothesis that image $x$ represents character $c_i \in \mathcal{A}$. If a crisp decision is needed we can use the maximum membership rule for soft outputs (3), viz.

$$\Psi_C(x) = arg\left(\max_i d_i(x)\right). \tag{4}$$

We have applied MLP-based classifier on this level. The vector of support values in (3) is the normalized output of MLP obtained by clipping network output values to $[0, 1]$ range and by normalizing their sum to 1.0.

Independently of nature of classifier $\Psi_C$, support vector (3) is usually interpreted as an estimate of *posterior* probabilities of classes (characters) provided that observation $x$ is given ([4], [9], [10]), i.e. in next considerations we adopt:

$$d_i(x) = P(c_i \mid x), \quad c_i \in \mathcal{A}. \tag{5}$$

## 3.2   Data Field Recognition on the Word Level

Let the length $\mid W \mid$ of currently recognized word $W \in \mathcal{L}$ be equal to $n$. This fact defines the probabilistic sublexicon $\mathcal{L}_n$

$$\mathcal{L}_n = \{(W_k, q_k)_{k=1}^{N_n} : W_k \in \mathcal{L}, \mid W_k \mid = n\}, \tag{6}$$

i.e. the subset of $\mathcal{L}$ with modified probabilities of words:

$$q_k = P(W_k / \mid W_k \mid = n) = \frac{p_k}{\sum_{j:|W_j|=n} p_j}. \tag{7}$$

The sublexicon (6) can be also considered as a soft classifier $\Psi_L$ which maps feature space $\{\mid W_k \mid : W_k \in \mathcal{L}\}$ into the product $[0,1]^{N_n}$, i.e. for each word length $n$ produces the vector of supports to words from $\mathcal{L}_n$, namely

$$\Psi_L(n) = [q_1, q_2, ..., q_{N_n}]^T. \tag{8}$$

Let suppose next, that classifier $\Psi_C$, applied $n$ times on the character level, on the base of character images $X_n = (x_1, x_2, ..., x_n)$, has produced the sequence of character supports (3) for the whole recognized word, which can be organized into the following matrix of supports, or matrix of *posterior* probabilities (5):

$$D_n(X_n) = \begin{pmatrix} d_{11}(x_1) & d_{12}(x_1) & \dots & d_{1L}(x_1) \\ d_{21}(x_2) & d_{22}(x_2) & \dots & d_{2L}(x_2) \\ \vdots & \vdots & \dots & \vdots \\ d_{n1}(x_n) & d_{n2}(x_n) & \dots & d_{nL}(x_n) \end{pmatrix}. \tag{9}$$

Now our purpose is to built soft classifier $\Psi_W$ (let us call it *Combined Word Algorithm* - CWA) for word recognition as a fusion of activity of both lexicon-based $\Psi_L$ and character-based classifier $\Psi_C$:

$$\Psi_W(\Psi_C, \Psi_L) = \Psi_W(D_n, \mathcal{L}_n) = [s_1, s_2, ..., s_{N_n}]^T, \tag{10}$$

which will produce support vector for all words from sublexicon $\mathcal{L}_n$.

Let $\mathcal{N} = \{1, 2, ..., n\}$ be the set of numbers of character positions in a word $W \in \mathcal{L}_n$ and $\mathcal{I}$ denotes a subset of $\mathcal{N}$. In the proposed fusion method with "interleaving" first the algorithm $\Psi_C$ applied for recognition of characters on positions $\mathcal{I}$ on the base of set of images $X^{\mathcal{I}} = \{x_k : k \in \mathcal{I}\}$, produces matrix of supports $D^{\mathcal{I}}$ and next - using these results of classification - the lexicon $\mathcal{L}_n$ (or algorithm $\Psi_L$) is applied for recognition of a whole word $W$.

The main problem of proposed method consists in an appriopriate division of $\mathcal{N}$ into sets $\mathcal{I}$ and $\bar{\mathcal{I}}$ (complement of $\mathcal{I}$). Intuitively, subset $\mathcal{I}$ should contain these positions for which character recognition algorithm gives the most reliable results. In other words division of $\mathcal{N}$ should lead to the best result of classification accuracy of a whole word. Thus, subset $\mathcal{I}$ can be determined as a solution of an appropriate optimization problem.

Let $W^{\mathcal{I}} = \{c_{i_k} : k \in \mathcal{I}, c_{i_k} \in \mathcal{A}\}$ be any set of characters on positions $\mathcal{I}$. Then we have following posterior probability:

$$P(W^{\mathcal{I}} \mid X^{\mathcal{I}}) = \prod_{k \in \mathcal{I}} d_{k\,i_k}(x_k). \tag{11}$$

The formula (11) gives conditional probability of hypothesis that on positions $\mathcal{I}$ of word to be recognized are characters $W^{\mathcal{I}}$ provided that set of character images $X^{\mathcal{I}}$ has been observed.

Applying for remaining part of the word sublexicon $\mathcal{L}_n$, we can calculate conditional probability of the whole word $W_j \in \mathcal{L}_n$, which constitutes the support (10) for word $W_j$ of soft classifier $\Psi_W$:

$$s_j = P(W_j \mid X^{\mathcal{I}}) = P(W^{\mathcal{I}} \mid X^{\mathcal{I}})\, P(W_j \mid W^{\mathcal{I}}). \tag{12}$$

The first factor in (12) is given by (11) whereas the second one can be calculated as follows:

$$P(W_j \mid W^{\mathcal{I}}) = \frac{q_j}{\sum_{j:W_j\,contains\,W^{\mathcal{I}}}\ q_j}. \tag{13}$$

Since the support vector (12) of the rule $\Psi_W$ strongly depends on the set $\mathcal{I}$ hence we can formulate the following optimization problem:

It is neccesary to find such a subset $\mathcal{I}^*$ of $\mathcal{N}$ and such a set of charcters $W^{\mathcal{I}^*}$ which maximize the maximum value of decision supports dependent on sets $\mathcal{I}$ and $W^{\mathcal{I}}$, namely

$$Q(\Psi_W^*) = \max_{\mathcal{I},W^{\mathcal{I}}} \ \max_{j=1,2,\dots,N_n} s_j(\mathcal{I}, W^{\mathcal{I}}). \tag{14}$$

The detailed description of suboptimal solution of the problem (14) which was applied in further experimental investigations can be find in [8].

### 3.3   Complete Form Recognition on the Semantic Level

For recognition of the whole form (object) on the semantic level we propose procedure called *Combined Semantic Algorithm* (CSA), which is fully analogous to the approach applied on the word level, i.e. relation between word classifier $\Psi_W$ and probabilistic lexicon (2) is exactly the same as relation between the character recognizer $\Psi_C$ and word lexicon (1). In other words, the form lexicon is treated as a special kind of classifier producing the vector of form supports (probabilities)

$$\pi = (\pi_1, \pi_2, \dots, \pi_M), \tag{15}$$

which next are combined with soft outcomes (10) of word classifier $\Psi_W$.

Let suppose that form to be recognized contains $K$ data fields. Recognition of $k$th data field containing word from sublexicon (6) which length is equal to $n_k$, has provided the vector of supports (10)

$$\Psi_W(\Psi_C, \Psi_L) = \Psi_W(D_{n_k}, \mathcal{L}_{n_k}) = (s_1^{(k)}, s_2^{(k)}, ..., s_{N_{n_k}}^{(k)})^T, \qquad (16)$$

given by formula (12).

Now, repeating recognition method from section 3.2 and optimization of fusion procedure with "interleaving" for support vectors (16) for $k = 1, 2, ..., K$ instead of matrix (9) and probabilities (15) instead of (8) we get according to (11), (12) and (13) support vector which soft classifier on the semantic level $\Psi_B$ gives to forms from the lexicon (2)

$$\Psi_B = (\sigma_1, \sigma_2, ..., \sigma_M). \qquad (17)$$

As previously, that $\sigma_i$ can be interpreted as an estimate of *posterior* probability of the object described by form $b_i \in \mathcal{L}_\mathcal{B}$ provided that observation of character images of all data fields $X_B = (X_{n_1}, X_{n_2}, ..., X_{n_K})$ are given and both lexicons (1) and (2) are available, viz.

$$\sigma_i = P(b_i \mid X_B, \mathcal{L}, \mathcal{L}_\mathcal{B}). \qquad (18)$$

The crisp decision is possible by selection the object $b^*$ from (2) for which support value (probability) $\sigma^*$ is the greatest one.

In application like the one described here, probabilistic lexicons are derived from the contents of database, where previously recognized and verified forms are stored, It may happen that the object described by a form is not registered in the database yet. Forcing to always recognize one of registered objects would be unreasonable. In particular at the early stages of the recognition system operation, when the database contains few records it cannot be used as a reliable objects lexicon. In our approach, the database is periodically tested in order to estimate the probability $P_{new}$. $P_{new}$ is the probability that the verified form being entered describes the object not registered in the database yet. Before the new record is entered to the database, it is tested if the object described by the record is already in the database. The new record is appropriately flagged depending on the test result. By analyzing the flags associated with certain number of recently entered records we can estimate $P_{new}$. If the probability $\sigma^*$ is greater than $P_{new}$ then $b^*$ found by CSA is the final recognition. Otherwise CSA result is rejected and it is assumed that the object described by the form is not contained in the data base.

## 4    Application of Three-Level Form Recognition Concept in the Laboratory Orders Form Recognition Module

The concept described in previous sections has been applied to laboratory orders form recognition module in a hospital information system (HIS). In some cases specimens for laboratory test are taken in locations distant from central laboratory, e.g. in outpatient clinic. It is convenient to transfer information about ordered tests in paper form together with specimens. To improve operation of central laboratory, which processes hundreds of orders daily, the method of fast and reliable entering of test orders data into the information system controlling automatic laboratory devices is necessary.

For each specimen, individual order form is filled by the medical assistant in outpatient clinic where specimen is taken. The form for particular specimen is identified by its symbol represented by barcode label stick both to specimen container and to the corresponding form. In the central laboratory the forms are scanned, their contents is recognized and entered into HIS database. Next, each recognized form contents is manually verified by operator by comparing recognized data with the image of the form displayed on the screen. Finally, verified data are used to register test order record in HIS database, where it is later used to control bi-directional laboratory devices. The system architecture and data flow is presented on Fig.1. It consists of the following modules:

**form design module** – allows system administrator to design various form variants containing required subsets of tests being supported by the laboratory,

**scanning module** – controls the farm of scanners connected to the system and manages the form images repository,

**lexicon extraction module** – updates periodically probabilistic lexicons both for word and semantic level using actual contents of HIS database,

**recognizer module** – performs soft recognition of form images, fetches results of soft recognition of isolated data fields as well as the results of soft identification of patient, for which the form has been prepared,

**manual form verification module** – provides user interface for thorough verification of form recognition results. Support vectors that are results of soft recognition in recognizer module are used to build ordered list of alternative values for isolated fields and for the whole patient data identification section. They are used as combo boxes connected with corresponding fields in case where manual correction is needed.

The test order form is presented on Fig. 2. It consists of three sections: ordering institution/physician data, patient identification data and ordered tests list. Two-level recognition is applied to ordering institution/physician data because there are no clear relations between data fields in this section and hence third level cannot be defined. Full three-level concept is applied to patient identification data. The patient data contain: name, first name, sex, birth date, social security identifier. Probabilistic lexicons for all data fields in patient section are derived from HIS database contents using lexicon extraction module. In case of date field, probabilistic dictionary is applied only to 4-digit year section. The module updates lexicons periodically (every night) using current database contents.

The ordered laboratory tests are identified by marking check boxes in the lower section of the form. The count of check boxes is limited by the area of test

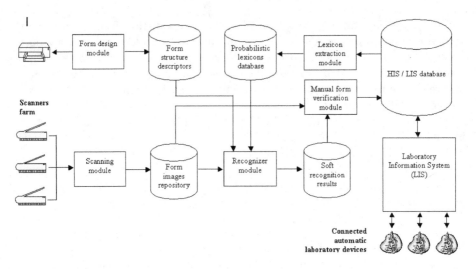

**Fig. 1.** Form recognition module architecture and data flow

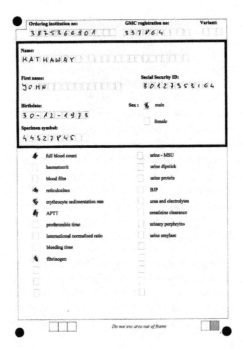

**Fig. 2.** Laboratory test order form image

selection section on the form and it is significantly lower that the total number of tests, carried out by the laboratory. To assure flexibility of the system,
many variants of forms can be designed in the system. Particular form variant
can be used to order a subset of tests, usually related each to other, e.g. test

in hematology, urine tests, clinical biochemistry etc. The user can define any number of form variants containing tests subsets using form design module. In different form variants the checkbox in given position represents different laboratory tests. Assignment of tests to particular check box positions is stored in variant description data structure used during final form contents interpretation. The form variant is defined by numeric field printed on the form. Correct recognition of form variant is absolutely essential for system usability and even for patient safety. To assure maximal accuracy and human-readability of form variant identification, the numerical variant symbol is pre-printed using fixed font.

## 4.1 Character Classification and Features Extraction

MLP has been used as the character level soft classifier. Feature extraction and MLP architecture is based on methods described in [1]. Directional features set has been selected as the basis for character classification due to its superior efficiency reported in literature and ease of implementation. The directional features describe the distribution of stroke directions in 25 evenly spaced subareas of the character image. The set of eight direction planes is created. Direction plane is an array of resolution equal to image resolution. Each plane corresponds to one of eight direction sections as shown on Fig. 3. According to the concept described in [1] for each image pixel the Sobel operator on image brightness $g(i, j)$ is calculated giving image brightness gradient. The brightness gradient vector is decomposed into two components $g_k$ and $g_{k+1}$ parallel to lines surrounding its section $k$. The lengths of components are then added to cells of corresponding direction planes. Finally, each plane is filtered using Gaussian filter resulting in 5x5 grid of values. 200 elements feature vector is built of all of grid values calculated for all plane arrays. MLP with 200 inputs and number of outputs corresponding to count of classes was used as character classifier. Data fields in test order form can be divided into purely numerical or alphabetical.

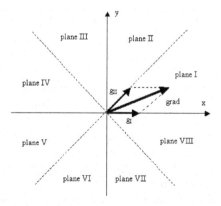

**Fig. 3.** Laboratory test order form image

We applied three independently trained classifiers for: printed numerical fields (ordering institution ID, form variant), handwritten numeric fields and handwritten alphabetical fields. The count of nodes in hidden layer was determined experimentally by maximizing character recognition accuracy. MLP for handwritten letters recognition contains 80 units in hidden layer while hidden layer of MLP for numerals recognition consists of 60 units.

## 5 Experiment Results

The aim of experiments was to assess the increase of form recognition accuracy resulting from application of described methods on word and semantic levels.

Character classifiers were trained using sets of character image database collected from almost 300 individuals. The training set for letters contained 9263 images. For numerals classifier training the set of 3130 images was used. For assessment of isolated characters classification accuracy, the characters extracted from real laboratory test order forms were used. The images extracted from 376 manually verified forms were used. Achieved accuracy in recognizing isolated handwritten characters level was: 90.7% for letters and 98.1% for digits. Probabilistic lexicons on word and semantic levels were derived from real data in HIS system database containing 53252 patient records.

In the system being described here, all automatically recognized forms are manually verified before they are entered into HIS database. To simplify the manual data correction it is expected that the system suggest alternate values for erroneous fields. The values are presented to the user as combo boxes filled with items ordered according to their support factors evaluated by soft recognition classifier on word level. It is expected that the correct value is located close to top of the list, so user can select it without typing but rather by simply clicking on the list. In the same way, the user can select the complete set of patient data from the list of suggested patients already registered in HIS database, ordered according to support values produced by recognition algorithm on the semantic level. In assessing the recognition algorithm it is therefore not essentially important if the actual class is the one with highest value of support factor, but rather if it is among $k$ ($k = 1, 2, 5$) classes with highest support values. The recognizer performance was therefore evaluated in three ways, using as the criterion the number of cases where actual class is among $k = 1, 3, 5$ classes with highest support factors.

On the word level the approach described in this article has been compared to two simple approaches. The first one is based only on the results of soft recognition on the character level. Support factor for a word is calculated as a product of support factors for subsequent letters. Only 5 words with highest values calculated in this way are taken into account. The second simple approach calculates support values in the same way, but the set of allowed words is defined by the lexicon. Probabilistic properties of the lexicon however are not used. Experiments have been performed for three levels of lexicon completeness: $p = 0.75$, $p = 0.90$ and $p = 1.0$, where $p$ is the probability that actual word belongs to

**Table 1.** Names recognition accuracy

| Criterion | S | SL p=0.75 | CWA p=0.75 | SL p=0.90 | CWA p=0.90 | SL p=1.00 | CWA p=1.00 |
|-----------|-----|-----|-----|-----|-----|-----|-----|
| 1 of 1 | 88.6% | 90.7% | 94.1% | 92.3% | 95.2% | 93.4% | 96.3% |
| 1 of 3 | 90.2% | 93.1% | 94.6% | 94.4% | 95.5% | 95.7% | 97.1% |
| 1 of 5 | 94.1% | 94.1% | 95.7% | 96.0% | 97.3% | 96.5% | 98.1% |

**Table 2.** Surnames recognition accuracy

| Criterion | S | SL p=0.75 | CWA p=0.75 | SL p=0.90 | CWA p=0.90 | SL p=1.00 | CWA p=1.00 |
|-----------|-----|-----|-----|-----|-----|-----|-----|
| 1 of 1 | 84.3% | 87.5% | 89.4% | 89.8% | 93.1% | 91.6% | 94.4% |
| 1 of 3 | 91.2% | 93.1% | 93.9% | 94.2% | 95.9% | 95.3% | 97.1% |
| 1 of 5 | 95.5% | 94.2% | 95.2% | 96.0% | 96.8% | 96.0% | 97.6% |

the lexicon. Results for names and surnames recognition are presented in tables below. S and SL denote here two described above simple reference algorithms. CWA denotes combined word algorithm described in section 3.2.

Similar experiment has been performed to assess the accuracy CSA algorithm on semantic level. Results are presented in table 3.

**Table 3.** Patient identification accuracy

| Criterion | S | SL p=0.75 | CSA p=0.75 | SL p=0.90 | CSA p=0.90 | SL p=1.00 | CSA p=1.00 |
|-----------|-----|-----|-----|-----|-----|-----|-----|
| 1 of 1 | 67.3% | 77.7% | 80.6% | 83.4% | 88.3% | 89.9% | 92.8% |
| 1 of 3 | 73.7% | 81.4% | 84.8% | 85.1% | 87.5% | 91.6% | 93.4% |
| 1 of 5 | 78.2% | 84.0% | 85.9% | 88.0% | 89.4% | 92.3% | 93.6% |

## 6   Conclusions

Experiments described in previous section have shown that application of proposed algorithms on both word and semantic levels significantly improves isolated data and patient recognition accuracy. In case of complete name and surname lexicons, average reduction of error rate on word level is 43% and 37% correspondingly. In case of patient identification on semantic level error is reduced by 23%. Obtained results, due to reduction of necessary corrections, contribute to making form verifier work more efficient, easier and less error prone.

Described here methods have been implemented in laboratory test order forms recognition subsystem cooperating with large hospital information system. Elimination of necessity of retyping of most data present on data forms reduced the average operator time needed for single form processing many times and in result reduced also laboratory operation costs.

## Acknowledgement

This work was financed from the State Committee for Scientific Research (KBN) resources in 2005–2007 years as a research project No 3 T11E 005 28.

## References

1. Liu C., Nakashima K., Sako H., Fujisawa H.: Handwritten Digit Recognition: Benchmarking of State-of-the-Art Techniques. Pattern Recognition, Vol. 36. (2003) 2271-2285
2. Lu Y., Gader P. Tan C.: Combination of Multiple Classifiers Using Probabilistic Dictionary and its Application to Postcode Generation. Pattern Recognition, Vol. 35. (2002) 2823-2832
3. Kuncheva L.: Combining Classifiers: Soft Computing Solutions. In: Pal S., Pal A. (eds.): Pattern Recognition: from Classical to Modern Approaches. World Scientific (2001) 427-451
4. Kuncheva L.I.: Using measures of similarity and inclusion for multiple classifier fusion by decision templates. Fuzzy Sets and Systems, Vol. 122. (2001) 401-407
5. Sas J., Kurzynski M.: Multilevel Recognition of Structured Handprinted Documents – Probabilistic Approach. In: Kurzynski M., Puchala E. (eds.): Computer Recognition Systems, Proc. IV Int. Conference, Springer Verlag (2005) 723-730
6. Sas J., Kurzynski M.: Application of Statistic Properties of Letter Succession in Polish Language to Handprint Recognition. In: Kurzynski M. (eds.): Computer Recognition Systems, Proc. IV Int. Conference, Springer Verlag (2005) 731-738
7. Sas J.: Handwritten Laboratory Test Order Form Recognition Module for Distributed Clinic. J. of Medical Informatics and Technologies, Vol. 8. (2004) 59-68
8. Kurzynski M., Sas J.: Combining Character Level Classifier and Probabilistic Lexicons in Handprinted Word Recognition – Comparative Analysis of Methods. In: Proc. XI Int. Conference on Computer Analysis and Image Processing, LNCS Springer Verlag (2005) (to appear)
9. Devroye L., Gyorfi P., Lugossi G.: A Probabilistic Theory of Pattern Recognition. Springer Verlag, New York (1996)
10. Duda R., Hart P., Stork D.: Pattern Classification. John Wiley and Sons (2001)
11. Vinciarelli A. et al.: Offline Recognition of Unconstrained Handwritten Text Using HMMs and Statistical Language Models. IEEE Trans. on PAMI, Vol. 26. (2004) 709-720

# Data Management and Visualization Issues in a Fully Digital Echocardiography Laboratory

Carlos Costa[1], José Luís Oliveira[1], Augusto Silva[1],
Vasco Gama Ribeiro[2], and José Ribeiro[2]

[1] University of Aveiro, Electronic and Telecommunications Department / IEETA,
3810-193 Aveiro, Portugal
{ccosta,jlo,asilva}@det.ua.pt
[2] Centro Hospitalar V. N. Gaia, Cardiology Department, 4434-502 V. N. Gaia, Portugal
{vasco,jri}@chvng.min-saude.pt

**Abstract.** This paper presents a PACS solution for echocardiography laboratories, denominated as *Himage*, that provides a cost-efficient digital archive, and enables the acquisition, storage, transmission and visualization of DICOM cardiovascular ultrasound sequences. The core of our approach is the implementation of a DICOM private transfer syntax designed to support any video encoder installed on the operating system. This structure provides great flexibility concerning the selection of an encoder that best suits the specifics of a particular imaging modality or working scenario. The major advantage of the proposed system stems from the high compression rate achieved by video encoding the ultrasound sequences at a proven diagnostic quality. This highly efficient encoding process ensures full online availability of the ultrasound studies and, at the same time, enables medical data transmission over low-bandwidth channels that are often encountered in long range telemedicine sessions. We herein propose an imaging solution that embeds a Web framework with a set of DICOM services for image visualization and manipulation, which, so far, have been traditionally restricted to intranet environments.

## 1 Introduction

In the last decade, the use of digital medical imaging systems has been increasing exponentially in the healthcare institutions, representing at this moment one the most valuable tools supporting the medical decision process and treatment procedures. Until recently, the benefits of digital technology were still confined to the medical imaging machinery and the examples of image data migration to a centralized (shared) archive were reduced.

The medical imaging digitalization and implementation of PACS (Picture Archiving and Communication Systems) systems increases practitioner's satisfaction through improved faster and ubiquitous access to image data. Besides, it reduces the logistic costs associated to the storage and management of image data and also increases the intra and inter institutional data portability. The most important contribution to the exchange of structured medical image between equipment, archive systems and information system was the establishment of DICOM (Digital Imaging and Communications in Medicine) Standard [1], in 1992. Currently, almost all medical imaging equipment manufacturers are including DICOM (version 3) digital output in their products.

J.L. Oliveira et al. (Eds.): ISBMDA 2005, LNBI 3745, pp. 13–21, 2005.

The digital medical imaging imposes new challenges concerning the storage volume, their management and the network infrastructure to support their distribution. In general, we are dealing with huge volumes of data that are hard to keep "online" (in centralized servers) and difficult to access (in real-time) outside the institutional broadband network infra-structure.

Echocardiography is a rather demanding medical imaging modality when regarded as digital source of visual information. The date rate and volume associated with a typical study poses several problems in the design and deployment of telematic enabled PACS structures, aiming at long-range interaction environments. For example, an uncompressed echocardiography study size can typically vary between 100 and 500Mbytes, depending on equipment technical characteristics, operator expertise factors and procedure type [2].

Several institutions are dealing with problems concerning the permanent historic availability of procedure images in network servers, as well their integration with other clinical patient information detained by the HIS (Healthcare Information Systems), in a unique access platform application. On other hand, current clinical practice reveals that patient related data are often generated, manipulated and stored in several institutions where the patient is treated or followed-up. In this heterogeneous and complex scenario, the sharing and the remote access to patient information is of capital importance in today's health care practice.

In these circumstances, compression techniques appear as an enabling tool to accommodate the source data rate to the communications channel and to the storage capacity. The definition of an adequate trade off between compression factor and diagnostic quality is a fundamental constraint in the design of both the digital archive and the telecommunications platform.

In the following sections we will present the architecture of a digital video archive and communication system, which, supporting and extending some DICOM services, is able to preserve the diagnostic image quality, and provides on-line access to whole echocardiographic datasets.

## 2  Materials and Method

The development scenario is a Cardiology Department PACS infra-structure that was been developed at the Central Hospital of V.N.Gaia, since 1997. This healthcare unit is supported by two digital imaging laboratories: - the Cardiac Catherization Laboratory (*cathlab*) [3] produces about 3000 procedures/year and; - the Cardiovascular Laboratory (*echolab*) with 6000 procedures/year.

The first *echolab* PACS module, installed in 2001, allowed the storage of 14-16 months of US procedures making use of a image compressed format in DICOM JPEG-lossy 85 (10 Mbytes/exam). This volume of information is difficult to handle, namely if one aims at permanent availability and cost-effective transmissions times for remote connections. Also in 2001, a transcontinental project started to develop a telematic platform capable of establishing cooperative telemedicine sessions between the Cardiology Departments of the Central Hospital of Maputo (Mozambique) and the CHVNG, to cardiovascular ultrasounds [4]. This telecardiology project was the catalyst factor for the development of a new PACS software solution (the Himage) to support the local storage, visualization and transmission of coronary US.

## 2.1  Image Compression

Digital video compression is a technology of the utmost importance when considering storage space and time transmission problems, associated to the huge number of US procedures/year realized in the CHVNG. The preceding Himage approach of using the existent DICOM JPEG lossy format [5], with above expressed study volumes, resulted in an unfeasable solution. As result, our research was initially concentrated in developing a new approach attempting to achieve maximum storage capacity and minimum download times without ever compromising image diagnostic quality.

The methodology to find the "best" compression factor (and other settings) was based on successive evaluations of compression codec/factor versus desired diagnostic image quality for each specific modality and corresponding utilization scenario [6].

Given the spatio-temporal redundancy that characterizes the echocardiographic video signals, there are important gains by choosing a compression methodology that copes with both intra-frame and inter-frame redundancies, like the MPEG family of encoders [7]. The novelty of our compression approach starts by embedding each storage server with highly efficient MPEG4 encoding software. If the JPEG explores exclusively the intra-frame information redundancy, the MPEG4 encoding strategy takes full advantage of object texture, shape coding and inter-frame redundancy ending up with best results.

Since MPEG4 is not a DICOM native coding standard, subsequent image transmission, decoding and reviewing is accomplished through a DICOM *private transfer syntax* mechanisms enabled between storage servers and diagnostic clients terminals equipped with the echocardiography viewing software. In order to achieve full compliance all the other DICOM information elements and structure were kept unchanged.

## 2.2  DICOM Private Transfer Syntax

As matter of ensuring flexibility, we decided not to insert the MPEG4 directly in the TLV (Tag Length Value) DICOM data structure, i.e. solve the US "specific problem". Instead, as Fig.1 points out, a multimedia container that dynamically supports different encoders was developed. The container has a simple structure, including a field to store the encoder ID code. When it is necessary to decompress the images, the Himage-PACS solicits the respective decoder service to the operating system (like other multimedia application).

This approach represents an optimized high level software solution. If, for instance, in a 5-10 years period a more efficient codec appears, we just need to change one parameter in the "Himage Conversion Engine" (i.e. set the new codec ID) to setup.

Fig. 2 shows a DICOM partial overlap dump, "default transfer syntax" versus "private syntax", of a US 25 frames image sequence with a RGB 576*768 matrix. It is possible to observe two important aspects. First, the DICOM DefaultTransferSyntaxUID identifier (1.2.840.10008.1.2) is replaced by a private PrivateTransferSyntaxUID (1.2.826.0.1.3680043.2.682.1.4). This private UID is based on the root UID 1.2.826.0.1.3680043.2.682 requested by our workgroup. Second, the "PixelData" field size "is reduced" 120 times (33177600/275968).

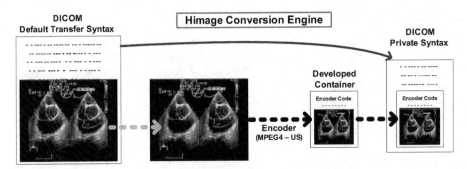

**Fig. 1.** Conversion of DICOM "default transfer syntax" in "private syntax"

| Tag | VR | Contents | length | Comments | |
|---|---|---|---|---|---|
| ..... | | | | | |
| (0002, 0010) | UI | [1.2.840.10008.1.2] | # 18, | DefaultTransferSyntaxUID | (before) |
| (0002, 0010) | UI | [1.2.826.0.1.3680043.2.682.1.4] | # 22, | PrivateTransferSyntaxUID | (after) |
| | | | | | |
| ..... | | | | | |
| (7fe0, 0010) | OW | 0000\0000\0000....... | # 33177600, | PixelData | (before) |
| (7fe0, 0010) | OW | 4952\4646\355c\0004\... | # 275968, | PixelData | (after) |

**Fig. 2.** DICOM dump: DICOM "default transfer syntax" vs "private syntax"

## 2.3  Echolab: Himage PACS Infra-structure

The clinical facility is equipped with 7 echocardiography machines including standard DICOM3.0 output interfaces. Daily outputs to the network reach about 800 images (still and cine-loops). All the acquisition units have a default configuration that ensures an automatic DICOM send service by the end of each medical procedure. Thus ultrasound image data is sent, uncompressed in a DICOM default transfer syntax mode, to an Acquisition Processing Unit (A.P.U) that acts as the primary encoding engine.

The received procedures are analyzed (the alphanumeric data is extracted from DICOM headers) to detect eventual exam/patient ID errors. As is represented in Fig. 3, after a consequent acknowledgement, the exam is then inserted in the database, i.e. the new exam is added to the Himage PACS. The DICOM image data is compressed and embedded in new DICOM private syntax file (Fig. 1). The result is stored in the "Storage Server", to be available to users. The original raw data images are saved and kept online during six months. However, the DICOM private syntax sequences are made available during all the system lifetime (in the "Storage Server"). The developed client application consists on a Web DICOM viewer (Fig. 4) that is available in the "Web Server" (Fig. 3). This software package allows the clinical staff to manipulate and visualize standard DICOM and also our private syntax exams available in the *Himage* system database. Because the Himage client solution is totally developed in web technology, the echolab procedures are accessible, through appropriate security measures, from any Internet point. The communication channel is encrypted with HTTPS following a requested username/password authentication procedure.

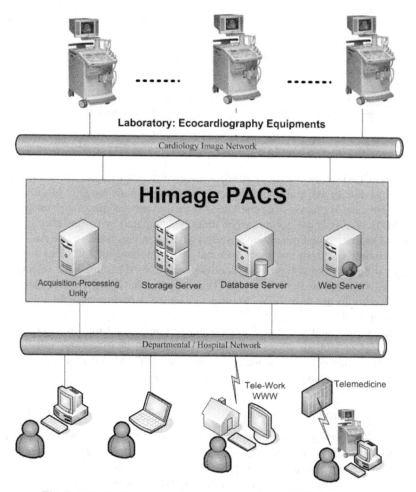

**Fig. 3.** CHVNG Cardiology Department: Himage PACS Infra-structure

## 3  Results

With this new Himage system, more than 12000 studies have been performed so far. For example, a typical Doopler color run (RGB) with an optimized time-acquisition (15-30 frames) and a sampling matrix (480*512), rarely exceeds 200-300kB. Typical compression ratios can go from 65 for a single cardiac cycle sequence to 100 in multi- cycle sequences. With these averaged figures, even for a heavy work-load Echolab, it is possible to have all historic procedures online or distribute them with reduced transfer time over the network, which is a very critical issue when dealing with costly or low bandwidth connections. According to the results of user's assessment, the achieved compression ratios do not compromise diagnostic quality, and reduce significantly the waiting time to download and display of images.

In Table 1 it is possible to observe some global statistics.

**Table 1.** Himage - three years of statistics

| Number of Procedures (Exams) | 12673 |
|---|---|
| N° of Images (Still + Cine-loops) | 253173 |
| Total of Space Volume | 48.934.047.237 Bytes (45,5GB) |
| Average Procedure Size | 3771 KBytes |
| Average File Size | 189 KBytes |
| Average number of Files/Procedure | 20 (majority of cine-loops) |

### 3.1  Image Quality Evaluation

The new DICOM-MPEG4 *private transfer syntax* offers an impressive trade off between image quality and compression factors spanning a wide range of bit-rates. Two studies were carried on assessing the DICOM cardiovascular ultrasound image quality of MPEG4 (768kbps) format when blindly compared with the uncompressed originals. Qualitative and quantitative results have been presented in [8]. An impressive fact coming out from this study was that, in a simultaneous and blind display of the original against the compressed cine-loops, 37% of the trials have selected the compressed sequence as the best image. This suggests that other factors related with viewing conditions are more likely to influence observer performance than the image compression itself.

The quantitative and qualitative assessment of compressed images shows us that quality is kept at high standards without ever impairing its diagnostic value. Using compression factors of the same magnitude in other DICOM coding standards JPEG [9] MPEG1 [10] will lead to a severe decrease on image quality.

### 3.2  Data Access and Management

Information Technologies (IT) dissemination and availability are imposing new challengers to this sector. The healthcare professionals are demanding new Web-based access platforms for internal (Intranet) consumption but also to support cooperative Internet based scenarios. Consequently, the digital medical images must be easily portable and enable integration with actual multimedia Web environments [8, 11]. The Himage client software answers to theses requirements: - first, all client modules are available in Web environment; - second, the image size volume (mean: 189 Kbytes) has acceptable download times, even to a remote access like, for instance, a telework session using an ADSL Internet connection; - third, easily integration with departmental Healthcare Information System.

As suggested in Fig. 4, the Web client module includes several facilities, besides the customization to handle DICOM 3.0 and our DICOM *private syntax*. Graphically, the main window includes a grid box with the patient's in the Himage database and a simultaneous cine preview of three sequences. The user can "search" procedures by different criterions: patient name; patient ID, procedure ID/type/date and provenience Institutions/equipment. A second graphical application layer provides the communications, DICOM viewer (Fig. 5) and report modules (Fig. 6).

In the Himage DICOM viewer window (Fig. 5) it is possible to visualize the image sequences (still and cine-loops), select image frames (from distinct sequences) that are sent to the report area. Other traditional specific functionalities have been included like, for example, the image manipulation tools (contrast/brightness), the printing capacities and the exportation of images in distinct formats (DICOM3.0, AVI, BMP, Clipboard...).

In the report module (Fig. 6), the user can arrange the images location or delete some frames with a drag-and-drop functionality. At the end, the output images matrix (2x3) is used jointly with the clinician report to generate a RTF file, compatible with any common text editor.

The export module makes possible to record a procedure in a CDROM/DVD-ROM in uncompressed DICOM *default transfer syntax* format and in AVI format.

The Himage client was developed in ASPX .NET and includes components in Visual Basic .NET, JavaScript and a binary object (ActiveX Viewer) that allows the direct integration of the private DICOM files in the Web contents.

### 3.2.1 Communication Platform

As referenced, the huge volumes of cardiac US medical image data are not easy to transfer in a time and cost-effective way. Healthcare professionals do not adopt telemedicine or telework platforms if they need to wait, for instance, 2-3 hours to receive/download a clinical image study with diagnostic quality.

The reduced image size of DICOM-MPEG4 private syntax US procedures prompts opportunities for new clinical telematic applications, especially in scenarios with low network bandwidth and/or reduced communications budget.

The Himage includes a communication platform that allows the creation of customized study packages, with more than one study, to be sent to predefined remote institutions. With this facility the physician can select images sequences, make annotations, append extra comments and send the examination over the network to the remote unit. In the main Himage interface exhibits a group of buttons ("on/off" and "send") that allows the (de)activation and the selection of sequences that are sent to the communication module. In this last area, the user just needs to choose the target institution and send the data following the email paradigm.

The first telemedicine project has established with Cardiology Department of the Central Hospital of Maputo (Mozambique) [4]. Both clinical partners were equipped with echocardiography machines including standard DICOM output interfaces, videoconference platforms. A network infrastructure was also installed with two server computer in the African partner. The sessions can be conference-based teleconsultations for analysis of real-time clinical cases, or analysis of previously transferred exams files (DICOM *private transfer syntax*). In the last case the remote physician can select echo image sequences, make annotations, append extra comments and send to the Gaia Hospital. This information is stored in one dedicated server and is accessible by the Gaia cardiovascular ultrasound specialists.

Both methods can be used at same time, the physician in Gaia can open and examine a clinic record file previously sent by the remote place, and use videoconferencing equipment for a face to face consultation and comment about diagnosis and therapy. The communication relies on 2x64 Kbps ISDN channels (1 full US exam takes typically 2-5 min using a 64kb channel).

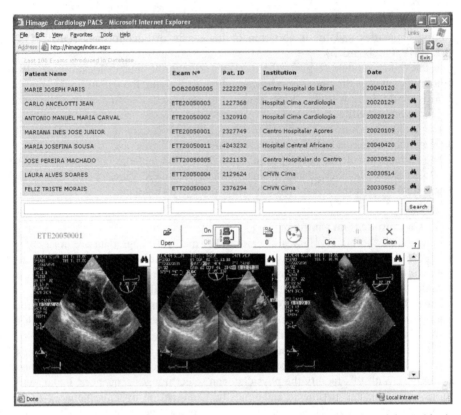

**Fig. 4.** Himage Web client: database, preview and buttons to second level modules (with virtual names)

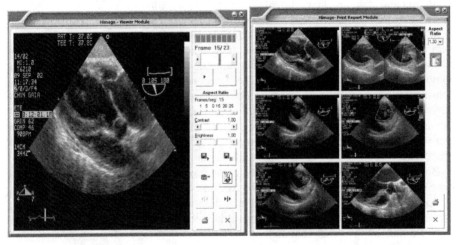

**Fig. 5.** Himage Visualization Module          **Fig. 6.** Himage - Report Module

# 4  Conclusions

This paper presents and describes a totally home made Web PACS solution that is based on a customized DICOM encoding syntax and compression scheme in a Cardiology medical image scenario. The developed software provides a seamless integration with the traditional DICOM entities, boosts the storage capacity up to the level of several years of online availability while preserving the image quality at high standards. The overall result is a cost-effective system with improved workflow and generalized user satisfaction.

The Himage development was the core element that allows us to implement the first fully digital Echocardiography Laboratory in Portugal. Moreover, the presented Himage-PACS characteristics make it particularly important in telemedicine communication scenarios with low network bandwidth and/or reduced communications budget like, for instance, the Maputo-Gaia project. Thus, as a final overview the following remarks about the Himage-PACS software environment are worth to be pointed out: a) Excellent trade-off between compression ratio and diagnostic quality; b) Availability of all the recorded procedures in Web environment; b) Reduced download times that enables data access outside the institutions wall (telework); d) Huge storage space savings. Very reduced average storage cost per study.

# References

1. DICOM, *Digital Imaging and Communications in Medicine version 3.0*, ACR (the American College of Radiology) and NEMA (the National Electrical Manufacturers Association). http://medical.nema.org/dicom/2003.html.
2. Costa, C., *PhD Thesis: "Concepção, desenvolvimento e avaliação de um modelo integrado de acesso a registos clínicos electrónicos"*, in *DET*. 2004, Aveiro University. p. 250.
3. Silva, A., C. Costa, et al. *A Cardiology Oriented PACS*. in *Proceedings of SPIE*. 1998. San Diego - USA.
4. Costa, C., A. Silva, et al., *A Transcontinental Telemedicine Platform for Cardiovascular Ultrasound*. Technology and Health Care - IOS Press, 2002. **10**(6): p. 460-462.
5. Huffman, D.A., *A method for the construction of minimum redundancy codes*. Proceedings IRE, 1952. **40**: p. 1098-1101.
6. Maede, A. and M. Deriche, *Establishing perceptual limits for medical image compression*. Proceedings of SPIE - Image Perception and Performance, 2001. **4324**: p. 204-210.
7. Soble, J.S., *MPEG Digital Video Compression*. Digital Cardiac Imaging in the 21st Century: A Primer, 1997: p. 192-197.
8. Costa, C., A. Silva, et al. *Himage PACS: A New Approach to Storage, Integration and Distribution of Cardiologic Images*. in *PACS and Imaging Informatics - Proceedings of SPIE*. 2004. San Diego - CA - USA.
9. Karson, T., S. Chandra, et al., *JPEG Compression Echocardiographic Images: Impact on Image Quality*. Journal of the American Society of Echocardiography, 1995. **Volume 8 - Number 3**.
10. Thomas, J., S. Chandra, et al., *Digital Compression of Echocardiograms: Impact on Quantitative Interpretation of Color Doppler Velocity*. Digital Cardiac Imaging in the 21st Century: A Primer, 1997.
11. Costa, C., J.L. Oliveira, et al., *A New Concept for an Integrated Healthcare Access Model*. MIE2003: Health Technology and Informatics - IOS Press, 2003. **95**: p. 101-106.

# A Framework Based on Web Services and Grid Technologies for Medical Image Registration

Ignacio Blanquer[1], Vicente Hernández[1], Ferran Mas[1], and Damià Segrelles[2]

[1] Departamento de Sistemas Informáticos y Computación,
Universidad Politécnica de Valencia, Camino de Vera S/N, 46022 Valencia, Spain
{iblanque,vhernand,fmas}@dsic.upv.es
[2] Instituto de Aplicaciones de las Tecnologías de la Información y de las
Comunicaciones Avanzadas, Universidad Politécnica de Valencia,
Camino de Vera S/N, 46022 Valencia, Spain
dquilis@itaca.upv.es

**Abstract.** Medical Imaging implies executing complex post-processing tasks such as segmentation, rendering or registration which requires resources that exceeds the capabilities of conventional systems. The usage of Grid Technologies can be an efficient solution, increasing the production time of shared resources. However, the difficulties on the use of Grid technologies have reduced its spreading outside of the scientific arena.

This article tackles the problem of using Grid Technologies for the co-registration of a series of volumetric medical images. The co-registration of time series of images is a needed pre-processing task when analysing the evolution of the diffusion of contrast agents. This processing requires large computational resources and cannot be tackled efficiently on an individual basis. This article proposes and implements a four-level software architecture that provides a simple interface to the user and deals transparently with the complexity of Grid environment. The four layers implemented are: Grid Layer (the closest to the Grid infrastructure), the Gate-to-Grid (which transforms the user requests to Grid operations), the Web Services layer (which provides a simple, standard and ubiquitous interface to the user) and the application layer.

An application has been developed on top of this architecture to manage the execution of multi-parametric groups of co-registration actions on a large set of medical images. The execution has been performed on the EGEE Grid infrastructure. The application is platform-independent and can be used from any computer without special requirements.

## 1 Introduction and State of the Art

The current use of Internet as main infrastructure for the integration of information through web based protocols (i.e. HTTP, HTTPS, FTP, FTPS, etc.), opened the door to new possibilities. The Web Services (WS) [1] are one of the most consolidated technologies in web environments, fundamented on the Web Services Description Language (WSDL). WSDL defines the interface and constitutes a key part of the Universal Description Discovery and Integration

J.L. Oliveira et al. (Eds.): ISBMDA 2005, LNBI 3745, pp. 22–33, 2005.

(UDDI) [2], which is an initiative for providing directories and descriptions of services for e-business. WSs communicate through the Simple Object Access Protocol (SOAP) [3], a simple and decentralized mechanism for the exchange of typed information structured in XML (Extended Mark-up Language).

As is defined in [4] a Grid provides an abstraction of resources for sharing and collaborating through different administrative domains. These resources can be hardware, data, software and frameworks. The key concept of Grid is the Virtual Organization (VO) [5]. A VO is defined as temporal or permanent set of entities, groups or organizations that provide or use resources. The usage of Grid Computing is currently in expansion. This technology is being introduced in many different application areas such as Biocomputing, Finances or Image Processing, along with the consolidation of its use in more traditional areas such as High Energy Physics or Geosciences.

In this process of development, many basic middlewares such as the different versions of Globus Toolkit [6] (GT2, GT3, GT4), Unicore [7] or InnerGrid [8] have arisen. At present, Grid technologies are converging towards Web Services technologies. The Open Grid Services Architecture (OGSA) [9] represents an evolution in this direction. OGSA seems to be an adequate environment for obtaining efficient and interoperable Grid solutions, some issues (such as the security) still need to be improved. Globus GT3 implemented OGSI (Open Grid Service Infrastructure) which was the first implementation of OGSA. OGSI was deprecated and substituted by the implementation of OGSA by the Web Services Resource Framework (WSRF) [10] in GT4. WSRF is totally based in WSs.

Although there exist newer versions, Globus GT2 is a well-stablished batch basic Grid platform which has been extended in several projects in a different way in which GT3 and GT4 have evolved. The DATAGRID project [11], developed the EDG (European Data Grid) , a Middleware based on GT2, which improved the support of distributed storage, VO management, job planning and job submission. The EDG middleware has been improved and extended in the LCG (Large Hadron Collider Computing Grid) [12] and Alien Projects to fulfil the requirements of the High Energy Physics community. Another evolution of the EDG is gLite [13], a Grid Middleware based in WS and developed in the frame of the Enabling Grids for E-sciencE (EGEE) [14]. gLite has extended the functionality and improved the performance of critical resources, such as the security, integrating the Virtual Organisation Membership System (VOMS) [15] for the management of VOs. VOMS provides information on the user's relationship with his/her Virtual Organization defining groups, roles and capabilities.

These middlewares have been used to deploy Grid infrastructures comprising thousands of resources, increasing the complexity on the usage of the Grid. However, the maturity of these infrastructures in terms of user-friendliness is not sufficient yet. Configuration and maintenance of the services and resources or fault tolerance is hard even for experimented users. Programming Grid applications usually involve a non-trivial degree of knowledge of the intrinsic structure of the Grid. This article presents a software architecture that abstracts the users from the management of Grid environments by providing a set of simple services.

Although the architecture proposed is open to different problems, this article shows the use for the implementation of an application for the co-registration of medical images. This application is oriented to either medical end-users and researchers for the execution of large numbers of runs using different values for the parameters that control the process. Researchers can tune-up the algorithms by executing larger sets of runs, whereas medical users can obtain the results without requiring powerful computers.

The application does not require a deep knowledge of the Grid environments. It offers a high-level user-friendly interface to upload data, submit jobs and download results without requiring to know the syntax of commands, Job Description Language (JDL) data and resource administration or security issues.

## 2   Motivation and Objectives

The final objective is to provide the users with a tool that could ease the process of selecting the best combination of parameters that produces the best results for the co-registration, and to provide high-quality co-registered images from the initial set of data. This tool will use the Grid as a source of computing power and will offer a simply and user-friendly interface to deal with.

The tool enables the execution of large sets of co-registration actions varying the values of the different parameters, easing the process of transferring the source data and the results. Since Grid concept is mainly batch (and the co-registration is not an interactive process due to its long duration), it must provide with a simply way to monitor the status of the processing. Finally the process must be achieved in the shorter time possible, considering the resources available.

### 2.1   Pharmacokinetic Modelling

The pharmacokinetic modelling of the images obtained after a quick administration of a bolus of extracellular gadolinium chelates contrast can have a deep impact on the diagnosis and the evaluation of different pathogen entities.

Pharmacokinetic models are designed to forecast the evolution of an endogenous or exogenous component on the tissues. To follow-up the evolution of the contrast agent a sequence of MRI volumetric images is obtained at different times following the injection of contrast. Each of these images comprises a series of image slices that cover the body part explored. Since the whole process takes a few minutes, images are obtained in different break-hold periods. This movement of the patient produces artefacts that make images directly incomparable. This fact is even more important in the area of the abdomen, which is strongly affected by the breathing and the motility of the organs.

The study of pharmacokinetic models for the analysis of hepatic tumours is an outstanding example of the above. A prerequisite for the computation of the parameters that govern the model is the reduction of the deformation of the organs in the different images obtained. This process can be performed by co-registering all the volumetric images with respect to the first one.

## 2.2   Co-registration

The co-registration of images consists on aligning the voxels of two or more images in the same geometrical space by using the necessary transformations to make the floating images as much as possible similar to the reference image.

In general terms, the registration process could be rigid or deformable. Rigid registration only uses affine transformations (displacements, rotations, scaling) to the floating images. Deformable registration enables the use of elastic deformations on the floating images. Rigid registration introduces fewer artefacts, but it can only be used when dealing with body parts in which the level of internal deformation is lower (e.g. the head). Deformable registration could introduce unrealistic artefacts, but is the only one that could compensate the deformation of elastic organs (e.g. in the abdomen).

Image registration can be applied in 2D (individually to each slice) or in 3D. Registration in 3D is necessary when the deformation happens in the three axes.

## 2.3   Post-processing

Although the co-registration of images is a computationally complex process which must be performed before the analysis of the images, it is not the only task that needs high performance computing. Extracting the parameters that define the model and computing the transfer rates for each voxel in the space will require large computing resources. The platform implemented has been designed to cope with following post-processing in the same way.

# 3   Architecture

The basic Grid middleware used in this architecture is the LCG, developed in the LHC Computing Grid Project, which has a good support for high throughtput executions. A four-layered architecture has been developed to abstract the operation of this middleware. The registration application has been implemented on top of this architecture.

Medical data is prone to abuse and need careful treatment in terms of security and privacy. This is even more important when the data has to flow from different sites. It is crucial both to preserve the privacy of the patients and to ensure that people accessing the information are authorised to do so. The way in which this environment guarantees the security is explained in detail in Section 3.2.

## 3.1   Layers

As it has been mentioned in previous sections, the development of this framework has been structured into four layers, thus providing a higher level of independence and abstraction from the specificities of the Grid and the resources. The following sections describes the technology and the implementation of each layer. Figure 1 shows the layers of the proposed architecture.

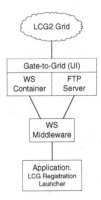

**Fig. 1.** The proposed architecture

**Grid Layer.** The system developed in this work makes use of the computational and storage resources being deployed in the EGEE infrastructure along a large number of computing centres distributed among different countries. EGEE currently uses LCG, although there are plans for migrating to gLite. This layer offers the "single computer" vision of the Grid through the storage catalogues and workload management services that tackle with the problem of selecting the rightmost resource. Figure 2 shows a sample LCG2 Grid structure.

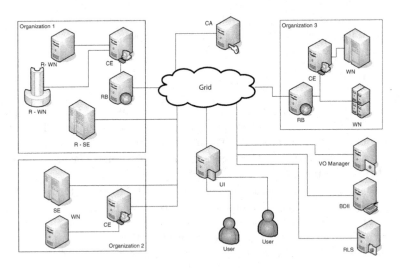

**Fig. 2.** A LCG2 Grid sample scheme

The LCG2 Grid middleware comprises the following elements:

– IS-BDII: Information Service - Berkeley Database Information Index, this element provides the information about the Grid resources and their status.
– CA: Certification Authority, the CA signs the certificates from both resources and users.

- CE: Computing Element, it is defined as a queue of Grid jobs. A Computing Element is a farm of homogeneous computing nodes called Worker Nodes.
- WN: Worker Node, is a computer in charge of executing jobs.
- SE: Storage Element, a SE is a storage resource in which a task can store data to be used by the computers of the Grid.
- RC, RLS: Replica Catalogue, Replica Location Service, are the elements that manage the location of the Grid data.
- RB: Resource Broker, the RB performs the load balancing of the jobs in the Grid, deciding in which CEs the jobs will be launched.
- UI: User Interface, this component is the entry point of the users to the Grid and provides a set of commands and APIs that can be used by the programs to perform different actions on the Grid.

The Job Description Language (JDL) is the way in which jobs are described in LCG. A JDL is a text file specifying the executable, the program parameters, the files involved in the processing and other additional requirements.

A description of the four layers of the architecture is provided along the following subsections.

**Gate-to-Grid Layer.** The Gate-to-Grid Layer constitutes the meeting point between the Grid and the Web environment that is intended to be used as user interface. In this layer there are WSs providing the interaction with the Grid similarly as if the user were directly logged in the UI. The WSs are deployed in a Web container in the UI which provides this mediation.

The use of the Grid is performed through the UI by a set of scripts and programs which have been developed to ease the task of launching executions and managing group of jobs. The steps required to execute a new set of jobs in the Grid are:

1. A unique directory is created for each parametric execution. This directory has separated folders to store the received images to be co-registered, the JDL files generated and the output files retrieved from the jobs. It also includes several files with information about the jobs, such as job identifiers and parameters of the registration process.
2. Files to be registered are copied to a specific location in this directory.
3. For each combination of parameters and pair of volumes to be registered, a JDL file filled-in with the appropriate values is generated.
4. Files needed by the registration process are copied in the SE and registered in the RC and the RLS.
5. The jobs are submitted to the Grid through an RB that selects the best available CE according to a predefined criteria.
6. Finally, when the job is done and retrieved, folders and temporal files are removed from the UI. The files registered in the SE that are no longer needed are also deleted.

The different Grid Services are offered through the scripts and programs aforementioned. These programs work with the UI instructions in order to ease the tasks for job and data management. The access to these programs and scripts is remotely available through the WSs deployed in the UI. The copying of the input files from the computer where the user is located to the UI computer is performed through FTP (File Transfer Protocol).

The most important WSs offered in the UI are:

**InitSession.** This service is in charge of creating the proxy from the Grid user certificates. The proxy is then used in the Grid environment as a credential of that user, providing a single sign-on for the access to all the resources.

**GetNewPathExecution.** As described before, jobs launched in a parametric execution (and not yet cleared) will have in the UI their own group folder. This folder has to be unique for each group of jobs. This service will get a unique name for each job group and it will create the directory tree to manage that job execution. This directory will store the image, logs, JDLs and other information files.

**Submit.** The submit call starts an action that carries on the registration of the files from the UI to the SE, creates the JDLs according to the given registration parameters and the files stored on the specified directory of the UI. It finally submits the jobs to the Grid using the generated JDL files.

**GetInformationJobs.** This service gets information about the jobs belonging to the same execution group. The information retrieved by this call is an XML document with the job identifiers and the associated parameters.

**CancelJob.** This call cancels a single job (part of an execution group). The cancellation of a job implies the removal of the registered files on the Grid.

**CancelGroup.** This service performs the cancellation of all the jobs launched to the Grid from a group of parametric executions. As in the case of CancelJob, the cancellation of jobs implies the removal of its associated files from the SEs. Moreover, in this case the temporal directory created on the UI is also removed when all the jobs are cancelled.

**GetJobStatus.** This service informs about the status of a job in the Grid, given the job identifier. The normal sequence of states of a job is: submitted, waiting, ready, scheduled, running, done and cleared. Other possible states are aborted and cancelled.

**PrepareResults.** This service is used to prepare the results of an execution before downloading them. When a job finishes, the resulting image, the standard output and the standard error files can be downloaded. For this purpose the *PrepareResults* service retrieves the results from the Grid and stores them in the UI.

The executable must exist in the UI system and it has to be statically compiled so that it can be executed without library dependencies problems in any machine of the Grid. The implemented registration of this project is based on the Insight Segmentation and Registration Toolkit (ITK) [16] software library. ITK is an Open Source software library for image registration and segmentation.

**Middleware Web Services Layer.** The Middleware Web Services Layer provides an abstraction to the use of the WSs. The abstraction of the WSs Layer has two purposes. On one hand, to create a unique interface independent from the application layer, and on the other hand to provide methods and simple data structures to ease the development of final applications.

The development of a separate software library for the access to the WSs will ease future extensions for other applications that share similar requirements. Moreover it will enable introducing optimizations in this layer without necessarily affecting the applications developed on top of this layer.

Moreover, this layer offers a set of calls based on the Globus FTP APIs to perform the data transferring with the Gate-to-Grid Layer. More precisely, the abstraction of the WSs lies, in first place, on hiding the creation and management of the necessary stubs for the communication with the published WSs. In second place, this layer manages the data obtained by the WSs by means of simple structures closer to the application. From each of the available WSs in the Gate-to-Grid layer there exists a method in this layer that gets the information in XML given by the WSs and returns that information in basic types or structured objects which can be managed directly by the application layer.

**Application Layer.** This layer is the one that offers the graphical user interface which will be used for the user interaction. This layer makes use of functions, objects and components offered by the middleware WS layer to perform any operation on the Grid.

The developed tool has the following features available:

– Parameter profile management. The management of the parameters allows creating, modifying and removing configurations of parameters for the launching of multi-parametric registrations. These registration parameters are defined as a rank of values, expressed as by three values: initial value, increment and final value. The profiles can be loaded from a set of templates or directly filled-in before submitting the co-registrations according to these parameters.
– Transferring of the volumetric images that are going to be registered. The first step is to upload the images that are going to be registered. The application provides the option to upload the reference image and the other images that will be registered. These files are automatically transferred to the UI to be managed by the Gate-to-Grid layer.
– Submission of the parametric jobs to the Grid. For each combination of the input parameters, the application submits a job belonging to the same job group. The user can assign a name to the job group to ease the identification of jobs in the monitoring window.
– Job monitoring. The application offers the option to keep track of the submitted jobs for each group.
– Obtaining the results. When a Grid execution has reached the state *done*, the user can retrieve the results generated by the job to the local machine. The results include the registered image, the standard output and standard error generated by the program launched to the Grid. The user can also download the results from a group of jobs automatically.

## 3.2  Security

When dealing with medical data, the security of information being managed is very important to guarantee the privacy of the patients. Among the personal data, medical data is especially critical due to the implications that can have the release of the information for its owner.

The use of Grid technologies in health implies actions that might put the privacy of medical data in risk. Processing medical data in a distributed environment implies the transference of the data to remote resources. Although transferences can be performed through secure protocols, the data could be accessed by users with the sufficient privileges in the remote systems. However, in the specific action of this work, the information related to the patient is removed from the headers of the images and only the raw images are sent, identified through a coded name that do not share any information with the patient data. So the privacy in not comprised by anonimizing the images.

Regarding the access to the system, the different layers of the architecture defined in the Section 3 have different approaches in the implementation of the security. Basically, it can be divided into two parts: One part is related to the WS environment and other part will deal with the Grid-LCG environment. In both cases secure protocols are used for the communication.

In relation to the scope of the security of WSs, the architecture defines a client layer (middleware WS) for interacting with the system. For WSs, the SOAP protocol is used on top HTTPS, which is based in Secure Sockets Layer (SSL). The HTTPS protocol guarantees the privacy of the data and the use of digital certificates guarantees authentication of user.

The Grid middleware used (LCG), provides a native infrastructure of security Grid Security Infrastructure (GSI). GSI is also based in SSL. Before accessing any resource of the Grid, a proxy must be created from the client certificate, which should have been. The certificate is duly signed by the Certificate Authority (CA). Each resource of the Grid establishes a mapping of the Distinguish Name (DN) obtained from the proxy generated, to a resource user. Finally each deployed resource in the Grid is certificated by a valid CA.

The security in the different environments of the architecture (Web and Grid) has already been defined. The following lines try to explain how the connection between the Grid and Web security environments is performed. In the described architecture, the Gate-to-Grid is the common point between the Web and Grid environment. A mapping system of the Web users (through Web user certificate) has been implemented in the Gate-to-Grid layer that associates the Web users with the Grid users. For each user, a Grid proxy is created from its Grid user certificate.

## 4  Results

The first result of this work has been the *LCG Registration Launcher* tool, which has been developed on top of the architecture described in this article. Figure 4 shows two screenshots of the application, one showing the panel for uploading

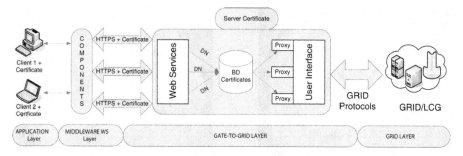

**Fig. 3.** Security scheme of the proposed architecture

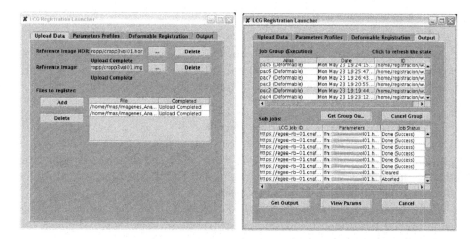

**Fig. 4.** Screenshots of the LCG registration launcher application

reference and floating volumes, and the other one showing the panel for the monitoring of the launched jobs.

The required time to perform a registration of a volumetric image in a PIII at 866 Mhz with 512 MB of RAM is approximately 1 hour and 27 minutes.

By one hand, when launching 12 simultaneous co-registrations using the EGEE Grid resources (for the biomed virtual organization) and giving the RB the ability to choose among the system resources, it took approximately 3 hours and 19 minutes. In this case it must be considered that the resources where shared by other Grid users. On the other hand, when just using resources with a lower degree of utilization, the total time for the 12 jobs was of 2 hours and 53 minutes (this case was using a CE with 20 WNs).

If a the same resources were used in a batch processing approach, using the same resources and running manually the jobs on the computing farm, the computing time would be a 8% shorter. The overhead of Grids is due to the use of secure protocols, remote and distributed storage resources and the scheduling overhead, which is in the order of minutes due to the monitoring policies which are implemented in a poll fashion.

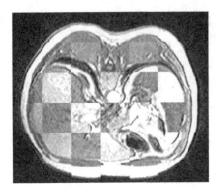 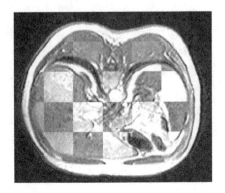

**Fig. 5.** Results of the co-registration before (left) and after (right)

Regarding the results obtained, Figure 5 shows a tiled composition of two images before the co-registration (left) and after the process (right). The figure clearly shows the improvement in the alignment of the voxels of the image.

## 5   Conclusions and Further Work

The final application developed in this work offers an easy-to-use high level interface that allows the use of the LCG2-based EGEE Grid infrastructure for image co-registration by Grid-unaware users.

With the use of the tool described in this work, the user achieves a large computational performance for the co-registration of radiological volumes and the evaluation of the parameters involved.

The Grid is an enabling technology that provides the clinical practice of processes with processes that, by its computational requirements, were not feasible with a conventional approach. It also offers a high throughput platform for medical research. The proposed architecture is adaptable to different platforms and enables the execution of different applications changing the user interface.

This work is a starting point for the realization of a middleware focused on the abstraction of the Grid to ease the development of interfaces for the submission of complex jobs to the Grid.

The application developed is a part of larger project. As introduced in Section 2, the co-registration application is a first step of the pharmacokinetic model identification. The next step to be treated will be the extraction of the pharmacokinetic model from a set of acquisitions. For this task, it was necessary to have the co-registration tool that has been developed in this work. Finally, the middleware WS layer will be enlarged to give support to new functionalities related with the extraction of the pharmacokinetic model. Finnally, the application currently supports the Analyze format [17], although the extension to other formats such as DICOM [18] is being considered among the priorities.

## Acknowledgements

The authors wish to thank the financial support received from The Spanish Ministry of Science and Technology to develop the project *Investigación y Desarrollo de Servicios GRID: Aplicación a Modelos Cliente-Servidor, Colaborativos y de Alta Productividad*, with reference TIC2003-01318. This work has been partially supported by the Structural Funds of the European Regional Development Fund.

## References

1. Scott Short. *Creación de servicios Web XML para la plataforma .NET*. Mc-Graw-Hill, 2002.
2. Universal Description, Discovery and Integratin (UDDI). http://www.uddi.org.
3. Simple Object Access Protocol (SOAP). http://www.w3c.org.
4. Expert Group Report. *Next Generation Grids 2*. European Commission, 2004. http://www.cordis.lu/ist/grids/index.htm.
5. Foster I. and Kesselman C. *The GRID: Blueprint for a New Computing Infraestructure*. Morgan Kaufmann Publishers, Inc., 1998.
6. I. Foster and C. Kesselman. Globus: A metacomputing infraetructure toolkit. *Intl J. Supercomputer Applications*, pages 115,128, 1997.
7. Forschungszentrum. *Unicore Plus Final Report*. Dietmar Erwin, 2003.
8. *InnerGrid Users' manual*. GridSystems, 2003.
9. I. Foster, C. Kesselman, J. Nick, and S. Tuecke. The physiology of the grid: An open grid services architecture for distributed systems integration, 2002.
10. The Globus Alliance. http://www.globus.org.
11. The DATAGRID project. http://www.eu-datagrid.org.
12. LHC Computing Grid. http://lcg.web.cern.ch/LCG.
13. gLite. Lightweight Middleware for Grid Computing. http://glite.web.cern.ch/glite.
14. Enabling Grids for E-sciencE. http://www.eu-egee.org.
15. Virtual Organization.
    http://hep-project-grid-scg.web.cern.ch/hep-project-grid-scg/voms.html.
16. Ibanez, Schroeder, Ng, and Cates. *The ITK Software Guide*. Kitware Inc.
17. Analyze 7.5 File Format. Mayo Clinic.
18. National Electrical Manufacturers Association. Digital Imaging and Communications in Medicine (DICOM). 1300 N. 17th Street, Rosslyn, Virginia 22209 USA.

# Biomedical Image Processing Integration Through INBIOMED: A Web Services-Based Platform

David Pérez del Rey[1], José Crespo[1], Alberto Anguita[1],
Juan Luis Pérez Ordóñez[2], Julián Dorado[2], Gloria Bueno[3],
Vicente Feliú[3], Antonio Estruch[4], and José Antonio Heredia[4]

[1] Biomedical Informatics Group, Artificial Intelligence Department,
School of Computer Science, Universidad Politécnica de Madrid
Campus de Montegancedo, s/n. 28660 Boadilla del Monte, Madrid
dperez@infomed.dia.fi.upm.es
[2] RNASA-IMEDIR (Lab. Redes de Neuronas Artificiales y Sistemas Adaptativos
Centro de Informática Médica y Diagnóstico Radiológico).Universidade da Coruña
julian@udc.es
[3] UCLM-ISA, E.T.S.I. Industriales, Ingeniería de Sistema y Automática
Universidad de Castilla-La Mancha – Av. Camilo José Cela, s/n – 13071 Ciudad Real
gloria.bueno@uclm.es
[4] Departamento de Tecnología, Universitat Jaume I, Castellón

**Abstract.** New biomedical technologies need to be integrated for research on complex diseases. It is necessary to combine and analyze information coming from different sources: genetic-molecular, clinical data and environmental risks. This paper presents the work carried on by the INBIOMED research network within the field of biomedical image analysis. The overall objective is to respond to the growing demand of advanced information processing methods for: developing analysis tools, creating knowledge structure and validating them in pharmacogenetics, epidemiology, molecular and image based diagnosis research environments. All the image processing tools and data are integrated and work within a web services-based application, the so called INBIOMED platform. Finally, several biomedical research labs offered real data and validate the network tools and methods in the most prevalent pathologies: cancer, cardiovascular and neurological. This work provides a unique biomedical information processing platform, open to the incorporation of data coming from other feature disease networks.

## 1 Introduction

INBIOMED is a cooperative research network on biomedical informatics, founded by the "Fondo de Investigación Sanitaria" (FIS) of the Institute of Health Carlos III. The research network is integrated by 13 groups of biomedical research, belonging to 11 research centres. INBIOMED provides storage, integration and analysis of clinical, genetic, epidemiological and image data oriented to pathology research. Its main objective is to develop analysis tools and to evaluate them in their corresponding research environments [1].

This paper is centred on the image data treatment within INBIOMED, and how a new developed, and so called INBIOMED platform, has been used to tackle these

J.L. Oliveira et al. (Eds.): ISBMDA 2005, LNBI 3745, pp. 34–43, 2005.

challenges in the biomedical image processing field. This platform permits the integration and reuse of image processing methods developed by different INBIOMED partners the widely adopted technology of web services [2].

In the last years, the World Wide Web growth has been increasing more and more, with a big amount of research and efforts to create models and tools for new applications. A new application concept has appeared together with this growth, web services. Web services are applications executed in remote server computers and they may be accessed by clients from any computer located in other place. These users can access the web services through a minimum software client or even none, since many of these applications are accessible using a regular web browser. In the Internet there exist many applications available in different areas that may be accessed by users. It is not necessary to code again these applications when this functionality is required for a new application. Therefore, it is important to be able to reuse these services to be integrated in other applications (at a lower scale, this was already present in the object-oriented approach). These challenges are common to many working areas and also to Biomedicine where the INBIOMED project is enclosed.

Image processing methods are one of the main application fields within Biomedical Informatics. Biomedical image processing is widely needed by professionals and researchers and may require important computation capabilities to reach satisfactory results. Therefore, the use of distributed applications based on web services is especially appropriate for these functionalities. Firstly, these systems are used by professionals working in centres spread over the country, and they use computers with different operating systems, capabilities, etc. and web services overcome these heterogeneities. In addition it is easier to maintain the application if it is centralized than if each user have to update its software when a modification is introduced. Nowadays the research in the image processing field is intensive and new techniques or modifications of existing ones appear periodically. Adding new modifications or even new methods using web services is straight forward, providing the users with the most advanced techniques instantaneously.

## 2  Background

Image processing usually require powerful systems with expensive hardware, because of the huge computation resources needed. In addition, many image processing problems had been solved and implemented before. Distributed image processing provides a framework to overcome both problems. On the one hand, time consuming processes can be performed using remote and powerful servers in a transparent way. On the other hand, these technologies may supply a system architecture to reuse image processing methods already implemented.

A study of the main technologies that support distributed processing has been accomplished within this work. Table 1 presents a list of some of these technologies together with a brief description, references and some advantages and drawbacks.

Although GRID follows a different approach than the rest of the technologies listed in Table 1, it has been included in the study due to its widespread among distributed computing. While GRID issues a task to a set of different computers (any node may run any part of the process), the other technologies can not perform that load balancing in a transparent way for the user.

**Table 1.** Distributed processing technologies

| Technology | Description | Advantages | Drawbacks |
|---|---|---|---|
| **Java RMI (Remote Method Invocation)** | Method invocation of remote java objects. Enables an object on one Java virtual machine to invoke methods on an object in another Java virtual machine [3] | Object oriented and very high level. Hides communication details. No need to learn a new IDL language [4], [5] | Constrained to java language, slow for complex tasks |
| **CORBA (Common Object Request Broker Architecture)** | Technology for creation of distributed object architectures. The core of its architecture is the Object Request Broker (ORB), the bus over which objects transparently interact with other objects [6] [7] | Object oriented. Allows communication between different operating systems and programming languages | You need to know the intermediate language of definition service (IDL). High complexity. Variations in vendor implementations. |
| **DCOM (Distributed Component Object Model)** | Distributed extension to COM (Component Object Model). Builds an object remote procedure call (ORPC) layer on top of DCE RPC [8] to support remote objects | Object oriented. Allows communication between different programming languages | Fully supported only in Microsoft Windows platforms. Architecture is not well-partitioned [9] |
| **RPC (Remote Procedure Calls)** | Protocol that enables programs to invoke procedures located in remote processes | Highly efficient | Not object oriented, low level. Different RPC systems can not communicate with each other [10] |
| **Web Services** | Service with a well defined and known interface that can be accessed through Internet [11] | Uses HTTP protocol for message transport. Communication based in XML and SOAP standards, avoids many incompatibilities | API is not fully standardized yet |
| **Mobile Agents** | Software entities capable of autonomous movement between hosts in a computer network [12] | Suitable for working with large amounts of data located in different places | Possible security issues. Code has to move to others computers. All computers involved must use the same language |
| **GRID Computing** | Distributed set of hardware and software resources that provided coordinated computing support [13] | Ideal for working with large amounts of data that require complex computing. Shows a set of computer as a single resource | Less useful working with small data. Requires complex software infrastructure. Big amount of bandwidth can be required [14] |

Distributed image processing systems may be implemented using any of these technologies, because all of them allow remote communication between clients and processing methods in different computers – e.g. systems using CORBA [15], or Web-Based Medical Image Retrieval and Analysis systems using Mobile Agents [16].

In this paper, we present an integrated approach, based on Web Services, for distributed image processing within the INBIOMED network. The reasons for this approach are based on the heterogeneity and different features of image data, computational resources needed and functionalities that users may require for this kind of project. Using this model, different users can benefit from integrated image processing tools remotely located.

## 3  The INBIOMED Platform

The INBIOMED platform [17] is a web services oriented architecture to provide a framework for sharing resources such as data sources or data processing algorithms. These resources can be located at different nodes that may be accessed through web services published in the platform catalogs.

The philosophy of the platform is to minimize the load of processing in complex client applications to convert the algorithms in data processing web services. Consequently, the platform applications only have to implement the user interface with their input forms and the presentation results, delegating in the platform the responsibility to query, store and process the information.

The communication between the client applications and the INBIOMED platform must be accomplished using the SOAP protocol, so the applications have to be implemented using a programming language that is able to be published using web services. There are not any more restrictions imposed to the development of INBIOMED platform client applications.

### Cooperative Model

Every INBIOMED network node collaborates with the other nodes in, at least, one of the following ways:

- Developing its own INBIOMED services
  In this way, it facilitates the sharing data access. Every node keeps its own site independently. Every node provides a user interface to the other nodes; that is used to send requests and will have the proper rights or privileges to access the data sources, data process service catalog and data repository.
  The client applications send all requests to the platform through this INBIOMED platform web service. This web service receives requests written in a language oriented to the batch execution called IQL (INBIOMED Query Language) [17]. The syntax of this language includes sentences for every function that can be performed within the Platform
  At this level is necessary to prepare the data to be accessible through the platform. The data warehouse technology plays a fundamental role, guaranteeing that the information stored in the databases is well consolidated and oriented to be queried, and is obtained from the internal or public data sources.
- Developing data processing services
  A node in the INBIOMED network, may not have information by itself, but can provide the development of a data processing web service implementing algorithms to analyze information for the other nodes.

The data processing web services are responsible to implement the methods necessary to analyze the data, process them and generate the results. The implementation of the algorithms can be made in the most suitable programming language for the resolution of the problem.

A common data processing web service needs to be built at the top of these algorithms to be included in the platform. The goal is to make available a wide catalog of services of this type that implement the processes needed for the different research groups.

• Developing client applications.

Using the resources available in the platform it is possible to implement client applications that solve a problem. These client applications may also be used by other members of the INBIOMED network.

Therefore, the topology of the INBIOMED network can vary from models completely centralized – where there is only one web service in a node, and all the other nodes receive and send data from this node – to federated models, where every node keeps its own information independently.

Among all the applications and data treated by the INBIOMED network, biomedical images and their corresponding processing algorithms are the main topic of this work.

## 4    Type of Images

The types of data that the INBIOMED Platform handles are molecular, functional and genetic images. These images group into three different medical applications: Pharmacogenetics, Cancer Epidemiology, and Molecular and Genetic Epidemiology. Table 2 shows the relationship between the image data and the medical application.

The data were collected from the user groups and an entity-relationship model, explaining the structures and characteristic for each image was created An *in-situ* observation of the tasks carry on by these groups was done in order to identify the image processing functions required. To this end, different image processing tools have been tailored to integrate them into INBIOMED's platform as explained in next section.

**Table 2.** Relationship between the image data and the medical application

| Image Type | Group |
|---|---|
| Electrophoresis Gels | Pharmacogenetics |
| ADN Gels | Pharmacogenetics |
| | Cancer Epidemiology |
| | Molecular and Genetic Epidemiology |
| Inmunofluorescency | Molecular and Genetic Epidemiology |
| Inmuno-histochemistry | |
| TMA | Epidemiology |
| Samples Sequencer | Cancer Epidemiology |
| CT | Cancer Epidemiology |

## 5    Image Processing Integration

The integrated image processing system included within the INBIOMED platform is based on distributed architecture with three different modules:

1. The image processing manager, responsible for image processing methods. This module is also responsible for publishing these methods and making them available to users.
2. A set of web services that implement the image processing methods.
3. The INBIOMED integration platform. It introduces and adapts the web services for remote use of data, provided by the shared databases within or outside the network.

The manager system contains a web application, used as user interface – that allows accessing the system from any computer connected to Internet and a web browser. It also includes an application server that is in charge of the image processing management methods and the communications with the remote applications that offer them. Thus, the application server incorporates a communication kernel that obtains the required information from local databases or from databases integrated in the INBIOMED platform – in the latter case, the use of a Data Access Object (DAO) is required. The application server included in the system also provides access to computation servers. Communication between user interface and application servers is accomplished through servlets, while communication with remote web services is performed using SOAP messages.

Together with web capabilities, the architecture offers the possibility of integrating algorithms developed in different computer languages, enabling the integration and linkage of already developed libraries. Some examples of image processing algorithms, integrated using the INBIOMED platform are described below.

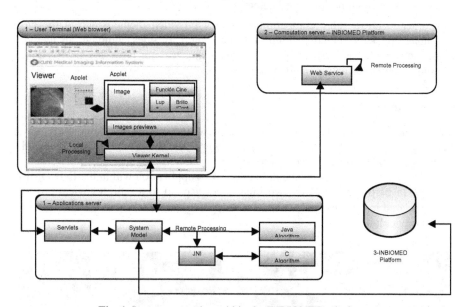

**Fig. 1.** Image processing within the INBIOMED platform

In the context of the INBIOMED research network different partners had been previously working on several biomedical image processing tools based on different methods. Some of these algorithms currently include:

- Morphological processing and analysis of images
  Morphological processing and analysis of images focus especially on the input image shapes, both in the filtering stage and in the segmentation stage [18] [19]. The Web services currently integrated in the platform provide morphological operators and filters, such as erosions, dilations, openings, closings, and openings and closings by reconstruction [20] [21]. In addition, analysis techniques are also provided, concretely, segmentation methods such as the watershed approach (with markers) and flat zone region merging techniques [22]. Such analysis techniques use a marker or mask image to indicate an initial set of particularly significant regions. This set can be computed in an automatic or semi-automatic manner, or previously by hand by a human user or specialist.
- Partial differential based models (PDE)
  PDE models are based on active contour models (ACM) or snakes [23]. The basic idea in the model is to evolve a curve, via minimization of an energy functional, subject to constraints from a given image, in order to detect region of interest within the image. Different models may be obtained, constrained to the resolution of the minimization problem. Thus, a geometrical and geodesic models where introduced by Caselles [24] and Malladi [25] to handle changes in the topology of the segmented shapes.
- Edge and region based detector
  Together with the previous image processing tools several pre-processing algorithms are included through the INBIOMED platform. These algorithms go from filters to edge and region based detectors.

In order to allow more compatibility, all the provided methods accept the most used image formats, including bmp, jpg, dic y pgm. Besides, although all methods return their results as grey scale images, input images can have any colour depth, avoiding the transformation in client application.

## 6   Use Cases

The integration image processing approach presented in this paper may have many applications within the biomedical area. Some of them, performed using the INBIOMED platform, are presented below.

Fig. 2 shows an example of the utilization of an application within the INBIOMED platform. In particular, the first two images correspond to a morphological erosion, and the second two visualizes a connected morphological filtering of the red band of a micro-array colour image.

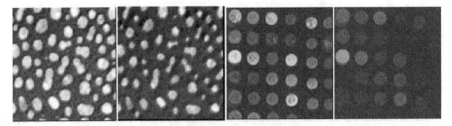

**Fig. 2.** Morphological processing through the INBIOMED platform

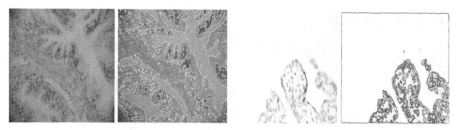

**Fig. 3.** Fuzzy C-Means clustering and PDE segmentation

**Fig. 4.** Edge and region based detectors image processing

Fig. 3 and 4 display the results of some image processing tools applied to histological images. Figure 3 shows an example of fuzzy c-mean clustering, a PDE segmentation [26]. Fig. 4 shows an edge and region based detectors to highlight the contour and the nucleolus of the cells into the images.

## 7   Conclusions

Every biomedical image process presented in the previous section are integrated within the INBIOMED platform and can be remotely executed. They are available to be used by any node of the INBIOMED network, but also open to contributions outside the network, speacially intended for academic training purposes [27].

This paper has presented a new web service-based approach to integrate and publish image processing algorithms within the INBIOMED. Previous studies have shown that web services features matched many needs of image processing algorithms in the field of biomedicine. These methods must be independently developed using any programming language, operating system, etc. They also needed to be available to reuse previous methods and to be easily used by others in the future. The integration of these algorithms into a common platform and format was a priority as well. And finally, the communication should be clear, passing through firewalls and methods must be remotely accessible with a regular web browser.

All these characteristics are supported by web services. As a result, this technology has proven to be especially suited for image processing integration within the biomedical field. In addition, the INBIOMED platform and its further experiments have confirmed the suitability of this approach to integrate biomedical image processing tools.

## Acknowledgements

This work was funded by the INBIOMED research network. We also would like to thank José Antonio López for its work within this project.

## References

1. Lopez-Alonso, V., Sanchez, J.P., Liebana, I., Hermosilla, I., Martin-Sanchez, F.: INBIOMED: a platform for the integration and sharing of genetic, clinical and epidemiological data oriented to biomedical research. Proceedings of Bioinformatics and Bioengineering, BIBE 2004. pp 222- 226 (2004)
2. Erl, T.: Service-Oriented Architecture: A Field Guide to Integrating XML and Web Services. Prentice Hall, New Jersey (2004)
3. Grosso, W.: Java RMI, O'Reilly and Associates, Inc., Sebastopol, CA, USA (2001)
4. Waldo, J.: "Remote procedure calls and Java Remote Method Invocation", IEEE Concurrency, pages 5--7, July (1998)
5. Vinoski, S.: Where is Middleware? IEEE Internet Computing 6(2): 83-85 (2002)
6. Felber, P., Garbinato, B., Guerraoui, R.: The Design of a CORBA Group Communication Service. Proceedings of 15$^{th}$ Symposium on Reliable Distributed Systems (1996)
7. Vinoski S. "CORBA: Integrating Diverse Applications Within Distributed Heterogeneous Environments", 1997, IEEE Communications Magazine, vol. 14, no. 2, February 1997.
8. Chung, Y. Huang, S. Yajnik, D. Liang, J. Shin, C. Y. Wang, and Y. M. Wang.: DCOM and CORBA: Side by Side, Step by Step, and Layer by Layer. C++ *Report,* 18-29 (1998)
9. Adamopoulos, D.X., Pavlou, G., Papandreou, C.A., Manolessos E.: Distributed Object Platforms in Telecommunications: A Comparison Between DCOM and CORBA. British Telecommunications Engineering 18, 43-49 (1999)
10. Liang, K.C., Chu, W.C., Yuan, S.M., Lo, W.: From Legacy RPC Services to Distributed Objects, Proceedings of the Sixth IEEE Computer Society Workshop on Future Trends of Distributed Computing Systems (1997)
11. Snell, J., Tidwell, D., Kulchenko, P.: Programming Web Services with Soap, O'Reilly and Associates, Inc., Sebastopol, CA, USA (2001)
12. D. Kotz, R. S. Gray.: Mobile agents and the future of the Internet. ACM Operating Systems Review, 33(3):7–13, 1999
13. Bote-Lorenzo, M., Dimitriadis, A., Gómez-Sánchez, E.: Grid Characteristics and Uses: a Grid Definition. Proceedings of the First European Across Grids Conference (ACG'03), Springer-Verlag LNCS 2970, pp. 291-298, Santiago de Compostela, Spain, Feb. 2004.
14. Ernemann, C., Hamscher, V., Schwiegelshohn, U., Yahyapour, R., Streit, A.: On Advantages of Grid Computing for Parallel Job Scheduling. Proceedings of the 2nd IEEE/ACM International Symposium on Cluster Computing and the Grid (CCGRID.02) (2002)
15. ImageProcess0.4 – Image Processing Tool, developed by sougeforge.net. Available in http://imageprocess.sourceforge.net/index.shtml (2004)
16. Smith, K., Paranjape, R.: Mobile Agents for Web-Based Medical Image Retrieval. Proceedings of the 1999 IEEE Canadian Conference on Electrical and Computer Engineering, Shaw Conference Center, Edmonton, Alberta, Canada May 9-12 1999.
17. Estruch, A., Heredia J.A., "Technological platform to aid the exchange of information and applications using web services" Lecture Notes in Computer Science, 3337, 458-468, Nov. 2004.
18. Serra, J.: Mathematical Morphology. Volume II: theoretical advances. Academic Press, London (1988)
19. Soille, P.: Morphological Image Analysis, Principles and Applications. Springer-Verlag, Heidelberg (2003)

20. Salembier P., Serra J.: Flat zones filtering, connected operators, and filters by reconstruction. IEEE Transactions on Image Processing, Vol 4, Num 8, pp 1153-1160 (1995)
21. Crespo, J., Maojo, V.: New Results on the Theory of Morphological Filters by Reconstruction. Pattern Recognition, Vol 31, Num 4, pp 419-429, Apr 1998
22. Meyer, F., Beucher, S.: Morphological segmentation. J. Visual Commun. Image Repres, Vol 1, Num 1, pp 21-45 (1990)
23. M. Kass, A. Witkin, D. Terzopoulos: Snakes: Active contour models. Int.J. Comput. Vis., 14 (26), pp. 321-331 (1988)
24. Malladi, R., Sethian, J.A., Vemuri, B.C.: Shape Modeling with Front Propagation: A Level Set Approach. IEEE Trans. on PAMI 17 (1995) 158–175
25. Caselles, V., Kimmel, R., Sapiro, G.: Geodesic Active Contours. Int. J. Comput. Vis. 22 (1) (1997) 61–79
26. Bueno G., Martínez A., Adán A.: Fuzzy-Snake Segmentation of Anatomical Structures. Lecture Notes in Computer Science, LNCS, Springer Pb. Volume 3212, pp. 33-42 (2004)
27. Maojo V, Kulikowski C.A.: Bioinformatics and medical informatics: collaborations on the road to genomic medicine. Journal of the American Medical Informatics Association. 10 (6) 515-22 (2003).

# The Ontological Lens:
# Zooming in and out from Genomic to Clinical Level

Domenico M. Pisanelli[1,2], Francesco Pinciroli[2], and Marco Masseroli[2]

[1] Laboratory for Applied Ontology, National Research Council,
Via Nomentana 56,00161 Rome, Italy
d.pisanelli@istc.cnr.it
http://www.loa-cnr.it/
[2] BioMedical Informatics Laboratory, Bioengineering Department,
Politecnico di Milano, Piazza Leonardo da Vinci 32, 20133 Milan, Italy
{marco.masseroli,francesco.pinciroli}@polimi.it
http://www.medinfopoli.polimi.it/

**Abstract.** Ontology is the talk of the day in the medical informatics community. Its relevant role in the design and implementation of information systems in health care is now widely acknowledged. In this paper we present two case studies showing ontologies "at work" in the genomic domain and in the clinical context. First we show how ontologies and genomic controlled vocabularies can be effectively applied to help in a genomic approach towards the comprehension of fundamental biological processes and complex cellular pathophysiological mechanisms, and hence in biological knowledge mining and discovery. Subsequently, as far as the clinical context is concerned, we emphasize the relevance of ontologies in order to maintain semantic consistency of patient data in a continuity of care scenario. In conclusion we advocate that a deep analysis of the structure and the concepts present at different granular level – from genes to organs – is needed in order to bridge this different domains and to unify bio-medical knowledge in a single paradigm.

## 1 Introduction

Ontology is the talk of the day in the medical informatics community. Its relevant role in the design and implementation of information systems in health care is now widely acknowledged. In the ever difficult and still unsatisfactory tuning between the old-and-new natural-language-described medical knowledge and the informatics methods-and-tools-and-systems, Medical Ontology gains the dignity to be given to one of the few hopes we have for making that tuning doing an effective positive jump. Practical results are not just around the corner. Appropriate works should be invested on both the knowledge and the informatics sides.

Starting from the spoken language side, it makes sense to remind that in the Webster's dictionary we read the following definitions of Ontology:

- a branch of metaphysics relating to the nature and relations of being,
- a particular theory about the nature of being and the kinds of existence.

We easily recognize that we deal with "objects" and "relations". Since the very old times, several philosophers - from Aristoteles (4th Century BC) to Leibniz (1646-1716), and more recently the 19th Century major ontologists like Bolzano, Brentano, Husserl and Frege - have provided criteria for distinguishing between different kind of objects (e.g. concrete vs. abstract) and for defining the relations among them.

J.L. Oliveira et al. (Eds.): ISBMDA 2005, LNBI 3745, pp. 44–50, 2005.

Coming to the informatics side, we see the word ontology – which pertained only the domain of philosophy – gaining new life in the realm of Computer Science. In the late 20th Century in facts, Artificial Intelligence (AI) adopted the term and began using it in the sense of a "specification of a conceptualization" in the context of knowledge and data sharing [1]. Guarino defined an ontology: "A set of logical axioms designed to account for the intended meaning of a vocabulary" [2]. Sowa proposed: "The subject of ontology is the study of the categories of things that exist or may exist in some domain. The product of such a study, called an ontology, is a catalogue of the types of things that are assumed to exist in a domain of interest D from the perspective of a person who uses a language L for the purpose of talking about D." [3]. According to Barry Smith: "An ontology is, in first approximation, a table of categories, in which every type of entity is captured by some node within a hierarchical tree." [4].

The relevance of ontology has been recognized in several practical fields, such as knowledge engineering, knowledge representation, qualitative modeling, language engineering, database design, information integration, object-oriented analysis, information retrieval and knowledge. Current applications areas include enterprise integration, natural language translation, mechanical engineering, electronic commerce, geographic information systems, legal information systems and, last but not least, biology and medicine.

Nowadays, Ontology for Medicine is no longer a pure research topic. It is the backbone of solid and effective applications in health care. Ontology can help to build more powerful and more interoperable information systems in healthcare. They can support the need of the healthcare process to transmit, re-use and share patient data and provide semantic-based criteria to support different statistical aggregations for different purposes. Ontology-based applications have also been built in the field of Medical Natural Language Processing.

An explicit formal ontological representation of entities existing at multiple levels of granularity (e.g. organs, cells, genes) is an urgent requirement for biomedical information processing. Kumar, Smith and Novotny discussed some fundamental principles which can form a basis for such a representation [5].

In this paper we present two paradigmatic case studies showing ontologies "at work" in the genomic domain and in the clinical context.

## 2   Zoom in: Genomics and Molecular Biology

In genomic and molecular biology domains, controlled vocabularies and ontologies are becoming of paramount importance to integrate and correlate the massive amount of information increasingly accumulating in heterogeneous and distributed databanks. Although at present they are still few and present some issues, they can effectively be used also to biologically annotate genes on a genomic scale and across different species, and to evaluate the relevance of such annotations.

In the current post-genomic era, biomolecular high-throughput technologies enable to study thousands of genes at a time generating a massive amount of experimental data at exponential rate. While in the past biologists studied single genes at a time, nowadays both the complete genomic sequences of many organisms (e.g. human, mouse, rat, and many other animals and plants), and the high-throughput technologies that allow investigating gene expressions and mutations on a whole genomic scale are available. Among the last, one of the most promising is the microarray technology, regarding either cDNA arrays or high-density oligonucleotide chips, which enables analyzing ten of thousand clones simultaneously. However, to biologically interpret high-throughput microarray results, it is paramount to annotate the selected identifier lists of candidate regulated clones with biological information about the correspondent genes. A great quantity of such information is increasingly accumulating in form of textual annotations within heterogeneous and widely distributed databases that are often publicly accessible through easy-to-use web-interfaces, e.g. Unigene [6,7], Entrez Gene [8,9], UniProt [10,11], KEGG [12,13], OMIM [14]. The gene annotations are composed of attributes that describe the name and the structural and functional characteristics of known genes, the tissues in which the genes are expressed, the genes' protein products and their protein domains, the known relations among genes, the genes' correlations with different pathologies, and the biochemical pathways in which they are involved. At present, the gene functional annotations are probably those carrying the most interesting information, and their analysis could highlight new biological knowledge such as the identification of functional relationships among genes or the involvement of specific genes in pathological processes.

In the genomic and molecular biology domains some controlled vocabularies are available, including the KEGG description of biochemical pathways, the PFAM list of protein domains [15], and the OMIM genetic disorders. The "Ontology Working Group" of the Microarray Gene Expression Data Society is carrying on the MGED project, which is charged with developing an ontology for describing samples used in microarray experiments [16].

In the genomic and molecular biology domains the most developed and widely accepted ontology is the Gene Ontology (GO) [17] (see also [18] for an analysis of its formal limits). Using the terms of the GO controlled vocabulary and their underlying semantic network, it is possible binding together several information relative to individual biological concepts, also with different meanings in different organisms, and perform cross-specie analyses. Moreover, using the classification of genes in GO categories, as provided by different sources (e.g. Entrez Gene), it is possible annotating a given gene list with GO terms and clustering the genes according to GO biological categories and concepts.

Lately, several web-accessible databases and tools are being implemented to enrich lists of candidate differentially expressed clones with the biological information available for the correspondent genes [19].

Masseroli and Pinciroli comparatively evaluated the three more advanced tools available, i.e. DAVID, GFINDer, and Onto-Tools [20]. Their aim was to assess the ability of these tools in using the Gene Ontology, and in case other genomic controlled vocabularies, to perform not only controlled annotations of gene lists from high-throughput experiments but also to statistically analyze the performed annotations and to help discovering biological knowledge. Results show the different degrees in which the evaluated tools facilitate a genomic approach in the biological interpretation of high-throughput experimental results and hence in the understanding of fundamental biological processes and complex cellular patho-physiological mechanisms.

Gene Ontology and genomic controlled vocabularies are very useful instruments to perform biological annotations of sets of genes on a genomic scale and across different species. Although some problems exist in the Gene Ontology, annotations provided by the GO vocabulary add semantic value to gene identifiers and allow binding together information of a single biological concept from different resources. Besides, through the GO tree-structure it is possible to represent a very wide range of biological specificity, from very general to very precise concepts, using their exact correspondent terms. Furthermore, through the annotations of the Gene Ontology and genomic controlled vocabularies, sets of genes can be clustered according to their biological characteristics, and the significance of these semantic biological classifications can be evaluated by applying specific statistical tests on them.

A few web tools are available to this aim. Comparative evaluation of these tools showed the different functionalities of each one in order to facilitate both the biological interpretation of high-throughput experiment results, and the understanding of which genes are more probably associated to which specific functional characteristics. The selected tools, the analysis of their functionalities, and the presented results are according to the time being. Other advanced tools and new releases of some of the analyzed Web tools are likely to become available in the near future. However, the methodological approach proposed by the selected tools is likely to remain significant and constitute the basis for more comprehensive analyses, as new controlled vocabularies and ontologies in the genomic and molecular biology domains will become available.

## 3   Zoom out: The Clinical Level

Web-based patient record management systems are becoming increasingly popular. Their aim is also to enable all professionals involved in the care process to provide a shared and continuous care.

The role of ontologies is essential in supporting interoperability and continuity of care.

In a complex scenario where multiple agents co-operate in order to allow continuity of care, ontologies are the glue needed to ensure semantic consistency to data and knowledge shared by the different actors involved in the process, including patients and their families.

Currently the management of patients requiring continuity of care presents some limitations, such as a variety of medical approaches and improper hospitalization. Undesired effect is a wasting of resources. These limitations are caused mainly by the lack of integration between hospital specialists and general practitioners, the lack of communication among hospital specialists and the lack of a well-defined care process to assure continuity of care.

Despite of the efforts to improve the communication between primary and secondary care (hospital and territory), there is not an homogeneous, shared approach focused on the patient. Several experiences have shown that the outcomes deriving from a proper management of heart failure patients - for example - can be more effective than the results obtained by drug therapies and reduce to the half the length of staying in the hospital [21].

An example will evidence how ontologies – reference ontologies in particular – are strategically relevant whenever continuity of care is an issue.

In this scenario – the typical e-Health scenario – semantic precision is essential in order to avoid possible misunderstandings among the different actors involved in the process of care.

Our example will consider a health-care structure using a patient-record designed with an openEHR architecture.

The openEHR architecture has evolved from the "Good Electronic Health Record" (GEHR) project. Its main feature consists in the splitting of the clinical content of a record in a two-level framework composed by an "information model" and a set of "archetypes" [22].

By using such an approach, the information and the knowledge level (represented by the archetypes) are clearly separated in a modular and scalable architecture.

Archetypes describe medical concepts, such as blood pressure, laboratory tests and diagnosis. They are defined by domain experts by means of an archetype editor and stored in a proper repository.

An archetype is a re-usable, formal model of a domain concept. Archetypes differ from ontologies in that the first model information as used in a patient record, whereas the second aim at representing reality. For example an archetype for "systemic arterial blood pressure measurement" is a model of the information that has to be captured for this kind of measurement: usually systolic and diastolic pressure, plus (optionally) patient state (position, exertion level) and instrument or other protocol information. On the other hand, an ontology would focus on describing in more or less detail what blood pressure is.

According to the openEHR point of view, archetypes themselves constitute an ontology, whose target domain happens to be information. On the other side, from a realist-ontology perspective – committed at depicting reality, independently from the information-system implementation – "blood pressure" is a physical phenomenon. Even without its measurement, without patient folders and all the actors and instruments involved in its assessment, it would still exist.

If we assume such a position, the archetype of blood pressure and the ontology of the physical phenomenon are clearly two different things. However, being realist does not mean to neglect the essential role that information systems, health-care actors and instruments must have. Nevertheless, archetypes are not ontologies, although someone might claim they are.

The ontology of "blood pressure", for instance, will account for the physical phenomenon, its measurement (a process), the outcome of such a process (data) and the participants to the process (physicians, instruments, etc.). An archetype has a different scope and does not feature such a semantic precision.

This is the reason why we need an evolution from a 'classic' openEHR architecture to an ontology-based patient record, whose information elements are mapped into a reference ontology of medicine like that designed at the Laboratory of Applied Ontologies of CNR in Rome. This reference ontology is mapped to the domain-independent DOLCE foundational ontology [23]. It currently consists of 121 concepts linked by 17 different relationships. It has been implemented by means of the RACER description-logic and can be browsed by the Protégé tool.

Without an ontological grounding, like that provided by ontologies, the same archetype may shift its sense according to the context in which it is placed and according to the tacit knowledge of the human agent who specifies its meaning. For example, how do we know if "blood pressure" is a value or a measurement in a given record? Humans understand the context and have no problems, but computers need ontologies.

## 4 Conclusions

In this paper we showed how ontologies and genomic controlled vocabularies can be effectively applied to help in a genomic approach towards the comprehension of fundamental biological processes and complex cellular patho-physiological mechanisms, and hence in biological knowledge mining and discovery.

As far as the clinical context is concerned, we emphasized the relevance of ontologies in order to maintain semantic consistency of patient data in a continuity of care scenario.

A deep analysis of the structure and the concepts present at different granular level – from genes to organs - is needed in order to bridge this different domains and to unify bio-medical knowledge in a single paradigm.

## References

1. Gruber, T.R.: A translation approach to portable ontologies, Knowledge Acquisition. 5(2) (1993) 199-220
2. Guarino, N.: Formal ontology and information systems. In: Guarino, N., (ed.): Formal Ontology in Information Systems. Proceedings of FOIS 1998. IOS Press, Amsterdam (1998)
3. Available on-line at: http://www.jfsowa.com/ontology/ontoshar.htm Last access: June 1st, 2005
4. Available on-line at: http://ontology.buffalo.edu/ontology[PIC].pdf Last access: June 1st, 2005
5. Kumar, A., Smith, B., Novotny, D.D.: Biomedical Informatics and Granularity, Available on-line at: http://ontologist.com Last access: June 1st, 2005
6. Schuler, G.D.: Pieces of the puzzle: expressed sequence tags and the catalog of human genes. J. Mol. Med. 75 (1997) 694-698
7. National Center for Biotechnology Information (NCBI), UniGene system, Available from: URL:http://www.ncbi.nlm.nih.gov/UniGene/ Last access: June 1st, 2005
8. Maglott, D., Ostell, J., Pruitt, K.D., Tatusova, T.: Entrez Gene: gene-centered information at NCBI. Nucleic Acids Research. 33 (2005) D54-D58
9. National Center for Biotechnology Information (NCBI). Entrez Gene database. Available from URL: http://www.ncbi.nlm.nih.gov/entrez/query.fcgi?db=gene Last access: June 1st, 2005
10. Apweiler, R., Bairoch, A., Wu, C.H., Barker, W.C., Boeckmann, B., Ferro, S., Gasteiger, E., Huang, H., Lopez, R., Magrane, M., Martin, M.J., Natale, D.A., O'Donovan, C., Redaschi, N., Yeh, L.S.: UniProt: the Universal Protein Knowledgebase. Nucleic Acids Research. 32 (2004) D115-D119
11. UniProt the universal protein resource. Available from URL: http://www.pir.uniprot.org/ Last access: June 1st, 2005
12. Kanehisa, M., Goto, S.: KEGG: Kyoto Encyclopedia of Genes and Genomes. Nucleic Acids Research. 28(1) (2000) 27-30

13. Kyoto Encyclopedia of Genes and Genomes – KEGG. Available from URL: http://www.genome.ad.jp/kegg/ Last access: June 1st, 2005
14. Online Mendelian Inheritance in Man – OMIM. Available from URL: http://www.ncbi.nlm.nih.gov/entrez/query.fcgi?db=OMIM Last access: June 1st, 2005
15. Sonnhammer, E.L.L., Eddy, S.R., Durbin, R.: PFAM: A comprehensive database of protein domain families based on seed alignments. Proteins. 28 (1997) 405-420
16. MGED NETWORK. Ontology Working Group (OWG). Available from URL: http://mged.sourceforge.net/ontologies/ Last access: June 1st, 2005
17. The Gene Ontology Consortium, Gene ontology: tool for the unification of biology, Nat. Gen. 25 (2000) 25-29
18. Smith, B., Williams, J., Schulze-Kremer, S.: The ontology of the gene ontology. AMIA Annu Symp Proc. (2003) 609-13
19. Guffanti, A., Reid, J.F., Alcalay, M., Gyorgy, S.: The meaning of it all: Web-based resources for large-scale functional annotation and visualization of DNA microarray data, Trends in Genetics. 18 (2002) 589-592
20. Masseroli, M., Pinciroli, F.: Using Gene Ontology and genomic controlled vocabularies to analyze high-throughput gene lists. To appear on Computers in Biology and Medicine
21. Task Force of the Working Group on Heart Failure of the European Society of Cardiology. The treatment of heart failure. European Heart J. 18 (1997) 736-53
22. Beale, T.: Archetypes: Constraint-based Domain Models for Future-proof Information System. Available at: http://www.openehr.org/downloads/archetypes/archetypes_new.pdf Last access: June 1st, 2005
23. Masolo, C., Gangemi, A., Guarino, N., Oltramari, A., and Schneider, L.: WonderWeb Deliverable D17. The WonderWeb Library of Foundational Ontologies. (2002)

# Dynamics of Vertebral Column Observed by Stereovision and Recurrent Neural Network Model

C. Fernando Mugarra Gonzalez[1], Stanisław Jankowski[2], Jacek J. Dusza[2], Vicente Carrilero López[1], and Javier M. Duart Clemente[3]

[1] Universitat de Valencia, Departament d'Enginyeria Electronica
Dr. Moliner 50, 46100 Burjassot (Valencia), Spain
{fernando.mugarra,vicente.carrilero}@uv.es
[2] Warsaw University of Technology, Institute of Electronic Systems
ul. Nowowiejska 15/19, 00-660 Warszawa, Poland
{sjank,jdusza}@ise.pw.edu.pl
[3] Hospital Universitario La Fe,
Av. Campanar sn, 46019 Valencia, Spain
javier.duart@uv.es

**Abstract.** A new non-invasive method for investigation of movement of selected points on the vertebral column is presented. The registration of position of points marked on patient's body is performed by 4 infrared cameras. This experiment enables to reconstruct 3-dimensional trajectories of displacement of marked points. We introduce recurrent neural networks as formal nonlinear dynamical models of each point trajectory. These models are based only on experimental data and are set up of minimal number of parameters. Therefore they are suitable for pattern recognition problems.

## 1 Introduction

Spinal problems such as scoliosis and kyphosis affect mainly young population, and require X-rays in their diagnoses. In order to reduce the use of the radiology in the diagnosis of such pathologies the stereoscopic vision system with four cameras and infrared light can be used. This system can analyze the stereoscopic images not only from a static point of view, but also those obtained from the movement of the patient under study. The results of data analysis by new computing algorithms based on adaptive methods, artificial neural networks and principal component analysis should be compared with radiological- clinical research.

Artificial vision techniques are widely used in multiple science and technological field [1]. Its application in the field of medicine is almost new and specifically in the spine pathology study is not a new topic, but with very few antecedents [2]. One antecedent of the proposed topic is a project exposed in the International Congress Innabis 2000 [3]. This project develops a study of static images of spine using stereovision, the system carries out the stereovision with 3 cameras in a black-lighted environment. The employment of this kind of light does not allow taking images with the cameras when the normal light is on, making it harder the acquisition of images.

This project overcomes the difficulties posed in the other project by means of:

- Sixteen points in dynamical tracking against ten points in static images
- Four TV cameras against three TV cameras.
- Infrared light in a normal luminosity environment against black light in a low luminosity environment.

J.L. Oliveira et al. (Eds.): ISBMDA 2005, LNBI 3745, pp. 51–60, 2005.

Stereovision recordings of 16 images per second produces large data sets, e.g. one minute experiment of 16 selected points contains 45 000 numbers. Therefore it is necessary to use advanced methods of data mining in order to compress data and to preserve the most relevant information. The efficient way is the construction of formal models of observed processes, as e. g. neural networks, fuzzy models, PCA etc. [11].

We decided to apply recurrent neural networks (RNN) as models of nonlinear dynamics of selected points on human body.

## 2  Hardware Equipment (Measurement Stand)

The stereovision system consists in four progressives scanning monochromatic CCD cameras filtered to be sensitive in near infrared spectra.

An infrared light canyon has been designed and adapted to each camera in order to increase the infrared light

The cameras are connected with the PC by a Matrox Genesis-LC frame-grabber who allow fifteen times a second simultaneously capture data from the images.

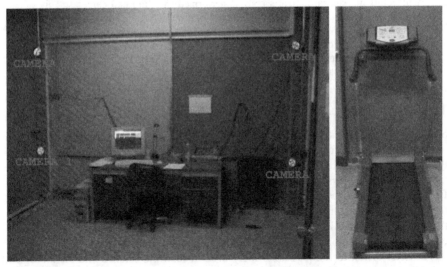

**Fig. 1.** Stereovision system composed by four cameras, frame-grabber and PC, and the conveyor belt

The image of the figure 1 shows the stereovision system. The image of each camera covers enough fields to track a person 190 cm tall, walking over the mechanical conveyor, inside the laboratory, Fig.1.

To pin down the points over the person, in the camera images, we put an infrared reflector disk in each of the sixteen points, Fig.2. Infrared light reflected by each point allows calculating its position in each frame.

As a person walk on the conveyor belt with his back looking to the cameras, cameras 1 and 2 can see all the time the reflecting infrared points positioned on the vertebral spine (C7, T3, T5, T8, T10, L1, L3 and L5) and the four reflecting infrared points on the right side of the back (Acromion, hip, perineum head and astragalus). Cameras 3 and 4 can see all the time the reflecting infrared points positioned on the vertebral spine and the four reflecting infrared points on the left side of the back.

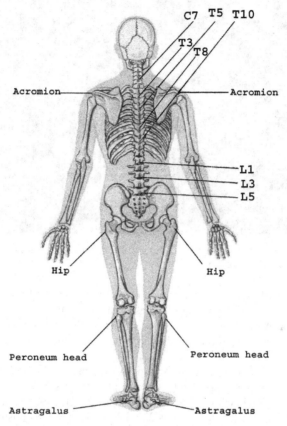

**Fig. 2.** Points to be pin down by the stereovision system

Different calibration algorithms have been used [4]. We obtained the best results with the calibration algorithm proposed by Heikkila [5]. We use a flat panel with a rectangular array of infrared reflecting disks to do the camera calibration for this algorithm. We place the panel, with the minimum possible error, in four different positions; these positions must to cover the space of the conveyor belt with a person walking on it. This last requirement is very important to minimize the error in the 3D position obtained, by the stereovision system, of the 16 points.

## 3   Experiment

The stereovision system works getting data from the images acquired simultaneously from the four cameras, fifteen times a second. The data are stored in a text file. The Fig.3 shows the image from the four cameras rebuilt from the data obtained from a frame.

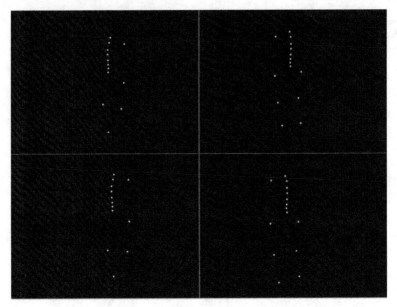

**Fig. 3.** View of the points from the four cameras: camera 2 top left, camera 4 top right, camera 1 bottom left and camera 3 bottom right

The data of each frame in the cameras are processed to identify: four points on the right side of the back in the camera 1, these four point and the eight points on the vertebral column in the camera 2, the four points on the left side of the back in the camera 3 and these four points and the eight points on the vertebral column in the camera 4.

We use three different combinations of pairs of cameras to obtain the 3D position of the different points: cameras 1 and 2 to obtain the 3D position of the four points on the right side of the back, cameras 3 and 4 to obtain the 3D position of the four points on the left side of the back, and cameras 2 and 4 to obtain the 3D position of the eight points on vertebral column.

The table 1 shows a test of the measures obtained by the stereovision system, the best results are obtained with the pair cameras two and four; we use this pair of cameras to obtain the 3D position of eight vertebral column points.

## 4  3D Trajectory of Displacement

The coordinates of displacement of any point indicated on the human body are registered by stereoscopic system of cameras, and the measurement of time intervals between successive images allows us to the present the signal graphically as shown in Fig.4. The functions $x(t)$, $y(t)$ and $z(t)$ were calculated for T8 in the reference system with centre located e.g. in the centre of gravity (CG) point. Usually it is not possible to mark CG point, so the referent point is the midpoint of the line segment which links two markers placed on the left and right coxa of examined patient. In this way the disturbance of signal produced by walking-tape shift irregularity and non-rectilinearity of movement was significantly reduced.

**Table 1.** Error in measures obtained by the stereovision system. We compare the distance measured by the stereovision system, with three different pair of cameras, among a reflecting disk and other four, all five disks located over a straight ruler, against the real distances

|  | Real distances (mm) | Stereovision measures (mm) | Error (mm) |
|---|---|---|---|
|  | 193,50 | 193,48 | -0,02 |
| 3D Camera 1-2 | 160,00 | 159,05 | -0,95 |
|  | 116,00 | 117,14 | 1,14 |
|  | 70,00 | 70,39 | 0,39 |
|  | 193,50 | 191,85 | -1,65 |
| 3D Camera 3-4 | 160,00 | 160,31 | 0,31 |
|  | 116,00 | 114,56 | -1,44 |
|  | 70,00 | 68,37 | -1,63 |
|  | 193,50 | 194,03 | 0,53 |
| 3D Camera 2-4 | 160,00 | 160,98 | 0,98 |
|  | 116,00 | 117,00 | 1,00 |
|  | 70,00 | 70,42 | 0,42 |

The trajectory of a point displacement in 3-dimensional space can be easily obtained from x(t), y(t) and z(t) time-plots registration, as presented in Fig. 5. This trajectory consists of 21 closed loops corresponding to 21 double steps of a patient on the walking track. These loops are not identical due to natural irregularity of human walk as well as to some external disturbances.

Formally, it is possible to treat the observed trajectory as a trajectory of a non-linear dynamical system. The reconstruction of this hypothetical dynamical system can be performed in the form of a recurrent neural network.

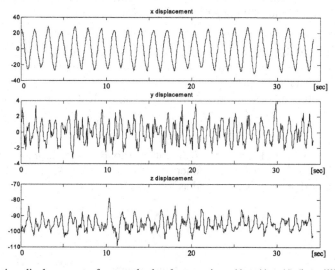

**Fig. 4.** Relative displacement of a vertebral column point x(t), y(t), z(t) (in millimeters) obtained by stereoscopic vision system.

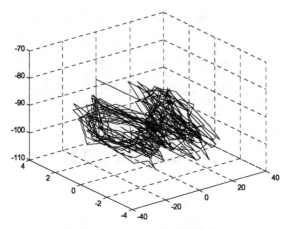

**Fig. 5.** Trajectory of a point displacement in 3D space (all coordinates scaled in mm)

## 5   Approximation of the Trajectory by Recurrent Neural Network

Recurrent neural network is an adaptive model of any non-linear dynamical system [6, 7, 8]. The calculation of the network parameters is called training. The required information about an object under study is a set of measurement of input and output variables – this training set should be representative with respect to the properties in real conditions. The physical knowledge can be implemented, e.g. as a prior choice of the network architecture.

The notion "recurrent network" means that a subset of network inputs consists of delayed outputs calculated by this network. If the inputs are equal to the observed delayed outputs the network is feedforward.

The recurrent neural network (RNN) can implement a NARMAX model (Nonlinear AutoRegressive and Moving Average with eXogeneous inputs) [9]:

$$y(t) = F[(x(t), x(t-1),..., x(t-M+1), y(t-1),..., y(t-N)]  \tag{1}$$

The most efficient RNN has a form of canonical network that is set up of two blocks: feedforward multilayer perceptron and feedback loops. The number of loops is equal to the order of dynamical system.

Using the notion of state variables, the RNN is equivalent to the usual object form:

$$\mathbf{s}(t+1) = f[\mathbf{s}(t), \mathbf{x}(t)] \quad state \quad equation$$
$$y(t) = g[\mathbf{s}(t), \mathbf{x}(t), w(t)] \quad output \quad equation  \tag{2}$$

where: $\mathbf{s}$ – state vector of dimension $N$ (number of loops), $y$ – output, $\mathbf{x}$ – control signal, $w$ – noise, $f$ and $g$ – nonlinear functions. The network is set up of discrete-time neurons. Each $i$-th neuron is described as follows:

$$z_i(t) = g_i[h_i(t)] = g_i[\sum_{j \in P_i} \sum_{\tau=0}^{T_{ij}} w_{ij;\tau} z_j(t-\tau)]  \tag{3}$$

where:
  $g$ – activation function (usually, hyperbolic tangent)
  $h$ – input potential

$z$ – output signal
$w$ – network parameter (calculated by using training method)
$T$ – maximal delay
$P$ – subset of neurons connected to a given unit
Thus the network output $y(t)$ is equal to:

$$y(t+1) = \sum_{j=1}^{K} w_j \tanh\left(\sum_{i=1}^{M}[w_{ji}x_i(t) + \sum_{l=M+1}^{M+N}w_{jl}y(t-l+1)] + w_{j0}\right) + b_1 \qquad (4)$$

During training the network weights are updated in order to minimize the mean square error between the network output and desired observed output.

Signals x(t), y(t), z(t) formally considered as orthogonal coordinates of a given point displacement are simulated by 3 recurrent neural networks. Their structures are shown in Fig. 6. The training (error backpropagation plus quasi-Newton method of gradient minimization) was performed by using specialized software toolbox: Neuro One 2.2 [10]. The coordinates x(t) and z(t) were reconstructed by a neural network of 3 hidden neurons with hyperbolic tangent activation function and 2 feedback loops while the coordinate y(t) can be reconstructed by a neural model of 5 hidden neurons and 2 feedback loops. Hence, it can be concluded that the point displacement can be modeled by second-order dynamical system. The external input excitation signal is a short step pulse.

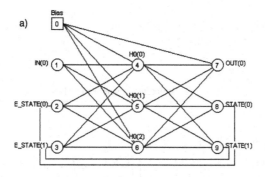

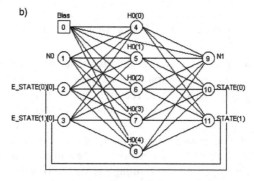

**Fig. 6.** Structure of recurrent neural network models for: a) x(t) and z(t) reconstruction, b) y(t) reconstruction

58    C. Fernando Mugarra Gonzalez et al.

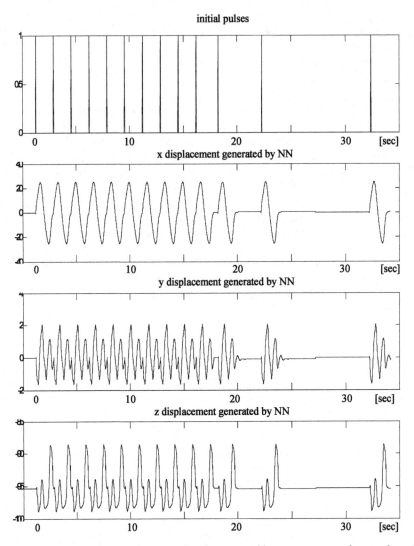

**Fig. 7.** Initial pulse excitations and output signals generated by recurrent neural network models

The signals generated by recurrent neural network models are presented in Fig. 7.

The trajectory of the recurrent neural network model is reconstructed from the signals presented in Fig. 7 by a system of recurrent neural networks as in Fig. 8. The resulting trajectory is shown in Fig. 9. The function of the RNN model of a point dynamics is the following.

Each pulse of excitation generates the output signals corresponding to one double step of patient's walk. The obtained trajectory is shown in Fig. 9 as bold dotted line.

The grey trajectories represent the training set corresponding to 21 double steps. Hence, the recurrent network output can be interpreted as an estimation of typical movement of examined point.

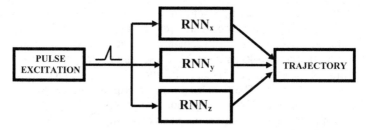

**Fig. 8.** Recurrent neural networks for the reconstruction of point displacement trajectory

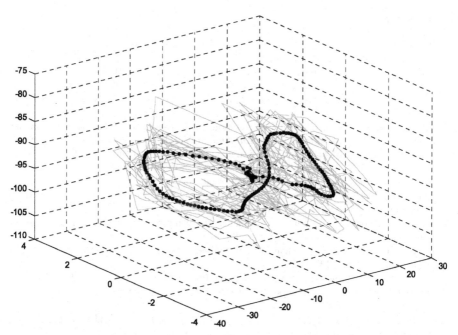

**Fig. 9.** Recurrent neural network model of 3D trajectory of selected point T8 (see Fig. 2) displacement (all coordinates scaled in mm)

## 6   Conclusions

In this paper a new non-invasive experimental technique is presented for the study of dynamical behaviour of human vertebral column. A set of selected points placed on human body is observed and registered in infrared light by a system of 4 cameras giving rise to stereoscopic vision. Therefore, each point displacement can be considered as a 3-dimensional trajectory.

The recognition of pathology of vertical column dynamics requires to define an appropriate set of quantitative descriptors corresponding to each experimental trajectory.

The most suitable approach to this problem is to construct a recurrent neural network as a non-linear dynamical system that can simulate the displacement of each observed point. The structure and the set of network parameters can thus be used as formal description of the vertebral column point dynamics.

The formal model of one point displacement is thus described by 3 recurrent neural networks that are defined by 80 parameters. These parameters provide the most relevant information of the experiment. Hence, the information of 60 seconds recording is compressed nearly 30 times. The reduced set of descriptors enables to use statistical classifiers for discrimination of selected pathological cases.

The studies of pathology recognition based on experiments obtained by our method will be performed in collaboration with medical staff.

# References

1. Faugueras, O.: Three-Dimensional Computer Vision, MIT Press (1993)
2. Sotoca, J.M., Buendía, M., Iñesta, J.M., Ferri, F.: Geometric properties of the 3D spine curve, *Lecture Notes in Computer Science*, Vol. 2652, Springer (2003) 1003-1011
3. Mugarra, C.F., Llorens, I., Gutierrez, J., E. M. Perez, E.M.: A low cost stereo vision system for vertebral column measures, *6th Internet World Congress for Biomedical Sciences* (2000)
4. Tsai, R.:A versatile camera calibration technique for high-accuracy 3D machine vision metrology, *IEEE Journal of Robotics and Automation*, Vol. RA-3 (1987) 323-344
5. Heikkilä, J.: Geometric camera calibration using circular control points, *IEEE Transactions on Pattern Analysis and Machine Intelligence*, Vol. 22 (2000) 1066-1077
6. Nerrand O., Roussel-Ragot P., Urbani D., Personnaz L., Dreyfus G.: Training recurrent neural networks: why and how? An illustration in dynamical process modeling, IEEE Transactions on Neural Networks, Vol. 5 (1994) 178-184
7. Rivals I., Personnaz L.: Nonlinear Internal model using neural networks, *IEEE Transactions on Neural Networks*, Vol. 11 (2000) 80-90
8. Dreyfus G., Idan I.: The canonical form of nonlinear discrete-time models, *Neural Computation*, Vol. 10 (1998) 133-164
9. Soderstrom T., P. Stoica: System Identification, Prentice Hall International, 1994
10. NeuroOne ver. 2.2, documentation, Netral S.A., France, available from: http://www.netral.com/
11. Becerikli Y.: On three intelligent systems: dynamic, neural, fuzzy, and wavelet networks for training trajectory, *Neural Comput. & Applic.* 13 (2004), 339-351

# Endocardial Tracking in Contrast Echocardiography Using Optical Flow

Norberto Malpica[1], Juan F. Garamendi[1],
Manuel Desco[2], and Emanuele Schiavi[1]

[1] Universidad Rey Juan Carlos, Móstoles, Madrid, Spain
norberto.malpica@urjc.es, jf.garamendi@alumnos.urjc.es,
emanuele.schiavi@urjc.es
[2] Hospital General Universitario Gregorio Marañón de Madrid
desco@mce.hggm.es

**Abstract.** Myocardial Contrast Echocardiography (MCE) is a recent technique that allows to measure regional perfusion in the cardiac wall. Segmentation of MCE sequences would allow simultaneous evaluation of perfusion and wall motion. This paper deals with the application of partial differential equations (PDE) for tracking the endocardial wall. We use a variational optical flow method which we solve numerically with a multigrid approach adapted to the MCE modality. The data sequence are first smoothed and a hierarchical-iterative procedure is implemented to correctly estimate the flow field magnitude. The method is tested on several sequences showing promising results for automatic wall tracking.

## 1 Introduction

The analysis of the motion of the heart wall is a standard technique for studying myocardial viability [1]. The measurement of cardiac perfusion using ultrasound imaging has recently become available with contrast agents. Contrast agents provide information about the degree of perfusion and the speed of reperfusion of the myocardium. Tracking of the myocardium would allow simultaneous quantification of wall motion and perfusion. Caiani [3] et al. proposed an interactive tracking method that segmented each frame independently.

In this work, we consider a new optical flow algorithm proposed recently in [4] where real time performances are reported when test sequences of synthetic images are considered. Our aim is to evaluate the effectiveness of the Combined Local-Global approach (CLG) for endocardial tracking where noisy corrupted image sequences are analysed. The algorithm combines local and global regularization of the flow field. A hierarchical implementation has been designed that allows to capture large inter-frame motion present in clinical sequences acquired with low frame rate. To filter the high degree of speckle present in the image, we evaluate two different smoothing schemes.

This paper is organized as follows: Section 2 introduce to the basic material and definitions. The algorithms used in the model cases are detailed, the reconstruction steps and the numerical implementation are presented. In Section 3

J.L. Oliveira et al. (Eds.): ISBMDA 2005, LNBI 3745, pp. 61–68, 2005.

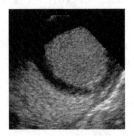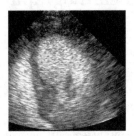

**Fig. 1.** Frames of two myocardial contrast echocardiography sequences. Short axis view (left) and apical four-chamber view (right)

we show the results we have obtained with different pre-processing steps as well as a parametric study relating the optical flow problem and the segmentation paradigm. Section 4 is devoted to the discussion and future work.

## 2    Material and Methods

The use of PDE's is becoming a standard tool in digital images processing and reconstruction [8]. On the other hand multigrid methods represent a new computational paradigm which can outperform the typical gradient descent method.

### 2.1    Preprocessing

Before applying an optical flow algorithm a pre-processing smoothing step is necessary ([2]), specially when dense flow fields and noisy images are analised. Otherwise a mismatch in the direction field appears. This step, usually, amounts to a convolution of the original sequences with a Gaussian kernel of standard deviation $\sigma$. As a result we get a smoothed new sequence of images which can be considered as the initial data of the CLG algorithm. This filtered version can suffer from several drawbacks when tracking is performed. In fact the resulting convolved images have a blur in the boundary of the cardiac wall and this can complicate its tracking. As an alternative, we have also pre-processed the original sequence with a Total Variation scheme because of its well known properties ([8]) as regards to edge preservation.

### 2.2    CLG Approach for Optical Flow

In this section, following Bruhn et al. ([4]), we shall introduce the basic notation and equations. Let $g(x,y,t)$ be the original image sequence where $(x,y) \in \Omega$ denotes the pixel location, $\Omega \subset \mathbb{R}^2$ is a bounded image domain and $t$ denotes time. Let $f(x,y,t)$ be its smoothed version which represents the initial data of the CLG algorithm.

The optical flow field $(u(x,y), v(x,y), t)^T$ at some time $t$ is then computed as the minimum of the energy functional

$$E(u,v) = \int_{\Omega} \left( \omega^T J_\rho(\nabla_3 f)\omega + \alpha(|\nabla u|^2 + |\nabla v|^2) \right) dxdy$$

where the vector field $\omega(x, y) = (u(x, y), v(x, y), 1)^T$ is the displacement, $\nabla u = (u_x, u_y)^T$ and $\nabla_3 f = (f_x, f_y, f_t)^T$. The matrix $J_\rho(\nabla_3 f)$ is given by $K_\rho * (\nabla_3 f \, \nabla_3 f^T)$ where $*$ denotes convolution, $K_\rho$ is a gaussian kernel with standard deviation $\rho$ and $\alpha > 0$ is a regularization parameter. More details of CLG method can be found in [7]. As usual in the variational approach the minimum of the energy $E(u, v)$ corresponds to a solution of the Euler-Lagrange equations

$$\alpha \Delta u - [J_{11}(\nabla_3 f)u + J_{12}(\nabla_3 f)v + J_{13}(\nabla_3 f)] = 0 \tag{1}$$

$$\alpha \Delta v - [J_{12}(\nabla_3 f)u + J_{22}(\nabla_3 f)v + J_{23}(\nabla_3 f)] = 0 \tag{2}$$

where $\Delta$ denotes the laplacian operator. This elliptic system is complemented with homogeneous Neumann boundary conditions.

As reported in Bruhn et al. [7] this approach speeds up the computation when compared with the clasical gradient descent method and we shall follow his indication here.

**Discretisation.** Optical flow seeks to find the unknown functions $u(x, y, t)$ and $v(x, y, t)$ on a rectangular pixel grid of cell size $h_x \text{x} h_y$. We denote by $u_{ij}$ and $v_{ij}$ the velocity components of the optical flow at pixel $(i, j)$. The spatial derivates of the images have been approximated using central differences and temporal derivatives are approximated with a simple two-point stencil.

The finite difference approximation to the Euler-Lagrange equations (1) and (2) is given by

$$0 = \frac{\alpha}{h_x^2}(u_{i,j-1} - 2u_{ij} + u_{i,j+1}) + \frac{\alpha}{h_y^2}(u_{i-1,j} - 2u_{ij} + u_{i+1,j}) - \tag{3}$$

$$-(J_{11,ij}u_{ij} + J_{12,ij}v_{ij} + J_{13,ij})$$

$$0 = \frac{\alpha}{h_x^2}(v_{i,j-1} - 2v_{ij} + v_{i,j+1}) + \frac{\alpha}{h_y^2}(v_{i-1,j} - 2v_{ij} + v_{i+1,j}) - \tag{4}$$

$$-(J_{21,ij}u_{ij} + J_{22,ij}v_{ij} + J_{23,ij})$$

where $J_{nm}, ij$ is the component $(n, m)$ of the structure tensor $J\rho(\nabla_3 f)$ in the pixel $(i, j)$.

## 2.3 Numerical Implementation

The system of equations 3 and 4 has a sparse system matrix and may be solved iteratively with a Gauss-Seidel scheme [5].

**System Resolution.** Iterative solvers of equation systems, such as Gauss-Seidel, have an great initial convergence, however, after the initial iterations the convergence slows down significantly. Multigrid algorithms, take this idea and combine it with a hierarchy model of equations systems that come from differents levels of detail in the problem discretisation. The main idea of multigrid methods is to first aproximate one solution in a level of discretisation, then calculate the error of the solution from a coarser level and correct the aproximate solution with the error.

Suppose that the system $Ax = b$ has arisen from the above discretization. We can obtain a approximate solution, $\tilde{x}$, with a few iterations of an iterative method, say the Gauss-Seidel method. It is possible to calculate the error $e$ of the solution and correct the solution: $x = \tilde{x} + e$. We calculate $e$ from the residual error

$$r = b - A\tilde{x}$$

As $A$ is a linear operator, we can find $e$ solving the equation system

$$Ae = r$$

Solving this equation requires the same complexity as solving $Ax = b$, but we can solve $Ae = r$ at a coarser discretisation level, where the problem is much smaller and will be easier to solve [6]. Once we have $e$ at he coarser level, we calculate $\hat{e}$ interpolating $e$ from the coarser level to the finest level and suddenly we do the correction $x = \tilde{x} + \hat{e}$.

The multigrid method takes this idea of approximate-correction and yields it to several levels of discretisation. The detailed algorithm is: (the superscripts $h_1$, $h_2...h_n$ indicates the level of discretisation: $h_1$ the finest level and $h_n$ the coarsest level)

– We start at the finest level ($h_1$). With a few iterations of Gauss-Seidel method we aproximate a solution $\tilde{x}^{h_1}$ of the system

$$A^{h_1} x^{h_1} = b^{h_1}$$

We calculate the residual error $r^{h_1}$

$$r^{h_1} = b^{h_1} - A^{h_1} \tilde{x}^{h_1}$$

– In this step we solve the equation $Ae = r$ in the next coarser level of discretisation:

$$A^{h_2} e^{h_2} = r^{h_2}$$

Restrict $r^{h_1}$ to $r^{h_2}$, $A^{h_1}$ to $A^{h_2}$. Now, we recall $e^{h_2}$ such as $x^{h_2}$, and $r^{h_2}$ such as $b^{h_2}$. The equation is now

$$A^{h_2} x^{h_2} = b^{h_2}$$

At level $h_2$, we aproximate a solution $\tilde{x}^{h_2}$ with a few iterations, and repeat the process throw levels until reach the level $h_n$.

– At level $h_n$, the equation $A^{h_n} x^{h_n} = b^{h_n}$ is exactly solved. Be $\tilde{x}^{h_n}$ to the exact solution in this level.

Now we start the correction of the solutions calculated at the levels.

– Interpolate $\tilde{x}^{h_n}$ to level $h_{n-1}$ and add to $\tilde{x}^{h_{n-1}}$ to obtain $\hat{x}^{h_{n-1}}$

- Starting from $\hat{x}^{h_{n-1}}$, run a few iterations of Gauss-Seidel over

$$A^{h_{n-1}}x = b^{h_{n-1}}$$

- Interpolate the new calculated aproximation $\tilde{x}^{h_{n-1}}$ to the level $h_{n-2}$ and repeat the process throw levels until the level $h_1$ is reached.
- The solution obtained in the $h_1$ is the solution of the original system.

Two operators are needed to move between levels in the multigrid. The restrictor operator used is a arithmetic mean and the interpotalion operator is a constant interpolation.

**Large Displacements: Hierarchical Approach.** The CLG method, as all variational methods, assumes that the movement of the objects in two consecutive images is small. This is necessary because CLG is based on the a linearisation of the grey value constancy assumption and the movements must be small for holds the linearisation. However, this is not in MCE imaging where acquisition frame rate can be small and heart motion can be non-linear.

This limitation can be overcome by calculating optical flows in coarser scales, where the displacements are smaller. To calculate the optical flow of two consecutive images $I1$ and $I2$, we scale the two images into several levels, $L_1$ to $L_n$, obtaining $I1_1, I1_2, .., I1_n$ and $I2_1, I2_2, .., I2_n$, where $L1$ is the finest level (the original images), $Ln$ is the coarsest level and $I1_i$ $I2_i$ correspond to the scaled images. The detailed process is:

- Compute the optical flow, $u_n$ and $v_n$, at the level $L_n$ between the images $I1_n$ and $I2_n$.
- Interpolate the optical flow to the next level $L_{n-1}$. Compute a new image $I1_{n-1}^w$ warping $I_{n-1}$ with the interpolated optical flow, then compute a new optical flow among $I1_{n-1}^w$ and $I2_{n-1}$. Correct the interpolated optical flow with the new optical flow adding the two fields.
- Repeat the process until level $L1$. The final optical flow is the corrected optical flow at level $L1$.

**Improving the Optical Flow: Iterative Hierarchical Approach.** In the ultrasound test sequences we noticed that the algorithm computes a correct field direction but it subestimates the magnitude of the displacement. To overcome this handicap we repeat the computation of the optical flow, warp the initial image and calculate a new optical flow. This process is repeated iteratively until the highest displacement is less than $\| h_x, h_y \|$. The final optical flow is the sum of all previously obtained optical flow values. This iterative process is applied only at scale levels $L_2$ to $L_n$.

# 3   Results

Algorithms were evaluated using images provided by Gregorio Marañón Hospital in Madrid. The algorithm has been tested on two type of sequences, to take

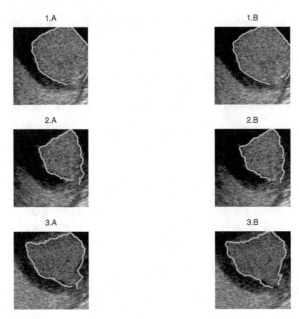

**Fig. 2.** Results of endocardial tracking with different preprocessing methods. Figures 1.A, 2.A, 3.A have been obtained with TV filtering while figures 1.B, 2.B and 3.B have been obtained with a gaussian ($\sigma = 3$)

into account different signal to noise ratios and different acquisition views. The first are short-axis views obtained during experimental surgery. The second are clinical images obtained from patients in a four-chamber view.

A first set of experiments was carried out to evaluate the two different presmoothing filters. The role of the smoothing process is to improve the results of the tracking. Therefore, the evaluation was carried out by comparing the result of the tracking with both filters on the same sequences. Fig. 2 shows the result on three frames with both filters. We did not appreciate any substantial improvement using the TV filter. As the gaussian filter is computationally more efficient, all further tracking results are obtained using this filter. To evaluate the optical flow algorithm the endocardial wall was manually segmented in the first frame of the sequence and the wall was automatically tracked in the remaining frames using the CLG algorithm with $\alpha = 200$ and $\rho = 3$. Fig. 2 shows examples of automatic tracking on short axis views. Fig. 3 shows the results on several frames of a four-chamber view sequence. Notice the different degrees of perfusion of the heart wall.

## 4 Discussion and Future Work

The results obtained in our study indicate that the CLG algorithm can be succesfully applied to myocardial contrast echocardiography sequences. We have designed an iterative-hierarchical approach that allows to capture large displace-

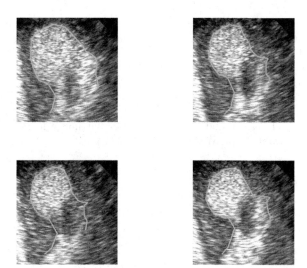

**Fig. 3.** Results of automatic tracking on several frames of a clinical sequence

ments accurately. The cardiac wall is tracked using only the information provided by the dense optical-flow. Post-processing of the curve to include curvature or smoothness assumptions would improve the current results. Also notice that the model does not assume any hypothesis about the specific noise of this specific imaging modality. Future work will consist in evaluating a denoising step combined with nonlinear regularization as preprocessing.

## Acknowledgements

Images were kindly provided by the department of Cardiology, Hospital Gregorio Marañón. This work is partially supported by the Spanish Health Ministry (Instituto de Salud Carlos III), Red Temática IM3 (G03/185) and also by project MTM2004-07590-C03-01.

## References

1. García-Fernández M.A., Bermejo J., Pérez-David E., López-Fernández T., Ledesma M.J., Caso P., Malpica N., Santos A., Moreno M., Desco M. New techniques for the assessment of left ventricular wall motion Echocardiography, 20 (2003), 659-672
2. Barron J., Fleet D., Beauchemin S. Performance of optical flow techniques. Int. J. Comput. Vision 12 (1994) 43-77.
3. Caiani EG, Lang RM, Caslini S et al. Quantification of regional myocardial perfusion using semiautomated translation-free analysis of contrast-enhanced power modulation images. J. Am. Soc. of Echocardiog 16(2003) 116-23.
4. Bruhn A., Weickert J., Feddern C., Kohlberger T., and Schnörr C. Variational optical flow computation in real time. IEEE T. Image Process, 14 (2005) 608-615

5. Bruhn A., Weickert J., Feddern C., Kohlberger T., and Schnörr C. Real-Time Optic Flow Computation with Variational Methods. Lecture Notes in Computer Science, Vol 2756. Springer-Verlag, Berlin Heidelberg New York (2004)
6. Golub G., Ortega J.M. Scientific Computing an Introduction with Parallel Computing. Academic Press Inc. 1993
7. Bruhn A., Weickert J., Schnörr C. Lucas/Kanade Meets Horn/Schunck: Combining Local and Global Optic Flow Methods. Int. J. Comput. Vision 61(2005) 211-231
8. Chan T., Shen J. Variational restoration of nonflat image features: Models and algorithms. SIAM J. Appl. Math. 61 (2000) 1338-1361

# Unfolding of Virtual Endoscopy Using Ray-Template

Hye-Jin Lee, Sukhyun Lim, and Byeong-Seok Shin

Inha University, Dept. of Computer Science and Information Engineering
253 Yonghyun-dong, Nam-gu, Inchon, 402-751, Rep. of Korea
{jinofstar,slim}@inhaian.net, bsshin@inha.ac.kr

**Abstract.** Unfolding, one of virtual endoscopy techniques, gives us a flatten image of the inner surface of an organ. It is more suitable for a diagnosis and polyp detection. Most common unfolding methods use radial ray casting along with pre-computed central path. However, it may produce false images deformed and lost some information because adjacent ray planes cross when the organ's curvature is relatively high. To solve it, several methods have been presented. However, these have severe computational overhead. We propose an efficient crossing-free ray casting for unfolding. It computes ray-cones according to curvature of the path. Then in order to avoid intersection between ray-cones, it adjusts direction of ray-cones detected while testing intersection. Lastly, it determines direction of all rays fired from sample points between control points by simple linear interpolation. Experimental results show that it produces accurate images of a virtually dissected colon and takes not much time.

## 1 Introduction

Virtual endoscopy is a non-invasive inspection of inner structure of the human cavities using tomography images (e.g., CT and MRI). While an optical endoscopy is invasive and contains a certain degree of risk for patients, virtual endoscopy avoids the inconvenience of an optical endoscopy and improves the accuracy of diagnosis. However, virtual endoscopy cannot provide entire view of organ surface due to limited field-of-view. Also some polyps may be hidden from view because of extremely complex structure of organs and folds. So, we have to devise a variety of the visualization methods that do not have physical limitation in the virtual world.

Recently, new methods to visualize human cavities have been proposed [5],[9],[10]. These methods based on the efficient way to inspect the inner surface of organ would be to open and unfold it. The virtual dissection of organs is analogies to the real anatomical dissection. We can easily and intuitively recognize special features and pathologies of the organ. However, unfolding methods have a problem of missing or duplicating polyps and take a lot of time to solve the problem. Therefore, they are not adequate for real-time application.

In this paper, we propose an efficient unfolding method that solves the problem of missing or duplicating significant features and produces a flatten image in short time. It determines direction of rays using ray-templates, the set of pre-computed direction vectors according to curvature of the path. In order to avoid intersection between adjacent ray-cones, it adjusts the normal vector of crossing ray-plane. Then it determines direction of all rays fired from sample points between control points using simple linear interpolation. Experimental results show that it does not produce erroneous images and takes less time compared with the previous methods.

J.L. Oliveira et al. (Eds.): ISBMDA 2005, LNBI 3745, pp. 69–77, 2005.

In the next section, we review problems of conventional unfolding methods. Sect. 3 presents the main steps of the method in detail. Experimental results are presented in Sect. 4 and conclusions and future work are described in Sect. 5.

## 2   Problems of Conventional Unfolding Methods

In general, unfolding methods compute ray directions using radial ray casting along with a central path of an organ. To transform a curved path into a straight one, computed ray planes are arranged in a straight line. A flatten image is generated by searching intersection points of organ surface with rays, and computing color of the points.

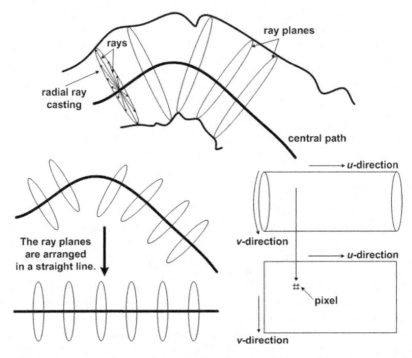

**Fig. 1.** Basic concept of unfold rendering. At first, it casts radial rays along with a central path and performs color composition along the ray directions. Then, it arranges ray planes on a straight line and applies mapping function to transform 2D image

These methods allow visualizing the complete surface of organs at once. Main problem of them is intersection of ray planes in high-curvature areas of the central path. In consequence, a polyp can appear twice in the flatten image or it can be missed completely (see Fig. 2).

To solve the intersection of ray plane, several methods have been presented. Wang et. al used electrical field lines generated by a locally charged path [5]. Electrical field lines that have the same polarity do not intersect each other. However, this method requires a lot of computations to calculate capacity for all electrical charges in the electrical field and to simulate the electrical force line.

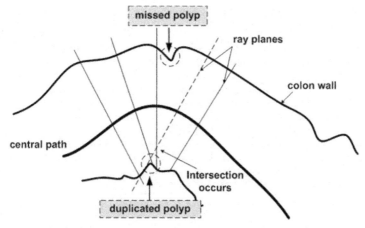

**Fig. 2.** Illustration of the missing and duplicating polyp in the conventional unfolding method due to intersections of the ray planes in high-curvature area. The ray plane represented as a dashed line produces a double polyp

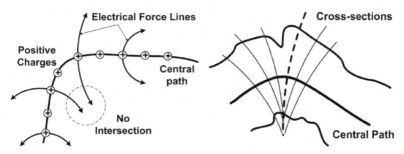

**Fig. 3.** Ray models proposed by Wang and Vilanova. Wang used electrical field lines generated by a locally charged path (left). Vilanova used a non-linear ray casting using the minimum distance of distance-map and vector field (right)

Vilanova et. al proposed a non-linear ray casting that prevents ray intersections by traversing non-linear rays using the minimum distance of distance-map and vector field[10]. It requires a lot of storages and long preprocessing time because it creates a distance-map and a vector field for traversing the minimum distance.

## 3   Unfold Rendering Using Ray-Template

We propose an efficient method to produce a flatten image without duplicating or missing features. In order to obtain the image of the virtually dissected organ, four steps have to be performed as shown in Fig. 4. We are given the central path of an organ cavity that represented as sample points. Control points are the part of sample points for taking shape of the path. Firstly, it classifies the path into two regions, curved regions and straight regions, according to the curvature of the central path. Then, it defines ray-cones only on control points. Secondly, it performs intersection test between two adjacent ray-cones and adjusts normal vector of a ray-cone when it recognizes the occurrence of intersection. Then, it computes direction of rays fired from sample points using simply interpolation method. Lastly, it generates the entire unfold image by traversing rays with pre-calculated ray directions. It cannot offer an accurate image that identifies exact shapes of organ features. However, it can quickly generate images of excellent quality reflected important features.

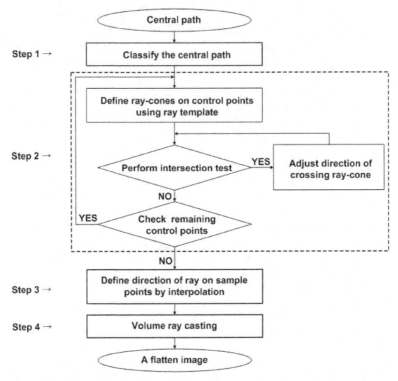

**Fig. 4.** A flow chart presents our procedure

## 3.1 Classification of the Central Path

A central path is composed of sampling points that correspond to $u$-coordinate. We classify it into curved regions and straight regions according to its curvature. After calculating dot product of a tangent vector $T$ of control point $P_i$ and that of adjacent control point $P_{i-1}$, it determines type of region. If the result of dot product is greater than pre-defined threshold($\varepsilon_1$), the type of region is regarded as straight. Otherwise, it is curved one.

*Ray template (RT)* is a set of ray vectors and its $k$-th ray is denoted as $RT(k)$. Radial and conic ray template mean that rays are fired in the shape of circle and cone form control points respectively. We apply a radial ray template to straight regions and a conic ray template to curved regions. Fig. 5 shows comparison of a radial template and a conic template. The ray plane given by a radial template **s** is crossing with adjacent ray. However, a conic template **t** can avoid intersection because its effect is similar to searching points that have minimum distance using distance-map. Fig. 6 shows a section of a ray-cone.

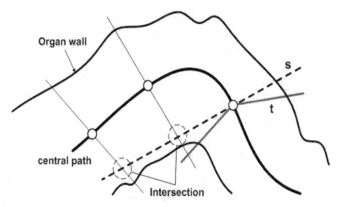

**Fig. 5.** Illustration of applying radial and conic templates on the path: **s** is a radial template, **t** is a conic template

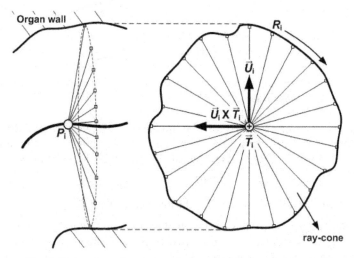

**Fig. 6.** Illustration of a section of a ray-cone $R_i$ on control point $P_i$

We assume that a central path $C(P_i)$ has control points $P_i$. It finds new basis that has a tangent vector of control point $P_i$ like as $z$-axis. Then it calculates direction of rays fired from control points by multiplying ray template $RT$ with transformation matrix. The advantage of basis transform is faster processing speed because it applies a transformation matrix to a lot of vectors in a lump. The point where a ray meets with the first non-transparent voxel is called *a boundary point*. A set of boundary points consisted of the outline of ray-cone. It derived from control point $P_i$ is $R_i$, the $k$-th point is $R_i(k)$.

### 3.2  Intersection Avoidance

Under occurrence of intersections, severe artifacts such as duplicating or eliminating features may come in a final image. So we have to check whether two consecutive ray-cones are intersected and adjust one of ray-cones when they are intersected.

We propose an intersection test using spatial coherence of shape of organ. At First, it computes vectors between boundary points that have identical index in adjacent ray-cones. It calculates dot product of the first vector $\vec{v}_1$ and the other vectors respectively. If results of dot product for all points are more than threshold ($\varepsilon_2$), the ray-cones do not intersect.

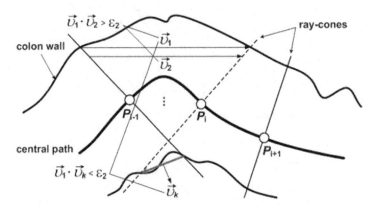

**Fig. 7.** Illustration of intersection test: Vector $\vec{v}$ is defined as direction vector that has identical index in two consecutive ray-cones

When intersection occurs, it adjusts the direction of current ray-cone $R_i$. The new normal vector of $R_i$ is determined by simply interpolating normal vectors of crossing ray-cones. It performs intersection test repeatedly until they do not intersect.

$$T'_i = \omega \times T_{i-1} + (1-\omega) \times T_i$$

$$\omega = \alpha \times k, \ 0 \le \omega < 1, \ k \text{ is the number of iteration.} \tag{1}$$

In equation 1, if $\omega$ is zero, the new normal vector is the same as the normal vector of neighboring ray-cone. Consequently, two consecutive ray-cones do not intersect because they are parallel to the each other.

### 3.3 Determination of Ray Direction Using Linear Interpolation

After determining ray-cones on control points, it computes direction of ray fired from sample points. It does not produce ray-cones for sample points instead of determining direction of rays using linear interpolation of boundary points.

As shown in Sect. 3.2, it computes vector $\vec{v}$ between boundary points that have the same index in adjacent ray-cones. Then, the direction of ray is computed using linear interpolation. We can compute the direction of each ray on sample point $S_n$ as following equation.

$$R_{S_n} = R_{i-1} + \vec{v} \times \frac{n}{m+1} - S_n \tag{2}$$

In above equation, $m$ is the total number of sample points between adjacent control points. The $n$ is current position of sample points between $P_i$ and $P_{i-1}$.

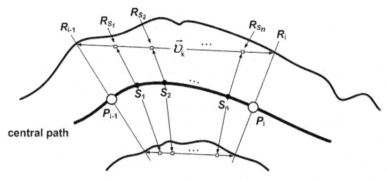

**Fig. 8.** Example of linear interpolation for direction of rays on sample points using ray-cones on control points. White circles are control points and black ones are sample points

### 3.4  Unfolding Image Generation

After determining ray directions, it performs color composition along the ray directions using ray casting.

As shown in Fig. 9, compositing color of each ray corresponds to each pixel on final image. The $u$-coordinate of an image is the index of sample points and the $v$-coordinate of an image is the index of rays.

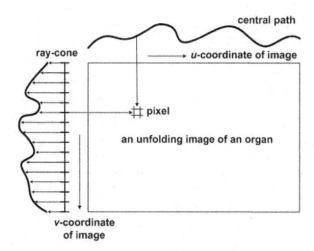

**Fig. 9.** Producing a flatten image by mapping ray-cones on 2D image

## 4  Experimental Results

In order to estimate the image quality and processing time of final image, we implemented basic radial ray casting and our method. These methods are implemented on a PC equipped with Pentium IV 3.0GHz CPU, 1GB main memory. The volume dataset is obtained by scanning a human abdomen with a multi-detector CT of which the resolution is $512 \times 512 \times 684$. We measure the processing time in a human colon. It

uses ray-templates consist of 360 vectors at the interval of one degree and produces
an image of N×360 resolution where N is the number of sample points.

**Table 1.** Comparison of the time of generating unfold images with basic radial ray casting and
our method. To obtain unfold image, we use volume ray casting. Although our method takes
some more time to avoid intersection, it takes time less than basic radial ray casting and image
quality is much better

| # of sample points | Our method(sec) | Original method(sec) | Efficiency (%) |
|---|---|---|---|
| 1992 | 0.9484 | 4.3284 | 456 |
| 1972 | 0.9406 | 4.2222 | 449 |
| 1320 | 0.7359 | 2.9963 | 407 |
| 996 | 0.5861 | 2.2668 | 387 |
| 334 | 0.1828 | 0.7641 | 418 |

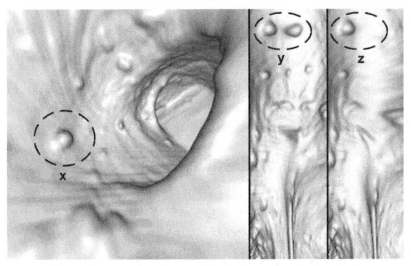

**Fig. 10.** Comparison of virtual endoscopy image (left) and unfold rendering images (middle,
right) for the same region. We can find the fact that single feature is represented twice in the
image produced by basic radial ray casting without intersection test (middle). However, a polyp
represented in our unfold image is the same as that of the endoscopy image (right)

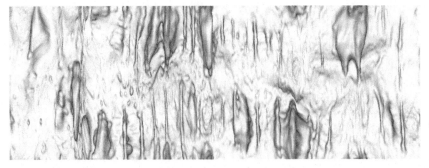

**Fig. 11.** Practical unfold image for human color in two different regions with 996×360 resolu-
tions

# 5  Conclusions

We propose an efficient unfolding method to inspect the inner surface of the colon. It computes ray-cones using ray-templates according to curvature of the path, and adjusts ray-cones that do not meet with adjacent ray-cones using a simple intersection avoidance method. Lastly, it determines direction of all rays fired from sample points by simple linear interpolation. The presented approach solves the problem of missing or duplicating polyps and produces a flatten image in less time compared with the previous methods. Scaling method to make virtual dissection image that is analogies to the real anatomical dissection is the subject of future work. Acceleration of processing time is also the theme of future study.

## Acknowledgement

This work was supported by INHA UNIVERSITY Research Grant.

## References

1. Wang, G., Vannier, M.W.: GI tract unraveling by spiral CT. In Proceedings SPIE, Vol. 2434. (1995) 307-315
2. Gröller, E.: Nonlinear Ray Tracing: Visualizing strange worlds, The Visual Computer, Vol. 11. (1995) 263-274
3. McFarland, E.G., Brink, J.A., Balfe, D.M., Heiken, J.P., Vannier, M.W.: Central axis determination and digital unraveling of the colon for spiral CT colography. Academic Radiology, Vol. 4. (1997) 367-373
4. Wang, G., McFarland, E.G., Brown, B.P., Vannier, M.W.: GI tract unraveling with curved cross-sections. IEEE Transactions on Medical Imaging, Vol. 17, No. 2. (1998) 318-322
5. Wang, G., Dave, S.B., Brown, B.P., Zhang, Z., McFarland, E.G., Haller, J.W., Vannier, M.W.: Colon unraveling based on electrical field: Recent progress and further work. In proceedings SPIE, Vol. 3660. (1999) 125-132
6. Vilanova, A., Gröller, E., Kőnig, A.: Cylindrial approximation of tubular organs for virtual endoscopy. In Proceedings of Computer Graphics and Imaging (2000) 283-289
7. Haker, S., Angenent, S., Tannenbaum, A., and Kikinis, R.: Nondistorting Flattening Maps and the 3-D Visualization of Colon CT Images. IEEE Transactions on Bio-medical Engineering, Vol.19, No.7. (2000) 665-671
8. Haker, S., Angenent, S., Tannenbaum, A., and Kikinis, R.: Nondistorting Flattening for Virtual Colonoscopy. MICCAI (2000) 358-366
9. Vilanova, A., Wegenkittl, R., Kőnig, A., Gröller, E., Sorantin, E.: Virtual colon flattening. In VisSym '01 Joint Eurographics – IEEE TCVG Symposium on Visualization (2001)
10. Vilanova, A., Wegenkittl, R., Kőnig, A., Gröller, E.: Nonlinear Virtual Colon Unfolding. IEEE Visualization (VIS) (2001) 411-579
11. Zhang, Z.: Methods to Visualize Interiors of Human Colons in Volumetric Datasets. Medinfo (2004) 1931
12. Zhu, L., Haker, S., Tannenbaum, A,.: Flattening Maps for the Visualization of Multibranched Vessels. IEEE Transactions on Medical Imaging, Vol. 24, No. 2. (2005) 191-206

# Integration of Genetic and Medical Information Through a Web Crawler System

Gaspar Dias[1], José Luís Oliveira[1],
Francisco-Javier Vicente[2], and Fernando Martín-Sánchez[2]

[1] Universidade de Aveiro, IEETA/DET, 3810 193 Aveiro, Portugal
[2] Instituto de Salud Carlos III (ISCIII), Madrid, Spain

**Abstract.** The huge amount of information coming from genomics and pro-teomics research is expected to give rise to a new clinical practice, where diag-nosis and treatments will be supported by information at the molecular level. However, navigating through bioinformatics databases can be a too complex and unproductive task.

In this paper we present an information retrieval engine that is being used to gather and join information about rare diseases, from the phenotype to the geno-type, in a public web portal – diseasecard.org.

## 1 Introduction

The decoding of the human genome and of other human beings has been promoting a better understanding of the evolution process of the species and the relation between diseases and genes. The integration of these massive amounts of genetic information in the clinical environment is expected to give rise to a new clinical practice, where diagnosis and treatments will be supported by information at molecular level, i.e., molecular medicine [1-3].

The generalization of molecular medicine requires an increased exchange of knowledge between clinical and biological domains [4]. Several databases already exist, covering part of this phenotype-to-genotype connection. A major hindrance to the seamless use of these sources, besides the quantity, is the use of ad-hoc structures, providing different access modes and using different terminologies for the same enti-ties. The specificity of these resources and the knowledge that is required to navigate across them leave its plain usage just to a small group of skilled researchers.

This problem is even more important if we deal with rare disease where the rela-tion between phenotype and genotype is typically strong (around 80% have genetic origins). Rare diseases are those affecting a limited number of people out of the whole population, defined as less than one in 2,000. Despite this insignificant rate, it is esti-mated that between 5,000 and 8,000 distinct rare diseases exist today, affecting be-tween 6% and 8% of the population in total. In the European Community this repre-sents between 24 and 36 million of citizens.

Many rare diseases can be difficult to diagnose because symptoms may be absent or masked in their early stages. Moreover, misdiagnosis, caused by an inadequate knowledge of these diseases, is also common. In this scenario, it is a major goal to maximize the availability of curate information for physicians, geneticists, patients, and families.

Consider a patient with Fabry disease, a rare disease with a prevalence of less than 5 per 10,000. So far, the main sources for finding biomedical information about this disease are basically two: bibliography and the Internet.

J.L. Oliveira et al. (Eds.): ISBMDA 2005, LNBI 3745, pp. 78–88, 2005.

The Web is a valuable source of information provided one knows where and how to look for the information. The easiest and most direct manner of making a search is to use the traditional search engines. To this date, a search engine such as Google indexes more then 8 billion web pages turning the search for a rare disease into an adventure. Just to cite an example, Fabry disease produces about 105,000 entries. To be able to deal with such a number of resources would be impossible if they are not previously filtered.

In this paper we present the DiseaseCard portal, a public information system that integrates information from distributed and heterogeneous public medical and genomic databases. In [5] we have present a first version of DiseaseCard, a collaborative system for collecting empirical knowledge disseminated along research centers, organization and professional experts. In this paper we present an automatic information retrieval engine that is now the computational support for DiseaseCard. Using a pre-defined navigation protocol, the engine gathers the information in real-time and present it to the user through a familiar graphic paradigm.

## 2  Information Resources Selection

The selection of the sources for biomedical information is crucial for DiseaseCard. *Nucleic Acids Research* (NAR) publishes annually "*The Molecular Biology Database Collection*", a list of the most important databases hierarchically classified by areas in this field of interest. In the 2005 update [6], there were 719 databases registered in the journal, 171 more than the previous year, a number, although appreciably lower than the one obtained in Google, still unmanageable by a primary care physician. Each database has its own domain, its own architecture, its own interface, and its own way of making queries, i.e., resources are heterogeneous and distributed.

There are resources specialized in rare diseases accessible via the Internet and that can be queried for all the information available about a pathology. However, they are oriented towards clinical aspects of the disease, relegating genetic aspects to the background. For instance, IIER (iier.isciii.es/er/) or ORPHANET (www.orphanet.net) are websites that in Spain and France, respectively, are points of reference.

### 2.1  Public Databases on Biomedicine

Back to the practical case from which we started, the goal of DiseaseCard is the integration of heterogeneous and distributed genetic and clinical databases, under a common appearance and navigation method. Once Diseasecard does not generate information by itself it is crucial to look upon reliable, curate and frequently updated data sources. To achieve this, several databases registered in NAR have been selected. Guaranteed scientific reliability, exact and frequently updated data and public and free access are common characteristics shared by these databases.

Based on this list we can map several pathways along the web which we use to build a navigation protocol and obtain Diseasecard information.

1. **Orphanet** is a database dedicated to information on rare diseases and orphan drugs. It offers services adapted to the needs of patients and their families, health professionals and researchers, support groups and industrials.

2. **IIER** is the Spanish database of rare diseases, a 'meeting place' for professionals, patients, family, organizations, industry, media and society in general.
3. **ClinicalTrials** provides information about federally and privately supported clinical research in human volunteers. It gives information about a trial's purpose, who may participate, locations, and phone numbers for more details.
4. **OMIM** is the database of human genes and genetic disorders, compiled to support research and education on human genomics and the practice of clinical genetics. The OMIM Morbid Map, a catalog of genetic diseases and their cytogenetic map locations, is now available.
5. **GenBank** is a database that contains publicly available DNA sequences for more than 170000 organisms.
6. The NCBI **dbSNP** is database of genome variation that complements GenBank by providing the resources to build catalogs of common genome variations in humans and other organisms.
7. The **EMBL** Nucleotide Sequence Database is maintained at the European Bioinformatics Institute (EBI) in an international collaboration with the DNA Data Bank of Japan (DDBJ) and GenBank at the NCBI (USA). Data is exchanged amongst the collaborating databases on a daily basis.
8. **Entrez-Gene** provides a single query interface to curated sequence and descriptive information about genetic loci. It presents information on official nomenclature, aliases, sequence accessions, phenotypes, EC numbers, MIM numbers, UniGene clusters, homology, map locations, and related web sites.
9. **Swiss-Prot** is a protein sequence database. It provides a high level of annotation such as description of protein function, domains structure etc.
10. The **ProDom** database contains protein domain families automatically generated from the SWISS-PROT and TrEMBL databases by sequence comparison.
11. **KEGG** (Kyoto Encyclopedia of Genes and Genomes) is the primary database resource of the Japanese GenomeNet service for understanding higher order functional meanings and utilities of the cell or the organism from its genome information.
12. **GeneCards**, is an automated, integrated database of human genes, genomic maps, proteins, and diseases, with software that retrieves, consolidates, searches, and displays human genome information.
13. The **PubMed** database includes over 14 million citations for biomedical articles back to the 1950's. These citations are from MEDLINE and additional life science journals. PubMed includes links to many sites providing full text articles and other related resources.
14. **PharmGKB** is an integrated resource about how variation in human genes leads to variation in our response to drugs. Genomic data, molecular and cellular phenotype data, and clinical phenotype data are accepted from the scientific community at large. These data are then organized and the relationships between genes and drugs are then categorized into different categories.
15. **HGNC** (HUGO), the Human Gene Nomenclature Database is the only resource that provides data for all human genes which have approved symbols. It contains over 16,000 records, 80% of which are represented on the Web by searchable text files.

16. The goal of the Gene Ontology Consortium (**GO**) is to produce a dynamic controlled vocabulary that can be applied to all organisms even as knowledge of gene and protein roles in cells is accumulating and changing.
17. **EDDNAL** is the European Directory of DNA diagnostic Laboratories. It aims to disseminate information among medical genetics health-care professionals concerning the availability of DNA-based diagnostic services for rare genetic conditions in Europe. EDDNAL also seeks to promote the highest standards of genetic testing as well as to facilitate research into the development of new diagnostic tests.

## 3  Navigation Workflow

With all the databases filtered, we have developed a conceptual protocol that follows logical rules combining the two perspectives: genetic and clinical [5]. This protocol allows users to follow a specific pathway from the disease to the gene (Figure 1).

For a given disease, the protocol start search for its symptoms, for general information and related centers of reference (Pathology). This is the entry point of the protocol. A disease can be related to a disorder or a mutation in the genetic pattern of the patient (Disorder & Mutation), like a polymorphism (Polymorphism) for instance. This is due to a change in the sequence of nucleotides (Nucleotide sequence) causing a change in the corresponding protein (Protein: sequence) with a domain and belonging to a family (Domain & Families). As a result the 3D structure of the protein is altered (Protein: 3D structure) and therefore its biological function (Protein: function). The protein takes part in a metabolic pathway (metabolic pathway), carrying out enzymatic reactions (Enzyme). It is at this level where the drugs act, if available (Drug). Clinical trials are carried out with the drugs in Europe and the USA and also pharmacogenetics and pharmacogenomics research (Pharmacogenomics & Pharmacogenetics research). There is bibliography about the disease and there are centers where the genetic diagnosis is made (Genetic Test/Lab). There is also genetic information relevant for R&D such as the exact location in the chromosome, official name of the affected gene, genetic data integrated in an individual card for each gene and information relative to the "Molecular function, Biologic process and Cellular component" in which the gene(s) is (are) involved, included in Gene Ontology.

From the clinical perspective, we have divided the protocol into user profiles fitting the different types of users, according to their information needs. In the top part of the protocol, area 1 (in white), there are the resources that provide information useful to primary care physicians such as information about bibliography, centers for genetic diagnosis and available treatments. The patient asks the primary care physician for the information about the disease and the doctor gets the information by querying DiseaseCard from his own office PC connected to the Internet.

Generally, the hospital specialist, to whom the patient was referred by the primary care physician, does the follow up in this type of pathologies. The specialist looks for more specific and detailed information than the primary care doctor, the areas 1 and 4. Next to the specialist is the hospital geneticist, which uses the information available in the areas 2 and 3 of the protocol.

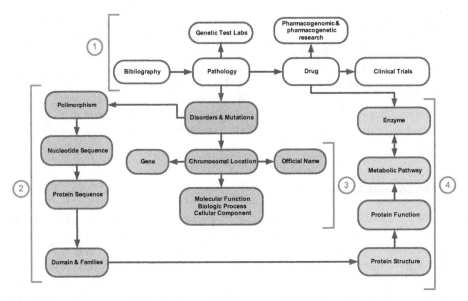

**Fig. 1.** Map of the concepts/databases used in Diseasecard. This illustration shows a navigation path, showing different areas associated with different clinical users: Area 1 – Primary care physician; Areas 2 and 3 – Geneticist; Areas 1 and 4 – Specialist, Pharmacologist

Due to their highly specialized training these professionals require exact genetic information, a need addressed by DiseaseCard by offering the possibility of designing ad hoc individualized cards for each pathology and user.

## 4   An Information Retrieval Engine

In a human-based navigation scheme each user would have to follow a predefined procedure in order to build one disease card manually. He has to navigate, in a given pathway, through several web pages and get web links (URLs) to fill diseasecard's concepts. Figure 2 describes the resource pathways that have to be followed in order to collect all diseasecard information. The "core" databases in this protocol are Orphanet, OMIM, Entrez-Gene, HGNC and Expasy from which most of the navigation nodes can be reached.

Since the cards are built based on a predefined template [5], the data sources to be explored are always the same so the respective queries/URLs are equal except their query ids. Exploring this commonality we can map the building task into a single protocol and construct an automatic retrieval engine.

With this aim, we develop a module (Cardmaker) integrated in Diseasecard System which automatically builds cards based on this single protocol. The protocol is described previously in XML using a specific schema (XML Protocol Descriptor, or XPD). With the protocol descriptor approach, instead of having a hard coded search engine we achieve a dynamic and more comprehensive system to assemble each card information.

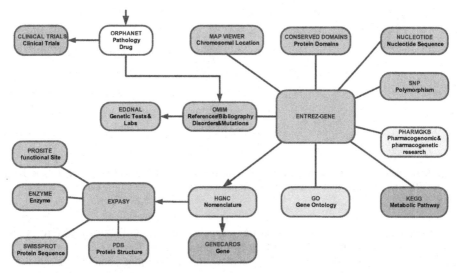

**Fig. 2.** Data sources network. Each box represents a data source containing respective retrievable concepts. Some of these concepts lead to other data sources and other concepts

Through this protocol, the expert user chooses the sources to be consulting during the querying process. This protocol is then used in general queries and interpreted by a parser which converts the XML syntax into search/extract actions (dynamic wrappers) used to explore and retrieve information from web resources. Once this file is defined, it is "plugged" into the system and it is then ready to generate cards.

Our goal is to focus on the development of a more flexible integration system that helps diseasecard's users to gather information from heterogeneous databases in an automated and transparent way.

A main characteristic of the diseasecard portal is that it only manages URL links to the relevant information. If a particular resource is mapped in a XPD file, to extract these links the system has to execute the following steps.

1. Download the web page specified in the protocol entry;
2. Search for data associated to diseasecard concepts;
3. Store extracted URL in a concept item;
4. Step into the next protocol entry;

This procedure is repeated until the end of the protocol.

Figure 3 illustrates an example of part of a protocol. Each block is a protocol entry that represents a data source with respective retrievable card concepts. Some of these concepts are also key parameters to navigate into other data sources. As an example, in the *Orphanet* data source the system can retrieve *pathology*, *Drug* and *Omim* concepts. The first two items are stored in a diseasecard instance and the third is used to step into OMIM site. Now, in OMIM, the system searches for *Reference*, *Disorders and mutations* and *Entrez Gene*'s links and repeats the process until the end of protocol.

The protocol is stored in an "xpd" file that is divided in *wrapper* elements, each associated to a single database. Inside each element it is specified the name of the

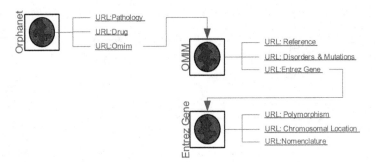

**Fig. 3.** A protocol preview with three data sources and respective retrievable card concepts. Each URL link is stored in a concept item and in some cases it is the entry point to access other sites

resource, respective URL and filtering terms to search and extract the relevant information. These filtering or matching terms are based on *Regular Expressions*[1] [7]. Each *wrapper* element will be interpreted by the *parser* that generates a run-time retrieval component. The following text is a piece of the protocol showed in Figure 3.

```
<wrapper>
   <resource-name>Orphanet</res    ource-name>
   <resource-url key-origin="_ext"><![CDATA[http://orphanet.url?KEY]]>
      </resource-url>
   <search-for>
      <regex for pathology><search-for>
      <regex><![CDATA[<regex for omim code>]]></regex>
      <put-into>_omim</put-into>
   </search-for>
</wrapper>
```

Once this process is mapped in a XML protocol descriptor, we need an engine to interpret the protocol language and to execute and control the automated card construction. To do this work we have developed and integrated into the Diseasecard System the XPD Engine, which is illustrated in Figure 4. Note that this engine works inside the *Cardmaker* tool.

This engine generates diseasecard instances based on the user request and on an XPD file. A *Parser* converts the XML language into a set of tasks called the *Task Plan*. These tasks will be used afterward to create web wrappers dynamically when requested by the *Controller*. Based on the *Task Plan*, the *Controller* manages these wrappers which generate card concepts that are finally assembled into a new card and stored in the Diseasecard's Database.

---

[1] A regular expression (abbreviated as regexp, regex or regxp) is a string that describes or matches a set of strings, according to certain syntax rules. Regular expressions are used by many text editors and utilities to search and manipulate bodies of text based on certain patterns

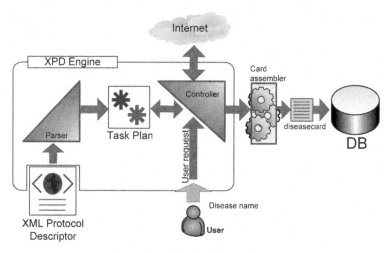

**Fig. 4.** The XPD Engine architecture

While this XPD engine is enough generic to be applied in any field of knowledge where information retrieval from public database must be performed, we show here its functionality inside the framework of diseasecard portal.

## 5 DiseaseCard Portal

DiseaseCard is a web portal publicly available that provides an integrated view over medical and genetic information with respect to genetic diseases (www.diseasecard. org). The end-user will typically start the investigation by searching for a disease or providing its name. Upon the identification of the disease, the portal will present a structured report, containing the details of the pathology and providing entry points to further details/resources either on the clinical and genetic domains.

Several aspects of older version of Diseasecard have been changed. We have decided to integrate the *Cardmaker/XPD Engine* tool into the querying process and this has implied a complete different usage paradigm – from a collaborative user-based annotation system (previous version) into an automatic information retrieval application (actual version). Although instead of having two different user operations, card querying and card creation, the system merges both tasks into a single one, providing a more useful and intuitive interface. *Cardmaker* works in the background if database doesn't have any card associated to the user request. Otherwise the system returns a card or a set of cards that matches with the user request (Figure 5).

Table 1 shows the main steps of Diseasecard evolution since it was created. The first goal that was behind the conception of Diseasecard was the development of a web tool where a group of experts could collaborate on-line in order to share, store and spread their knowledge about rare diseases (DC1.0). Since we have concluded that the majority of the diseasecard protocol could be automated we have developed an alternative way of creating cards through the *Cardmaker* tool. This utility was available only for authenticated users (DC2.0). After some improvements on *Cardmaker* we realized that this tool has a good time performance and it can create reliable diseasecard contents. So we decided to delegate this task totally to the system. With this approach, card creation is transparent to any user because the tool works in run-time when he asks for some card that isn't defined in database (DC3.0).

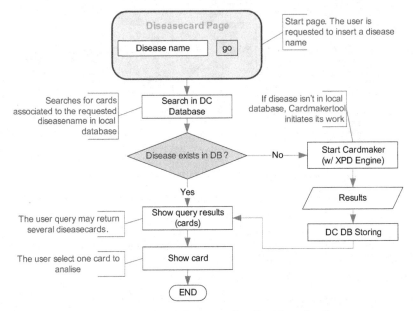

**Fig. 5.** Behavior of the Diseasecard querying operation. Based on the disease name requested by user, the Diseasecard checks for cards related to requested disease name in the database. If exists, the system shows the card contents. If not, the system launches the Cardmaker/XPD Engine. If this operation is succeeded, then the results are showed in the card view

**Table 1.** Diseasecard's versions

| Diseasecard Version | Main Features |
|---|---|
| Diseasecard 1.0 | Collaborative web tool |
| Diseasecard 2.0 | Collaborative web tool + Automated card creation |
| Diseasecard 3.0 | Automated card creation |

The next illustrations show two perspectives of the Diseasecard System. Our concern in the main page (Figure 6) is to provide a simple and intuitive interface in order to facilitate the user's search. The main page provides two different query modes which are "*search for disease*" through a single word or expression related to any disease or search from the "*list of available diseases*" where the user selects a disease from an alphabetic list.

Figure 7 appears after selecting a disease from the returned list originated by a user query. It shows on the left side the diseasecard structure which contains all disease-card's concepts and respective links to associated web resources. This example shows a rare disease called *Fanconi Anemia*. The selected concept in this case is *pathology* and it links to an *Orphanet* page which contains a description of this disease showed on the right side of the page.

**Fig. 6.** Diseasecard.org – The main page

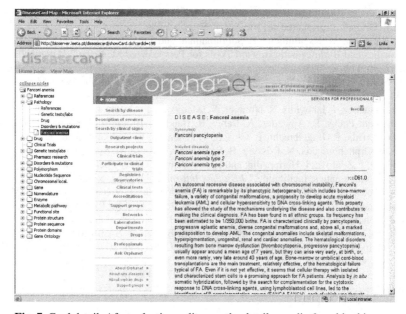

**Fig. 7.** Card detail. After selecting a disease, the details are displayed in this page

## 6   Conclusions

The recent advances on genomics and proteomics research bring up a significant grow on the information that is publicly available. However, navigating through genetic and bioinformatics databases can be a too complex and unproductive task for a primary care physician. Moreover, considering the rare genetic diseases field, we verify that the knowledge about a specific disease is commonly disseminated over a small group of experts.

The diseasecard is a web portal for rare diseases, based on an automatic information retrieval engine, that provides transparently to the user a virtually integration of distributed and heterogeneous information – using a pathway that goes from the symptom to the gene. With this system medical doctors can access genetic knowledge without the need to master biological databases, teachers can illustrate the network of resources that build the modern biomedical information landscape and general citizen can learn and benefit from the available navigation model.

## Acknowledgement

This portal was developed under the endorsement of the INFOGENMED project and of the NoE INFOBIOMED, both funded by the European Community, under the Information Society Technologies (IST) program.

## References

1. R. B. Altman, "Bioinformatics in Support of Molecular Medicine" Em AMIA Annual Symposium, Orlando.," presented at Proc AMIA Symp., 1998.
2. B. D. Sarachan, M. K. Simmons, P. Subramanian, and J. M. Temkin, "Combining Medical Informatics and Bioinformatics toward Tools for Personalized Medicine," *Methods Inf Med*, vol. 42, pp. 111-5, 2003.
3. D. B. Searls, "Data Integration: Challenges for Drug Discovery," *Nature Reviews*, vol. 4, pp. 45-58, 2005.
4. Martin-Sanchez, I. Iakovidis, S. Norager, V. Maojo, P. de Groen, J. Van der Lei, T. Jones, K. Abraham-Fuchs, R. Apweiler, A. Babic, R. Baud, V. Breton, P. Cinquin, P. Doupi, M. Dugas, R. Eils, R. Engelbrecht, P. Ghazal, P. Jehenson, C. Kulikowski, K. Lampe, G. De Moor, S. Orphanoudakis, N. Rossing, B. Sarachan, A. Sousa, G. Spekowius, G. Thireos, G. Zahlmann, J. Zvarova, I. Hermosilla, and F. J. Vicente, "Synergy between medical informatics and bioinformatics: facilitating," *J Biomed Inform*, vol. 37, pp. 30-42., 2004.
5. J. L. Oliveira, G. Dias, I. Oliveira, P. Rocha, I. Hermosilla, J. Vicente, I. Spiteri, F. Martin-Sánchez, and A. S. Pereira, "DiseaseCard: A Web-Based Tool for the Collaborative Integration of Genetic and Medical Information," presented at Biological and Medical Data Analysis: 5th International Symposium (ISBMDA'2004), Barcelona, Spain, 2004.
6. M. Y. Galperin, "The Molecular Biology Database Collection: 2005 update," *Nucleic Acids Research*, vol. 33, 2005.
7. Regular Expressions in Java. (n.d.) Retrieved June 17, 2005, from http://www.regular-expressions.info/java.html

# Vertical Integration of Bioinformatics Tools and Information Processing on Analysis Outcome

Andigoni Malousi, Vassilis Koutkias, Ioanna Chouvarda, and Nicos Maglaveras

Lab. of Medical Informatics, Faculty of Medicine, Aristotle University
P.O. Box 323, 54124, Thessaloniki, Greece
{andigoni,bikout,ioanna,nicmag}@med.auth.gr

**Abstract.** Biological sources integration has been addressed in several frameworks, considering both information sources incompatibilities and data representation heterogeneities. Most of these frameworks are mainly focused on coping with interoperability constraints among distributed databases that contain diverse types of biological data. In this paper, we propose an XML-based architecture that extends integration efforts from the distributed data sources domain to heterogeneous Bioinformatics tools of similar functionalities ("vertical integration"). The proposed architecture is based on the mediator/wrapper integration paradigm and a set of prescribed definitions that associates the capabilities and functional constraints of each analysis tool. The resulting XML-formatted information is further exploited by a visualization module that generates comparative views of the analysis outcome and a query mechanism that handles multiple information sources. The applicability of the proposed integration architecture and the information handling mechanisms was tested and substantiated on widely-known ab-initio gene finders that are publicly accessible through Web interfaces.

## 1 Introduction

The distribution of massive amounts of biological data through several multi-institutional databanks has prompted the development of numerous decentralized Bioinformatics tools that analyze, predict and attempt to interpret raw data into meaningful knowledge [1]. Currently, a wide range of analysis tools are freely accessible through Web interfaces implementing diverse algorithmic approaches on multiple Bioinformatics problems. The efficiency of these methods is increasingly inviting and the exported outcomes are quite valuable, in order to decipher and model biological processes. An even more challenging issue is to orchestrate multiple resources and accordingly interrelate their functionalities/outcomes through integration frameworks that are built upon globally accepted data exchange protocols [2]. Despite the complexity of the domain, systematic efforts gave rise to the establishment of structured and semi-structured data formats that facilitate advanced querying and interpreting capabilities, basically addressing requirements of diverse data that are deposited in heterogeneous databanks [3].

J.L. Oliveira et al. (Eds.): ISBMDA 2005, LNBI 3745, pp. 89–100, 2005.

In this work, we propose an XML-based architecture that a) integrates multiple, heterogeneous biological analysis tools of similar functionalities, i.e., vertical integration, b) structures the resulting outcomes in form of XML documents that follow common formatting instructions, and c) incorporates advanced information handling capabilities in terms of comparative graphical views and query processing on the analysis results. The applicability of the proposed approach was tested on a set of ab-initio gene prediction tools that are freely accessible through Web interfaces.

## 2    Technical Background

### 2.1    Bioinformatics Resources Profile

In general, the design and implementation requirements of an integration architecture are implied by physical and functional restrictions of the associated resources. In the Bioinformatics field most analysis tools share common characteristics as described below [4]:

- *Decentralization*: Typically, most Bioinformatics analysis tools are accessible through Web interfaces that span through geographically distributed servers. A specific functionality may be provided by multiple tools and the analysis outcome may diverse depending on the algorithmic approach followed.
- *Heterogeneity*: Diversities among Bioinformatics tools involve both their schematic and semantic representation. Interoperability issues are partially coped with the adoption of technologies such as XML, Web Services, etc. Currently, several biological data repositories have adopted XML and related technologies within their general data model, in order to offer structured views of non-structured data sources, facilitating this way flexible data handling [5]. Unlike databanks, most biological analysis tools do not offer machine-readable and processable data structures of their outcome.
- *Autonomy*: Most analysis tools are functionally autonomous, i.e., they perform analysis independently and their underlying design and implementation model may be changed/updated without prior public notification.
- *Multiple data formats*: The data types involved in an analysis range from simple, textual DNA sequences to complex 3D protein structures, according to domain-specific requirements.
- *Query capabilities, parameterization*: Analysis tools that typically offer similar functionalities may exhibit diverse query capabilities or even different configuration parameters, depending mostly on their underlying algorithmic approach.

Based on these remarks, it is evident that the development of integration architectures that interlink multiple types of biological analysis resources requires a thorough examination of the functional requirements, and the implied interoperability constraints.

## 2.2  XML and Bioinformatics

XML is the standard markup language for describing structured and semi-structured data over the Internet [6]. Together with other proposed and accepted standards, supervised also by the W3C Consortium, XML is considered an ideal means to provide the infrastructure for integrating heterogeneous resources. Several biological databases such as Entrez [7] and EMBL [8], support implementations of external modules that export or view contents as XML documents and plug them to their general relational schema. In addition, various XML-based markup languages have been introduced as standard formats for describing biological data such as BioML (Biopolymer Markup Language) [9], MAGE-ML (Microarray Gene Expression Markup Language) [10], and BSML (Bioinformatic Sequence Markup Language) [11].

With the advent of more sophisticated XML-based technologies, including RDF (Resource Description Framework) and Web Services, XML is intended to play a more important role in the future. Moreover, related technologies, such as XQuery [12], offer enhanced capabilities in performing complex queries and retrieving combined data from multiple XML data sources.

## 2.3  Integration Architectures in Bioinformatics

Several integration frameworks have been developed aiming to facilitate reusability capabilities among the incorporated resources, addressing diverse descriptive data models and transparency levels. Apart from the warehousing approach [13], mediator-based integration schemes, followed in the proposed architecture, handle on demand integration requests through tailored query formulation [14], [15]. For example, BioKleisli is an integration architecture relying on an object-oriented database system that takes advantage of the expressiveness of CPL (Collection Programming Language), in order to perform queries against multiple heterogeneous resources in pipelines of domain-specific processing nodes [16]. TAMBIS (Transparent Access to Multiple Bioinformatics Information Sources) follows a mediator-based integration scheme aiming to formulate queries against ontology-driven descriptions of Bioinformatics sources [17]. In addition, DiscoveryLink offers intermediary modules available to applications that compose requests on retrieving data from multiple biological sources [18]. Unlike TAMBIS, DiscoveryLink is a wrapper-oriented system that is intended to be used by other applications rather than end-users.

Most of these integration architectures focus on biological databases, rather than on analysis tools. Moreover, integration in the aforementioned architectures involves primarily the orchestration of complementary resources in form of workflows that execute predefined tasks in sequential order, aiming to reveal hidden dependencies/relations among different types of biological data. Unlike these horizontal integration architectures [19], this work focuses on a vertical integration scheme, in which multiple Bioinformatics analysis tools of similar functionalities are integrated, so as to provide machine-readable and comparative views of their outcomes.

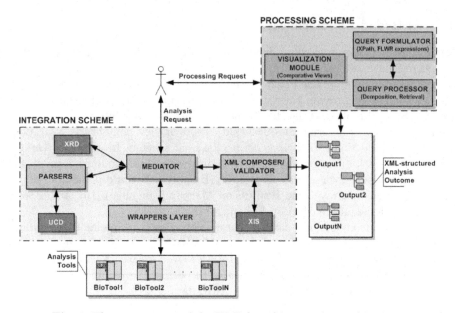

**Fig. 1.** The components of the XML-based integration architecture

## 3   Integration Scheme

As shown in Fig. 1, the proposed architecture involves two distinct parts: a) the Integration Scheme which follows the mediator/wrapper paradigm, and b) the Processing Scheme, which provides graphical representation and advanced query capabilities on the analysis outcome. It has to be noted that these two parts are not synchronized and may operate independently, since the XML documents generated by the former part are stored locally and, therefore, can be accessed by the processing modules on demand. The basic structural and functional components of the Integration Scheme are described in the following.

### 3.1   Structural Components

The specifications of the proposed integration architecture rely on particular features that describe Bioinformatics tools and distinguish them from the general physical and technical characteristics of the biological data sources. Unlike distributed databases, most Web-accessible biological analysis tools present their outcomes in unstructured or semi-structured forms, focusing more on human-readability, rather than machine-manageable formats. To cope with representational heterogeneities and potential accessibility incompatibilities of the associated resources, a set of structural components is included in the Integration Scheme as described below.

**Uniform Concept Description.** Even within the same class of Bioinformatics tools, there may be variations in the type of information provided, different metrics to encode quantitative parameters (e.g., probabilities and reliability scores), and grammatically diverse representations of common concepts. To increase the expressiveness of the extracted information, a Uniform Concept Description (UCD) is incorporated, which describes similar concepts, related to both input and output data, using unified terms that are stored in a global data index.

**Resource Description.** The XML Resource Description (XRD) module contains descriptions related to the accessibility parameters of each tool in machine-processable way that facilitates interoperation among relevant functional components. So far, most Bioinformatics tools that perform computational data analysis do not offer self-descriptive capabilities to allow direct access to their sources, thus, it is necessary to associate resource descriptions within the integration scheme. In addition, due to heterogeneities in the schematic representation, it is not feasible to incorporate a global schema that describes resources. Therefore, for each analysis tool the corresponding descriptive elements are defined in terms of XML-formatted tuples.

**Information Schema.** The information extracted from heterogeneous Bioinformatics analysis tools is structured in XML-formatted documents that are validated against a predefined XML-based Information Schema (XIS). The XIS provides a hierarchical view of the associated tagged elements, corresponding to the extracted data along with their relations. The exported XML-formatted outcome must comply with the formatting instructions of the XIS, in order to enable data handling by the query and visualization modules included in the Processing Scheme. The XIS elements are defined according to problem-specific requirements, and the selection is primarily based on the significance and relevance with the problem to be solved.

## 3.2   Functional Components

The functional components of the Integration Scheme provide transparent and coordinated access to the appropriate analysis tools, according to the requested analysis [15]. This way, the inherent complexities of submitting a request to the appropriate resource(s) and retrieving the analysis outcome(s) are hidden from the end-user. Furthermore, information extraction is applied on the obtained results, in order to select features that are relevant with the structural specifications, and, finally, compose XML-structured documents containing the information of interest. This functionality is supported by the following modules.

**Mediator.** Given a user request, the mediator decomposes the query, passes the relevant parameters to each one of the wrappers and coordinates the overall submission/retrieval procedure. To avoid inconsistencies among user requirements and the tools' specifications, the mediator excludes non-matching resources. The tools' compliance is validated by the mediator that matches the query parameters to the XRDs and simultaneously performs CGI and/or Java servlet requests on the matching resources, through the corresponding wrapping modules [15].

**Wrappers/Parsers.** For a set of analysis tools, a wrapping layer is incorporated, i.e., each tool corresponds to a relevant wrapper. Each wrapper accesses the XRD module of its assigned tool, through the mediator module, submits requests, and ships back the unstructured information obtained to the mediator [20]. After that, the mediator initiates the corresponding parsing modules. The outcome that is obtained by each wrapper is passed through the corresponding parsing module that filters the extracted information according to the XIS descriptions. Moreover, to overcome representational heterogeneities coming from semantically similar concepts that are encoded in different terms, the global XML schematic descriptions of the incorporated outcome are used to transform each tool's terminology into annotations of the predefined UCD.

**Document Composer/Validator.** The extracted information is processed by the Document Composer, which structures data within appropriate tagged elements, following the formatting rules of the predefined XIS. Accordingly, the Document Validator confirms the validity of the extracted XML documents and reports potential inconsistencies with the prescribed XIS definition. This process is iteratively applied and the resulting information is stored in XML documents.

## 4   Processing Scheme

In the proposed architecture, information handling is related with visualization of the generated XML documents, in terms of comparative views of the obtained outcomes, and an advanced query formulation/execution mechanism that enables feature extraction and interrelation among the information obtained from vertically integrated analysis tools. The basic modules contained in the Processing Scheme are shown in Fig. 1 and described below.

### 4.1   Visualization Module

The Visualization Module serves the need of combining and comparing evidence derived from multiple vertically integrated resources, in the form of user-friendly depictions that may help researchers get a more comprehensive view of the extracted information and even facilitate further investigation of the underlying biological processes. Apart from the XML documents' content, graphical representations may be applied on the resulting XML views generated by specific XQuery statements. In both cases, the information depicted may be configured with respect to the query criteria and the technical constraints implied by each analysis problem.

## 4.2   Query Processing

**Query Requirements.** The query mechanism of the Processing Scheme addresses comparative management and analysis of the information enclosed in the tuples of the validated XML documents. The query processing capabilities conform with specific structural and functional requirements such as:

- The XIS-compliant data that are dynamically extracted from the integration scheme are stored in valid and machine-readable documents.
- Query processing may involve more than one information sources.
- The tree nodes of the XML-formatted documents are a priori known, unlike the number of instances of each tuple which is dynamically determined.
- The resulting documents follow same schema definitions and incorporate semantically identical concepts.
- Information retrieval is performed using both simple XPath expressions, against a single source, and complex statements targeting multiple sources.
- The query processing may involve the construction of new tagged elements, generating aggregated views of single or multiple information sources.
- The query mechanism does not need to address error-handling, since the outcomes' well-formedness and validity are confirmed by the XML Validator.

The Query Processor is built on the XQuery data model [12]. XQuery considers sources as nesting structures of nodes, attributes and atomic values, and assigns XML-oriented descriptions in both query formulation and information retrieval. The implemented Query Processor constitutes the intermediary connecting the client interface with the XML data sources in an effort to further manage information derived from the vertically integrated tools and reveal hidden dependencies/relations among the incorporated I/O features.

**Types of Information Retrieval.** Three types of queries are supported:

1. *Join*: Generate aggregated views of node elements encoding similar concepts that span through multiple information sources. This type of query serves the vertical integration requirements, supporting customized views of selected tuples that are stored in same-schema information sources.
2. *Select-on-One*: Navigate through the tagged elements of a single source, in order to select node(s)/atomic value(s) that match user-defined criteria. This type of information is obtained by simple XPath/XQuery expressions.
3. *Select-on-Many*: Comparative exploration by defining selection rules against multiple information sources. To address this type of requests, the query mechanism has to formulate and execute complex FLWR (For-Let-Where-Return) expressions in the generic form of:

```
[LET <srcᵢ> :=doc(<URIᵢ>) <pᵢ>]
FOR [<vrᵢ> in <spᵢ>]
WHERE [<f(vrᵢ) >]
RETURN <atomic value(s)/tuple(s)> of <vrᵢ>
```

Assuming $N$ XML-formatted information sources, the LET clause locates the source documents by their $\{URI_i | i = 1 \ldots N\}$ definitions and binds each identifier to a variable $\{src_i | i = 1 \ldots N\}$. Expressions in square brackets may appear more than once, thus, the LET clause may be similarly repeated to include additional sources. $p_i$ and $sp_i$ correspond to simple path expressions that bind the target tuples to variables $src_i$ and $vr_i$ respectively (FOR clause). The WHERE clause contains the condition declarations that meet end-user requirements using logical, numerical operators and other functions among the predefined variables $vr_i$. The RETURN clause defines the resulting atomic values/tuples that match selection criteria.

# 5   Test Case: Ab-initio Gene Prediction

## 5.1   Problem Description

Computational gene prediction in eukaryotic organisms is performed by identifying coding features (*exons*) within usually larger fragments of non-coding regions (*introns*) [21]. Ab-initio methods rely exclusively on the intrinsic compositional and structural features of the query DNA sequence in order to build the most probable gene structure. Technically, these approaches implement various statistical methods, applied on species-specific gene models, and the analysis outcome is usually associated with probabilities/scores that reflect the reliability of the prediction [22]. The sensitivity/specificity of each prediction tool exhibits significant variations, depending mostly on the trained gene model and the underlying algorithmic approach. Moreover, various exhaustive assessments have shown that most ab-initio gene finders exhibit high accuracy prediction levels at nucleotide level; however the overall accuracy at gene level is still disputable, basically due to alternative expression patterns that change coding boundaries and therefore the identified gene assembly [23], [24].

Currently, there are 25 ab-initio gene prediction tools that are freely accessible through Web interfaces [22]. The applicability of the presented architecture was tested and substantiated on a subset of 5 widely known gene finders, namely, *Augustus*, *Fgenes*, *Fgenesh*, *Genscan*, and *HMMgene*. The following sections describe the implemented architectural components and illustrate the virtue of the approach adopted through examples of use.

## 5.2   Integrating Ab-initio Gene Finders

The gene finding integration scheme was implemented in accordance with the structural and the functional modules of the Integration Scheme shown in Fig. 1. Specifically, for each gene finder an XRD module was incorporated addressing individual accessibility requirements and descriptions of the associated I/O and configuration parameters. In addition, to cope with the representational heterogeneities of the extracted information, a common vocabulary was introduced that encodes the predicted features along with their positional descriptions into common terms, following the UCD specifications.

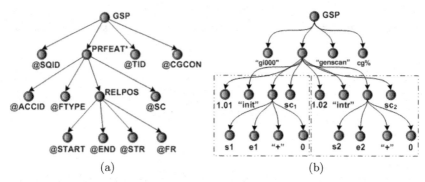

**Fig. 2.** (a) Hierarchical view of the XIS (* indicates potential repeated elements and @ denotes valued schema elements) and (b) an example well-formed XML document containing sample values

The associated information types along with their relations and accepted terminology were finally captured in an XIS description, as shown in Fig. 2(a). Each exonic feature $(PRFEAT)$ identified within the query DNA sequence $(@SQID)$ by a gene finder $(@TID)$ is further described by its type $(@FTYPE)$, its positional features $(RELPOS@[START, END, STR, FR])$, and the associated probability/score $(@SC)$. An example XML document that conforms to the XIS formatting instructions is shown in Fig. 2(b). The overall coordination of the processes involved in the integration scheme is taken by a mediator that was implemented according to the specifications of the generic Integration Scheme. Similarly, for each gene finder the corresponding wrapper and parser modules were developed resulting in uniform outcome descriptions that are finally structured in XML-formatted documents that comply with the predefined XIS rules.

### 5.3   Comparative Graphical Representation

In general, the Visualization Module is configured based on problem-specific requirements aiming to generate user-friendly and comparative depictions of the analysis outcome. In this case, the Visualization Module was used to dynamically formulate graphical representations of the identified coding features in horizontal panels that facilitate comparisons of the evidence derived from the incorporated gene finders. Figure 3 illustrates a comparative depiction of the exons identified within a query genomic sequence that is known to contain exons 2-9 of the human *p53* tumor suppressor gene. It is noteworthy that the identified exonic features are not identical (especially at 5' end region) in all predictions, resulting in alternative transcript forms.

### 5.4   Query Processing

According to the types of information retrieval described in Section 4.2, the query processing mechanism may be applied either on a single source, serving simple navigation/filtering requirements, or on multiple XML-based sources in

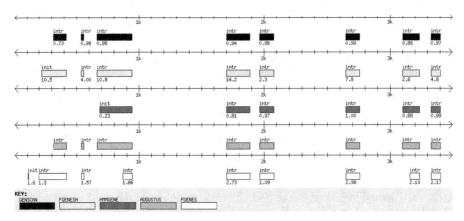

**Fig. 3.** Comparative graphical view of the exons identified in human chromosome 17, within 7,512,465:7,520,980 genomic region that is known to contain exons 2-9 of the *p53* tumor suppressor gene. The labels *init* and *intr* indicate an initial and internal exon respectively. Each exon (except those identified by *Augustus*) is associated with a probability (*Genscan, HMMgene*), or score (*Fgenesh, Fgenes*)

order to join, compare and/or select specific segments of the included tagged elements. An example *"Select-on-Many"* request that the query processor can answer is:

*"On the overall prediction outcome obtained for p53 human genomic sequence, identify exonic features that:*
1. *are predicted by both Genscan and Fgenesh, and*
2. *exhibit high probability scores (Genscan probability > 0.80), and*
3. *their relative start site is located within 100:1000 genomic region".*

The corresponding XQuery statement that retrieves the matching features of the declared XML sources is presented in the left column of Table 1. The tagged-elements in the right column contain the results obtained by executing the XQuery expression. As shown, the query is configured to retrieve specific elements (RETURN clause), providing a quantitative view of the matching features.

## 6   Conclusion

The architecture described in this paper addresses both vertical integration of biological analysis tools and information management on the resulting outcome. The mediator/wrapper approach was adopted for transparent and coordinated access to the incorporated resources and XML has been used to describe the structural components of the Integration Scheme, as well as the evidence derived from the analysis outcome. The information handling mechanisms are applied on the results obtained from the Integration Scheme, aiming to further exploit the information captured within the resulting XML documents. These handling capabilities are supported by a visualization module that generates comparative

**Table 1.** An example FLWR expression (left column) and the resulting tagged elements that match selection criteria (right column)

| Query | Results |
|---|---|
| `<QUERY>` | `<QUERY>` |
| `{` | `<VIEW>` |
| `let $h :=doc(<fgenesh source>)//RELPOS` | `<TYPE>intr</TYPE>` |
| `for $g in doc(<genscan source>)//PRFEAT` | `<BEGIN>535</BEGIN>` |
| `where (($g/RELPOS/START=$h/START)` | `<END>556</END>` |
| `and ($g/RELPOS/END=$h/END)` | `<SCORE>0.983</SCORE>` |
| `and ($g/RELPOS/BEGIN>100)` | `</VIEW>` |
| `and ($g/RELPOS/BEGIN<1000)` | `<VIEW>` |
| `and ($g/SC>0.80))` | `<TYPE>intr</TYPE>` |
| `return` | `<BEGIN>666</BEGIN>` |
| `<VIEW>` | `<END>944</END>` |
| `  <TYPE>{$g/FTYPE/text()}</TYPE>` | `<SCORE>0.982</SCORE>` |
| `  <BEGIN>{$g/RELPOS/START/text()}</BEGIN>` | `</VIEW>` |
| `  <END>{$g/RELPOS/END/text()}</END>` | `</QUERY>` |
| `  <SCORE>{$g/SC/text()}</SCORE>` | |
| `</VIEW>` | |
| `}` | |
| `</QUERY>` | |

graphical descriptions of the analysis outcome and a query processing mechanism that formulates and executes complex query expressions against multiple information sources, based on the expressiveness of the XQuery data model.

Following the need for interoperable biological resources, the proposed architecture may be adopted in various applications that address vertical integration problems. Furthermore, the modularity and extendability of the associated components enables incorporation of complementary resources that could help interlink diverse forms of biological data. Being able to co-relate different types of biological evidence and orchestrate distributed and heterogeneous resources is necessary, in order to increase our understanding and knowledge on the underlying biological processes.

# References

1. Stevens, R., Goble, C., Baker, P., Brass, A.: A Classification of Tasks in Bioinformatics. Bioinformatics 17(2) (2001) 180–188
2. Wong, R.K., Shui, W.: Utilizing Multiple Bioinformatics Information Sources: An XML Databases Approach. Proceedings of IEEE Symposium on Bioinformatics and Bioengineering IEEE Computer Society (2001) 73–80
3. Ng, S.-K., Wong, L.: Accomplishments and Challenges in Bioinformatics. IEEE IT Professional 6(1) (2004) 44–50
4. Hernandez, T., Kambhampati, S.: Integration of Biological Sources: Current Systems and Challenges. ACM SIGMOD Record 33(3) (2004) 51–60

5. Achard, F., Vaysseix, G. Barillot, E.: XML, Bioinformatics and Data Integration. Bioinformatics 17(2) (2001) 115–125
6. World Wide Web Consortium: Extensible Markup Language (XML) 1.0 W3C Recommendation (Third Edition) (2004) http://www.w3.org/TR/2004/REC-xml-20040204/
7. Entrez - Search and Retrieval System. http://www.ncbi.nlm.nih.gov/Entrez (2003)
8. Wang, L., Riethoven, J.-J., Robinson, A.: XEMBL: Distributing EMBL Data in XML Format. Bioinformatics 18(8) (2002) 1147–1148
9. Fenyö, D.: The Biopolymer Markup Language. Bioinformatics 15(4) (1999) 339–340
10. Spellman, P.T. et al.: Design and Implementation of Microarray Gene Expression Markup Language (MAGE-ML). Genome Biology 3(9) (2002)
11. BSML, XML Data Standard for Genomics: The Bioinformatic Sequence Markup Language http://www.bsml.org/
12. World Wide Web Consortium: XQuery 1.0: An XML Query Language (W3C Working Draft) (2005) http://www.w3.org/TR/xquery/
13. Hammer, J., Schneider, M.: Genomics Algebra: A New, Integrating Data Model, Language, and Tool for Processing and Querying Genomic Information. Proceedings of the 2003 CIDR Conference (2003)
14. Tork Roth, M., Schwarz, P.: Don't Scrap it, Wrap it! A Wrapper Architecture for Legacy Sources. Proceedings of the 23rd VLDB Conference Greece (1997) 266–275
15. Koutkias, V., Malousi, A., Maglaveras, N.: Performing Ontology-driven Gene Prediction Queries in a Multi-Agent Environment. In: Barreiro J. M., Martin-Sanchez, F., Maojo, V., Sanz, F. (eds.): Biological and Medical Data Analysis. LNCS, Vol. 3337. Springer-Verlag, Berlin Heidelberg New York (2004) 378–387
16. Davidson, S.B., Buneman, O.P., Crabtree, J., Tannen, V., Overton, G. C., Wong, L.: BioKleisli: Integrating Biomedical Data and Analysis Packages. In: Letovsky S. (Ed.): Bioinformatics: Databases and Systems. Kluwer Academic Publishers, Norwell, MA (1999) 201–211
17. Baker, P., Brass, A., Bechhofer, S., Goble, C., Paton, N., Stevens R.: TAMBIS: Transparent Access to Multiple Bioinformatics Information Sources. Proceedings of the 6th International Conference on Intelligent Systems for Molecular Biology (1998)
18. Haas, L., Schwarz, P., Kodali, P., Kotlar, E., Rice, J., Swope, W.: DiscoveryLink: A System for Integrated Access to Life Sciences Data Sources. IBM Systems Journal 40(2) (2001) 489–511
19. Sujansky, W.: Heterogeneous Database Integration in Biomedicine. Journal of Biomedical Informatics 34 (2001) 285–298
20. Aldana, J.F., Roldan, M., Navas, I., Perez, A.J., Trelles, O.: Integrating Biological Data Sources and Data Analysis Tools through Mediators. Proceedings of the 2004 ACM Symposium on Applied Computing Nicosia Cyprus (2004) 127
21. Zhang, M.Q.: Computational Prediction of Eukaryotic Protein-coding Genes. Nature 3 (2002) 698–710
22. Mathé, C., Sagot, M.F., Schiex, T., Rouzé, P.: Current Methods of Gene Prediction, their Strengths and Weaknesses. Nucleic Acids Research 30 (2002) 4103–4117
23. Rogic, S., Mackworth, A.K., Ouellette, F.: Evaluation of Gene-finding Programs on Mammalian Sequences. Genomic Research 11 (2001) 817–832
24. Mironov, A.A., Fickett, J.W., Gelfand, M.S.: Frequent Alternative Splicing of Human Genes. Genome Research 9 (1999) 1288–1293

# A Grid Infrastructure for Text Mining of Full Text Articles and Creation of a Knowledge Base of Gene Relations

Jeyakumar Natarajan[1], Niranjan Mulay[2], Catherine DeSesa[3],
Catherine J. Hack[1], Werner Dubitzky[1], and Eric G. Bremer[3]

[1] Bioinformatics Research Group, University of Ulster, UK
{j.natarajan,cj.hack,w.dubitzky}@ulster.ac.uk
[2] United Devices Inc, Austin, TX, USA
niranjan@ud.com
[3] Brain Tumor Research Program, Children's Memorial Hospital,
Feinberg School of Medicine, Northwestern University, Chicago, IL USA
egbremer@northwestern.edu, cdesesa@comcast.net

**Abstract.** We demonstrate the application of a grid infrastructure for conducting text mining over distributed data and computational resources. The approach is based on using LexiQuest Mine, a text mining workbench, in a grid computing environment. We describe our architecture and approach and provide an illustrative example of mining full-text journal articles to create a knowledge base of gene relations. The number of patterns found increased from 0.74 per full-text articles from a corpus of 1000 articles to 0.83 when the corpus contained 5000 articles. However, it was also shown that mining a corpus of 5000 full-text articles took 26 hours on a single computer, whilst the process was completed in less than 2.5 hours on a grid comprising of 20 computers. Thus whilst increasing the size of the corpus improved the efficiency of the text-mining process, a grid infrastructure was required to complete the task in a timely manner.

## 1 Introduction

Whilst the past decade has witnessed an inexorable rise in the volume, diversity and quality of biological data available to the life-science research community, arguably the more measured rise in natural language documents provides a richer source of knowledge (Figure 1).

With many documents such as research articles, conference proceedings, and abstracts being made available in electronic format they have the potential to be automatically searched for information. However, given the volume of literature available, a semi-manual approach which would require reading a large number of articles or abstracts resulting from a keyword-based search, is infeasible. This has prompted researchers to apply text-mining approaches to life-science documents. Text mining is concerned with identifying non-trivial, implicit, previously unknown, and potentially useful patterns in text [1]. For example identifying references to biological entities [2-5], molecular interactions (protein-protein, gene- protein, drug-protein/gene), interactions of complexes and entities [6-9], sub-cellular localization of proteins [10, 11], pathways and structure of biological molecules [12-14]. Text mining has also been

J.L. Oliveira et al. (Eds.): ISBMDA 2005, LNBI 3745, pp. 101–108, 2005.

used to complement other techniques to achieve more complex tasks such as assisting in the construction of predictive and explanatory models [15] and assisting the construction of knowledge-bases and ontologies [16-19].

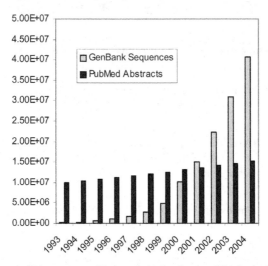

**Fig. 1.** The past decade has seen a rapid growth in databases of biological data such as Gen-Bank, whilst the growth in literature databases such as PubMed has remained linear

Each of these text mining applications employs one or more of a range of computationally expensive techniques, including natural language processing, artificial intelligence, machine learning, data mining, and pattern recognition. Predominantly these examples have used homogeneous and localized computing environments to analyze small data sets such as collections of subject specific abstracts. However, as the text collections, the domain knowledge, and the programs for processing, analyzing, evaluating and visualizing these data, increase in size and complexity the relevant resources are increasingly likely to reside at geographically distributed sites on heterogeneous infrastructures and platforms. Thus text mining applications will require an approach that is able to directly deal with distributed resources; grid computing is promising to provide the necessary functionality to solve such problems. Grid computing is a generic enabling technology for distributed computing, which can provide the persistent computing environments required to enable software applications to integrate instruments, displays, computational and information resources across diverse organizations in widespread locations [20]. In its basic form, grid technology provides seamless access to theoretically unlimited computing resources such as raw compute power and physical data storage capacity.

In this paper we describe the use of a grid infrastructure for conducting text mining over distributed data and computational resources. Full-text journal from the fields of molecular biology and biomedicine were used to create a knowledge base of gene annotations. It is envisaged that this knowledge base may be used to annotate or interpret high-throughput data-sets such as microarray data. The approach is based on extending LexiQuest Mine, a text mining workbench from SPSS (SPSS Inc, Chicago) [21] in a UD grid computing environment (United Devices Inc, Texas) [22].

## 2  Text Mining in LexiQuest Mine

LexiQuest Mine [21] employs a combination of dictionary-based linguistic analyses and statistical proximity matching to identify entities and concepts. Pattern matching rules are then used to identify the relationship between the named entities (Figure 2). The LexiQuest Mine natural language processing engine contains modules for linguistic, non-linguistic, named entity, events and relations extractor. Each module can be separately or consequently used, depending on the nature of the problem. User defined dictionaries for specific domains (e.g., gene, protein names in life science) may also be incorporated.

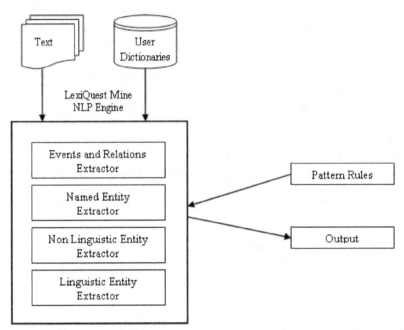

**Fig. 2.** Text mining in the LexiQuest Mine environment

The Pattern Rules component contains specific patterns or rules to extract relationships. The rules are regular expressions based on the arrangement of named entities (gene, protein names), prepositions and keywords that indicate the type of relationship between them, for example,

*gene[a-z]*(\s)interact\s[a-z]*(\s)+*gene[a-z]*(\s).

To perform the extraction of gene relations using pattern matching on gene names and relations, LexiQuest Mine relies on:

- Dictionaries of gene names
- Dictionaries of synonyms of gene names
- Linguistics tags following protein names which are used to automatically identify unknown genes (e.g. protein, kinase, phosphate )
- Dictionaries of gene relations (e.g., binds, inhibits, activates, phosphorylates)

## 3   Grid Deployment of Text Mining

Figure 3 illustrates the grid infrastructure Grid MP from United Devices Inc, Texas, USA [22]. Grid MP features a Linux based dedicated Grid server (labeled Grid MP Services here), which acts as a master or hub of the Grid. It balances compute demand by intelligently routing 'job' requests to suitable resources. The Grid MP Agent is software program that runs on each of the compute nodes, identifying the node and its capabilities to the Grid Server. The compute devices may run any Operating System (e.g. Windows, Mac, or Linux). The Grid MP Agent is responsible for processing work and for automatically returning the result files to the Grid MP Services. Users connect to the grid using a web interface or command line application-specific service scripts to deploy and run their applications.  Grid MP software transforms the execution of an application from a static IT infrastructure, where the execution of an application is always tied to specific machines, to a dynamic virtual infrastructure where applications execute on different machines based on resource availability, business priority, and required service levels. The common steps involved in grid deployment and running of an application include:

- Grid MP Services and Agents are installed to lay the foundation for the virtual infrastructure. Administrators define and install policies that govern the use of this infrastructure across multiple applications and users.
- Application Services (e.g. LexiQuest Mine) are created and deployed.
- Users may interact with their application which is transparently executed on the virtual infrastructure created by Grid MP
- Results are collected by Grid MP Services and passed back to the end user.

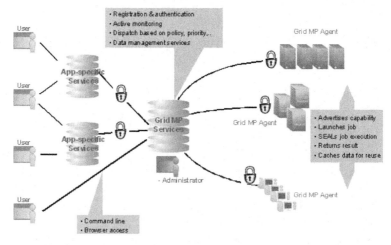

**Fig. 3.** Grid deployment of an application in UD Grid environment

## 4   Evaluation of Grid Based Text Mining for the Extraction of Gene Annotations

The aim of this work was to create a knowledge base of gene relationships which can be used to annotate and interpret microarray data. A corpus of approximately 125 000 full-text articles from 20 peer-reviewed journals in the field of molecular biology and biomedicine (1999-2003) (Table 1) was mined. The text mining methodology has been described previously [19] and comprises the following natural language processing (NLP) steps:

- *Sentence tokenization* to separate the text into individual sentences;
- *Word tokenization* to break pieces of text into word-sized chunks;
- *Part-of-speech tagging*   to assign part-of-speech information   (e.g., adjective, article, noun, proper noun, preposition, verb);
- *Named entity tagging* to find gene names and their synonyms and to replace the gene synonyms with unique gene identifier.
- *Pattern matching* to extract gene relations.

A sample output of final gene relations extracted using our system is illustrated in Table 2. In addition to the gene relations, we also extracted the PubMed ID and section ID corresponding to each gene relation using the pre-inserted XML tags in the corpus. This will help users to identify the source article and section from which the relations were extracted. The advantage here is that users can get corresponding gene annotations from full text articles for their genes of interest from their initial PubMed query results. The section tag helps users to identify the sections other than abstracts, which contains more gene relations for further research.

**Table 1.** List of downloaded journals and publisher's web sites

| Journal Name | URL |
| --- | --- |
| Biochemistry | http://pubs.acs.org/journals/bichaw/ |
| BBRC | http://www.sciencedirect.com/science/journal/0006291X |
| Brain Research | http://www.sciencedirect.com/science/journal/00068993 |
| Cancer | http://www3.interscience.wiley.com/cgi-in/jhome/28741 |
| Cancer Research | http://cancerres.aacrjournals.org |
| Cell | http://www.cell.com/ |
| EMBO Journal | http://embojournal.npgjournals.com/ |
| FEBS Letters | http://www.sciencedirect.com/science/journal/00145793 |
| Genes and Development | http://www.genesdev.org/ |
| International Journal of Cancer | http://www3.interscience.wiley.com/cgi-in/jhome/29331 |
| Journal of Biological Chemistry | http://www.jbc.org/ |
| Journal of Cell Biology | http://www.jcb.org/ |
| Journal of Neuroscience | http://www.jneurosci.org/ |
| Nature | http://www.nature.com/ |
| Neuron | http://www.neuron.org/ |
| Neurology | http://www.neurology.org/ |
| Nucleic Acid Research | http://nar.oupjournals.org/ |
| Oncogene | http://www.nature.com/onc/ |
| PNAS | http://www.pnas.org/ |
| Science | http://www.sciencemag.org/ |

Due to the high-through put computational tasks and huge amount of data (125 000 full-text articles from 20 different journals), we are unable to run the above application in a reasonable time frame using single computer. As an eventual solution to above situation, we have 'grid-enabled' the above application using UD Grid environment. The execution of this task on a single computer was compared with the grid-enabled application in terms of the computational efficiency and fidelity. The computational efficiency compares the time taken to complete the task, whilst fidelity provides a quantitative assessment of the quality of the results, i.e. whether the same results are returned from a single machine as on the grid, and whether repeated runs on the grid system produce consistent results.

**Table 2.** Sample output of gene relation extracted from full-text articles

| PubMed ID | Gene 1 | Gene 2 | Relation | Source |
|-----------|--------|--------|----------|--------|
| 12881431 | APOBEC2 | AICDA | Mediates | Abstract |
| 12101418 | NS5ATP13TP2 | P53 | Inhibits | Introduction |
| 15131130 | Map3k7 | nf-kappa b | Activates | Methods |
| 12154096 | Igf-1 | Pkb | Activates | Results |

The time taken to complete the task and the CPU time was calculated for a single machine and two grid infrastructures comprising of 10 computers and 20 computers respectively (Figure 4). Whilst the CPU time remains approximately constant for both the grid and the single computer, the advantage of the grid structure becomes apparent when looking at the total time to complete the task. Once the literature database contained more than 5000 full-text articles it was impractical to run on single machine.

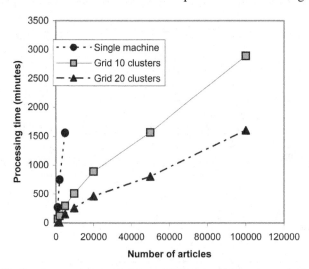

**Fig. 4.** Processing time to complete text mining of full-text articles using a single computer (●) and a grid of 10 computers (■) and a grid of 20 computers (▲)

Figure 5 illustrates the average number of patterns returned per article over 5 runs on a single machine and the two grid infrastructures described previously. The error bars show the maximum and minimum number of patterns found, illustrating that there was no significant difference observed on the different architectures. The number of patterns returned in repeated runs also remained constant. However as the size of the corpus increased from 1000 to 5000 articles, the number of patterns found per article increases from an average of 0.74 to 0.83, indicating the clear advantage of using a larger corpus.

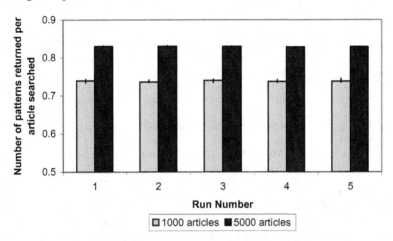

**Fig. 5.** The average number of patterns returned per article in the corpus using a single machine and two grid structures comprising of 10 and 20 cluster computers. The process was repeated over 5 runs and using a corpus of 1000 articles and a corpus of 5000 articles

## 5   Conclusion and Future Directions

Biological texts are rich resources of high quality information and the automatic mining of such repositories can provide the key to interpreting the large volumes of life-science data currently available. This work has demonstrated that as the size of the corpus increases, the text-mining process becomes more efficient in terms of extracting information from articles. However mining larger databases has significant implications in terms of processing time and computing resources; for example it was impractical to mine a corpus of more than 5000 full-text articles on a single machine. This paper has demonstrated that grid technology can provide the necessary middle-tier layer to allow the use of a distributed computing environment for such tasks. In addition to those described in this paper, we are also in the process of using the above infrastructure for developing other applications such as finding gene-disease relationship, disease-drug relation etc. using the same full-text articles. We plan to integrate these data with our in-house microarray data that integrates text and data mining applications in grid infrastructure. Incorporating this technique into the data mining pipeline of microarray analysis has the potential to effectively extract information and thus provide a greater understanding of the underlying biology in timely manner.

## References

1. Hearst M. A., Untangling text data mining, Proc. Of ACL, 37 (1999)
2. Fukuda, K., Tsunoda, T., Tamura, A., and Takagi, T., Towards Information Extraction: identifying protein names from biological papers, Pacific Symposium on Biocomputing, 707-718 (1998)

3. Eriksson, G., Franzen, K., and Olsson, F., Exploiting syntax when detecting protein names in text, Workshop on Natural Language Processing in Biomedical Applications, 2002, at http://www.sics.se/humle/projects/prothalt/
4. Wilbur, W., Hazard G. F. Jr., Divita G., Mork J. G., Aronson A. R., and Browne A. C., Analysis of biomedical text for biochemical names: A comparison of three methods, Proc. of AMIA Symposium, 176-180, (1999)
5. Kazama, J., Makino, T., Ohta, Y., and Tsujii, J., Tuning Support Vector Machines for Biomedical Named Entity Recognition, Proc. of the Natural Language Processing in the Biomedical Domain, Philadelphia, PA, USA (2002)
6. Ono, T., Hishigaki, H., Tanigami, A., & Takagi, T.: Automated extraction of information on protein-protein interactions from the biological literature, Bioinformatics, 17, 155-161 (2001)
7. Wong, L.: A protein interaction extraction system, Pacific Symposium on Biocomputing, 6, 520-531 (2001)
8. Yakushiji, A., Tateisi, Y., Miyao, Y., & Tsujii, J.: Event extraction from biomedical papers using a full parser, Pacific Symposium on Biocomputing, 6, 408-419 (2001)
9. Sekimizu, T., Park, H.S., & Tsujii, J.: Identifying the interaction between genes and gene products based on frequently seen verbs in Medline abstracts, Proceedings of the workshop on Genome Informatics, 62-71 (1998)
10. Craven, M., and Kumlien, J., Constructing biological knowledge base by extracting information from text sources, Proc. of the 7th International Conference on Intelligent Systems for Molecular Biology, 77-76 (1999)
11. Stapley, B. J., Kelley, L. A., and Strenberg, M. J. E., Predicting the sub-cellular location of proteins from text using support vector machines, Pacific Symposium on Biocomputing, 7, 374-385 (2002)
12. Gaizauskas, R., Demetriou, G., Artymiuk, P. J, and Willett, P., Protein structure and Information Extraction from Biological Texts: The PASTA system, Bioinformatics, 19:1, 135-143 (2003)
13. Rzhetsky, A., Iossifov, I., Koike, T., Krauthammer, M., Kra, P., Morris, M., Yu, H., Duboue, P. A., Weng, W., Wilbur, W. J., Hatzivassiloglou, V., and Friedman, C., GeneWays: a system for extracting, analyzing, visualizing, and integrating molecular pathway data, Jr of Biomedical Informatics, 37, 43-53 (2004)
14. Hahn, U., Romacker, M., and Schulz, S., Creating knowledge repositories from biomedical reports: The MEDSYNDIKATE text mining system, Pacific Symposium on Biocomputing, 7, 338-349 (2002)
15. Ideker, T., Galitski, T., and Hood, L., A new approach to decoding life: systems biology, Annu Rev Genomics Hum Genet 2:343-372 (2001)
16. Rzhetsky, A., etc.: GeneWays: a system for extracting, analyzing, visualizing, and integrating molecular pathway data, Jr of Biomedical Informatics, 37, 43-53 (2004)
17. Pustejovsky, J., etc.: Medstract: Creating large scale information servers for biomedical libraries, ACL-02, Philadelphia (2002)
18. Wong, L.: PIES a protein interaction extraction system, Pacific Symposium on Biocomputing, 6, 520-531 (2001)
19. Bremner,E.G., Natarajan, J. Zhang,Y., DeSesa,C., Hack,C.J. and Dubitzky,W.,: Text mining of full text articles and creation of a knowledge base for analysis of microarray data, LNAI, Knowledge exploration in Life Science Informatics, 84-95 (2004)
20. Foster I, and Kesselman C (eds), The Grid 2: Blueprint for a New Computing Infrastructure, Morgan Kaufmann (2004)
21. SPSS LexiQuest Mine available at http://www.spss.com
22. United Devices Grid MP Services available at http://www.ud.com

# Prediction of the Performance
# of Human Liver Cell Bioreactors by Donor Organ Data

Wolfgang Schmidt-Heck[1], Katrin Zeilinger[2], Gesine Pless[2],
Joerg C. Gerlach[2,3], Michael Pfaff[4], and Reinhard Guthke[1]

[1] Leibniz Institute for Natural Product Research and Infection Biology,
Hans Knoell Institute, Beutenbergstr. 11a, D-07745 Jena, Germany
{wolfgang.schmidt-heck,reinhard.guthke}@hki-jena.de
http://www.hki-jena.de
[2] Division of Experimental Surgery, Charité Campus Virchow, University Medicine Berlin,
Augustenburger Platz 1, D-13353 Berlin, Germany
[3] Depts of Surgery and Bioengineering, McGowan Institute for Regenerative Medicine,
University of Pittsburgh, PA, USA
{katrin.zeilinger,joerg.gerlach}@charite.de
[4] BioControl Jena GmbH, Wildenbruchstr. 15, D-07745 Jena, Germany
michael.pfaff@biocontrol-jena.com

**Abstract.** Human liver cell bioreactors are used in extracorporeal liver support
therapy. To optimize bioreactor operation with respect to clinical application an
early prediction of the long-term bioreactor culture performance is of interest.
Data from 70 liver cell bioreactor runs labeled by low (n=18), medium (n=34)
and high (n=18) performance were analyzed by statistical and machine learning
methods. 25 variables characterizing donor organ properties, organ preservation, cell isolation and cell inoculation prior to bioreactor operation were analyzed with respect to their importance to bioreactor performance prediction.
Results obtained were compared and assessed with respect to their robustness.
The inoculated volume of liver cells was found to be the most relevant variable
allowing the prediction of low versus medium/high bioreactor performance
with an accuracy of 84 %.

## 1 Introduction

Liver cell bioreactors are being developed and used for temporary extracorporeal
liver support [1, 2]. Primary human liver cells isolated from discarded human organs
are inoculated and cultured in these bioreactors. The 3D liver cell bioreactor investigated here consists of a system of interwoven capillaries within a special housing that
serve medium supply and removal as well as oxygenation of the cells that are culti-
vated in the inter-capillary space of the bioreactor. This bioreactor mimics conditions
close to those in the liver organ *in vivo*. It was shown that primary human liver cells
obtained from discarded human livers that were explanted but not suitable for transplantation reconstitute to liver tissue-like structures after inoculation into the bioreactor [3, 4].

The design of bioreactor operating conditions that support the long-term maintenance of liver cell functionality is of great importance with respect to the bioreactor's
clinical application. In previous work, data mining and pattern recognition methods
were applied to extract knowledge from bioreactor operation data in order to enable

the prediction of the long-term performance of the human liver cells in the bioreactor based on early culture data [5]. Using fuzzy clustering and rule extraction methods, the kinetics of galactose and urea over the first 3 culture days were found to be the best single predictors. In addition, kinetic patterns of the amino acid metabolism over the first 6 culture days and their relation to the long-term bioreactor performance were identified and described by different network models (correlation networks, Bayesian networks, differential equation systems) [6]. However, these results alone do not allow to draw conclusions with respect to the causes of the observed differences in the metabolic performance of the bioreactor cultures. These differences may in particular be due to donor organ properties and/or differences in organ preservation, cell isolation and cell inoculation prior to bioreactor operation. This paper presents results obtained by statistical tests and machine learning methods to quantify relations between donor organ and cell preparation characteristics and bioreactor performance.

## 2   Material and Methods

### 2.1   Cell Isolation and Bioreactor Culture

Cells for bioreactor inoculation were isolated from 70 human donor organs that were excluded from transplantation due to organ damage (steatosis, fibrosis, cirrhosis or other reasons). The organs were preserved at 4°C for varying time periods for the transport from the donor clinic to the Charité Virchow Clinic. Cell isolation from these organs was performed with the approval of the Deutsche Stiftung Organtransplantation (DSO) and the local ethics committee using a five-step enzyme perfusion technique as described elsewhere [4].

Cells were inoculated into the bioreactors immediately after isolation and cultured in the systems under standardized perfusion and oxygenation conditions. The culture period was one day to 60 days. The bioreactor culture performance was assessed on the basis of biochemical variables that were measured daily in the culture perfusate (see 2.2).

### 2.2   Data

A data set $x_{i,j}$ ($i = 1,\ldots, I; j = 1,\ldots, J$) for 21 metric and 4 categorical variables ($I = 25$) characterizing donor and organ properties as well as organ preservation, cell isolation and cell inoculation of $J = 70$ bioreactor runs was analyzed (Tables 1 and 2). For some metric variables $i$, a number of values was missing ($70$-$N_i$, Table 1).

Each run was labeled by $L_j \in$ {L, M, H} denoting 'low', 'medium' and 'high' performance, respectively, categorizing the long-term maintenance of the functionality of the liver cells in the bioreactor culture. 18, 34 and 18 runs were labeled L, M and H, respectively. This performance had been assessed by an expert based on the biochemical variables that were measured during the bioreactor operation quantifying enzyme liberation, glucose and lactate metabolism, galactose and sorbitol uptake, ammonia elimination, urea and albumin production and amino acid metabolism.

**Table 1.** Metric variables characterizing donors and organs, organ preservation, cell isolation and cell inoculation (BMI – body mass index, GGT – gamma glutamyltranspeptidase, LDH – lactate dehydrogenase, ALT – alanine aminotransferase, AST – aspartate aminotransferase, GLDH – glutamate dehydrogenase, AP – alkaline phosphatase, PS – preservation solution; Min – minimum value, Max – maximum value, $N_i$ – number of available values)

| Variable | Unit | Min | Max | $N_i$ |
|---|---|---|---|---|
| BMI (of the donor) | kg·m$^{-2}$ | 21 | 39 | 68 |
| Weight (of the donor) | kg | 55 | 140 | 70 |
| Height (of the donor) | cm | 155 | 195 | 68 |
| Age (of the donor) | a | 20 | 79 | 69 |
| GGT (in the donor plasma) | U·L$^{-1}$ | 6 | 1075 | 66 |
| LDH (in the donor plasma) | U·L$^{-1}$ | 71 | 2013 | 47 |
| ALT (in the donor plasma) | U·L$^{-1}$ | 5 | 647 | 70 |
| AST (in the donor plasma) | U·L$^{-1}$ | 3 | 405 | 70 |
| DeRitis (quotient AST/ALT) | - | 0.21 | 5.83 | 65 |
| Bilirubin (total bilirubin in the donor plasma) | μmol·L$^{-1}$ | 0.38 | 133 | 67 |
| Urea (in the donor plasma) | mmol·L$^{-1}$ | 1.40 | 91 | 64 |
| Preservation_Time (of the organ) | h | 2.50 | 27 | 69 |
| Organ_Weight | g | 997 | 3378 | 70 |
| LDH_PS (LDH in the preservation solution) | U·L$^{-1}$ | 11 | 5310 | 57 |
| AST_PS (AST in the preservation solution) | U·L$^{-1}$ | 3 | 2110 | 57 |
| GLDH_PS (GLDH in the preservation solution) | U·L$^{-1}$ | 0 | 29 | 57 |
| AP_PS (AP in the preservation solution) | U·L$^{-1}$ | 0 | 19 | 56 |
| Remaining_Mass (of the organ after cell isolation) | g | 179 | 1344 | 66 |
| Dissolved_Mass (of the cells) | g | 20 | 89 | 67 |
| Viability (of the cells) | % | 30 | 85 | 67 |
| Inoculated_Volume (of the cells) | % | 144 | 800 | 67 |
|  | mL |  |  |  |

**Table 2.** Categorical variables characterizing donors, organ preservation, cell isolation and bioreactor culture performance (f – female, m – male, UW – University of Wisconsin solution, HTK – histidine-tryptophane-ketoglutarate solution, Coll. – collagenase P, HSA – human serum albumin)

| Variable | Categories | Distribution |
|---|---|---|
| Gender (of the donor) | {f, m} | [33, 37] |
| Preservation_Solution | {UW, HTK, Celsior, -} | [36, 25, 3, 6] |
| Digestion_Enzyme | {Coll., Liberase, Serva} | [57, 11, 2] |
| Additives (used during cell isolation) | {DNAse, HSA, none} | [19, 1, 50] |
| Performance (of the bioreactor culture) | {low, medium, high} | [18, 34, 18] |

## 2.3  t-Test and Wilcoxon Test

The 21 metric variables averaged over the groups of runs that were assigned to the different bioreactor performance levels were compared by the two-sided t-test and Wilcoxon's rank sum test (Wilcoxon-Mann-Withney test, MATLAB Statistics Toolbox, The MathWorks, Natick, MA, USA). Tests were performed comparing the groups 'high' versus 'low or medium', 'low' versus 'high or medium', 'low' versus 'high', 'high' versus 'medium' and 'low' versus 'medium'. The variables were ranked according to the p-values as determined by the t-test, i.e. according to the probability that two samples with a normal distribution of unknown but equal variances have the same mean.

## 2.4  Contingency Table Analysis

For each variable a 2 by 2 table was generated determining the numbers of runs assigned to two clusters with respect to the variable values as well as to the bioreactor performance. The clustering of the metric variables was performed using the minimum variance criterion. For the categorical variables the following pairs were analyzed: 'female' versus 'male', 'University of Wisconsin solution' versus 'histidine-tryptophane-ketoglutarate solution', 'Collagenase P' versus 'Liberase' and 'DNAse' versus 'no additives'. In respect of the bioreactor performance the pairs 'high' versus 'low or medium', 'low' versus 'high or medium', 'low' versus 'high', 'high' versus 'medium' and 'low' versus 'medium' were analyzed. For each 2 by 2 table the two-sided $p$-values were calculated by Fisher's exact test [7].

## 2.5  Random Forest Analysis

The variables were ranked according to their importance as calculated by Breiman's Random Forest algorithm [8, 9] available in R [10]. Before starting the algorithm '*randomForest*' missing values were imputed using the proximity obtained from the random forest imputing algorithm '*rfImpute*' configured for 5 iterations and 2500 trees. Running '*randomForest*' in the supervised mode an ensemble of 5000 trees was generated using a *mtry* parameter of 3 (estimated by '*tuneRF*' with stepfactor = 2, ntreeTry = 5000, improve = 0.05) and default values for the other parameters. The ensemble of the 5000 trees generated was then analyzed with respect to the first and second level split variables of the decision trees.

## 2.6  Support Vector Machines

The Support Vector Machine algorithm [11, 12] with a linear kernel and c = 0.1 (cost of constraint violation) together with a leave-one-out cross-validation was used in order to find single variables as well as pairs and triplets of variables that can robustly predict the bioreactor performance. This was done comparing the performance levels 'low', 'medium', 'high', 'low or medium' and 'high or medium'. When running the algorithm for missing values $x_{i,j}$, the corresponding runs $j$ were ignored when the variable $i$ was involved. The prediction accuracy was determined as the quotient $Q$ dividing the number of correctly predicted runs by the total number of tests (which equals the number of runs $J$ minus the number of runs ignored when dealing with missing values). $Q$ characterizes the predictive strength of the respective variable set. Single variables, pairs or triplets of variables with maximum prediction accuracy $Q$ were then selected.

# 3  Results and Discussion

Tables 3 and 4 show the results obtained by the ranking of the variables according to the $p$-values as calculated by the t-test and the exact Fisher's test as well as according to the importance as calculated by the Random Forest algorithm (see also Fig. 2). The t-test can only be applied to the metric variables. Comparing the performances 'low' versus 'high or medium' and 'low' versus 'medium', the 'Inoculated_Volume' was found to be significantly correlated to the bioreactor performance ($p<0.002$, see also

Fig. 1) by all three statistical tests (t-test, Wilcoxon's rank sum test, Table 3; exact Fisher's test for the contingency table analysis, Table 4).

The three statistical tests applied assess the relevance of individual variables but not of their combinations. To test such combinations of variables, Random Forests (RF) and Support Vector Machines (SVM) were applied.

The RF algorithm [8, 9] combines two powerful concepts in machine learning: bagging and random feature selection. Bagging stands for bootstrap aggregating which uses resampling to produce pseudo-replicates in order to improve predictive accuracy. Random feature selection can considerably improve predictive accuracy, too. Fig. 2 shows the variables ranked by their importance as calculated by the RF algorithm. The RF out-of-bag (OOB) estimate of error rate obtained was 44 %.

Each binary decision tree generated by the RF algorithm contains one first level split variable and two second level split variables (or one or two leaf nodes instead of them). Looking at individual variables, the 'Inoculated_Volume' most frequently occurred as first level and as second level split variable, i.e. in 10 % and 7 % of the cases, respectively (508 times in the 5000 first level nodes, 671 times in the 10000 second level nodes; Table 5). Often, one of the two second level split variables does not appear, i.e. there exists a leaf node following the first split. There are 1054 trees among the 5000 generated ones (21 %) with a leaf node instead of one of the two second level split variables (Table 5). These 1054 leaf nodes at the second level stand 430 times for 'low', 155 times for 'medium' and 469 times for 'high' performance.

**Table 3.** Rankings of the metric variables with respect to their influence on the bioreactor performance as obtained by the two-sided t-test for the performances 'high' versus 'low or medium' (A), 'low' versus 'high or medium' (B), 'low' versus 'high' (C), 'high' versus 'medium' (D) and 'low' versus 'medium' (E), respectively (*: $p<0.05$, **: $p=0.0013$, ***: $p=0.0003$); significant results obtained by Wilcoxon's rank sum test are indicated by crosses (+: $p<0.05$, ++: $p=0.0046$, +++: $p<0.001$)

| Variable | A HvsL\|M | B LvsH\|M | C LvsH | D HvsM | E LvsM |
|---|---|---|---|---|---|
| BMI | *, + 1 | 6 | * 3 | 3 | 15 |
| Weight | 16 | 2 | 6 | 21 | 4 |
| Height | 7 | 10 | 17 | 2 | 3 |
| Age | + 3 | 4 | 4 | 6 | 9 |
| GGT | 9 | 20 | 14 | 10 | 16 |
| LDH | 18 | 13 | 21 | 14 | 7 |
| ALT | 17 | 12 | 18 | 15 | 8 |
| AST | 13 | 16 | 15 | 13 | 13 |
| DeRitis | 8 | 19 | 11 | 7 | 12 |
| Bilirubin | 11 | 11 | 7 | 12 | 17 |
| Urea | 5 | 5 | 19 | *, + 1 | *, + 2 |
| Preservation_Time | 14 | 14 | 13 | 16 | 19 |
| Organ_Weight | 19 | 8 | 12 | 19 | 5 |
| LDH_PS | 10 | 7 | 8 | 8 | 10 |
| AST_PS | + 12 | 3 | 9 | + 9 | 6 |
| GLDH_PS | 21 | 18 | 20 | 20 | 21 |
| AP_PS | 20 | 15 | 16 | 18 | 20 |
| Remaining_Mass | 6 | 9 | + 5 | 11 | 11 |
| Dissolved_Mass | * 2 | 17 | * 2 | 4 | 18 |
| Viability | 4 | 21 | 10 | 5 | 14 |
| Inoculated_Volume | 15 | ***, +++ 1 | *, ++ 1 | 17 | **, +++ 1 |

Searching for pairs of variables that most frequently appear in the set of generated decision trees, 'Urea' and 'Inoculated_Volume' were most often found jointly as first and second level split variables in 67 of the 5000 trees generated (1.3 %; Table 6). The occurrence of other pairs is less frequent, i.e. smaller than 1.3 % (Table 6). The 'pair' consisting of the variable 'Inoculated_Volume' (as the first level split variable) and a leaf node (instead of one of the two second level split variables) was found with the highest frequency, i.e. in 242 of the 5000 trees (5%; Table 6). Almost all, i.e. 239 of these 242 trees (99 %) can be expressed by the following rule:

IF *'Inoculated_Volume is smaller than B1'*, THEN *'Performance is low'*.    (1)

The split value $B1$ is different for the 239 trees. The distribution of the split values $B1$ is bimodal (Fig. 3): The split value $B1$ for the 'Inoculated_Volume' lies 166 times (i.e. 69 %) between 350 and 380 mL and 52 times (22 %) between 250 and 260 mL. 21 times the split values $B1$ lie outside these intervals.

**Table 4.** Rankings of the metric and the categorical variables with respect to their influence on the bioreactor performance as obtained by the two-sided exact Fisher's test of the contingency table analysis for the performances 'high' versus 'low or medium' (A), 'low' versus 'high or medium' (B), 'low' versus 'high' (C), 'high' versus 'medium' (D) and 'low' versus 'medium' (E), respectively (*: $p<0.05$, **: $p<0.002$) as well as the rankings according to the importance as calculated by the Random Forest algorithm (see also Fig. 2)

| Variable | A HvsL\|M | B LvsH\|M | C LvsH | D HvsM | E LvsM | F RF |
|---|---|---|---|---|---|---|
| BMI | 7 | 16 | 8 | 7 | 20 | 14 |
| Weight | 16 | 4 | 9 | 21 | 4 | 21 |
| Height | 12 | 22 | 16 | 11 | 19 | 17 |
| Age | * 2 | 12 | 5 | * 2 | 21 | 3 |
| GGT | 20 | 14 | 19 | 14 | 9 | 13 |
| LDH | 18 | 11 | 20 | 12 | 8 | 10 |
| ALT | 21 | 23 | 21 | 22 | 22 | 15 |
| AST | 22 | 24 | 22 | 23 | 23 | 6 |
| DeRitis | 8 | 15 | 7 | 8 | 16 | 8 |
| Bilirubin | 11 | 6 | 10 | 17 | 13 | 19 |
| Urea | 6 | 10 | 23 | 3 | 3 | 2 |
| Preservation_Time | 13 | 25 | 15 | 13 | 24 | 16 |
| Organ_Weight | 23 | 19 | 24 | 24 | 15 | 9 |
| LDH_PS | 15 | 8 | 11 | 20 | 11 | 7 |
| AST_PS | 24 | 13 | 13 | 25 | 10 | 4 |
| GLDH_PS | 19 | 18 | 25 | 15 | 12 | 11 |
| AP_PS | 17 | 20 | 17 | 18 | 25 | 20 |
| Remaining_Mass | 5 | 3 | 4 | 9 | 6 | 5 |
| Dissolved_Mass | * 1 | 7 | * 3 | * 1 | 18 | 12 |
| Viability | 10 | 21 | 14 | 6 | 14 | 18 |
| Inoculated_Volume | 14 | ** 1 | * 2 | 19 | ** 1 | 1 |
| Gender | 9 | 17 | 18 | 5 | 7 | 23 |
| Preservation_Solution | 25 | 5 | 12 | 16 | 5 | 24 |
| Digestion_Enzyme | 3 | 9 | 6 | 4 | 17 | 22 |
| Additives | 4 | * 2 | * 1 | 10 | 2 | 25 |

Searching for triplets of variables, those trees were most frequently found that have 'AST' as first level split variable with either 'Inoculated_Volume' (31 trees;

0.6 %) or 'Urea' (24 trees; 0.5 %) as one of the two second level split variables and a leaf node instead of the other second level split variable (Table 7).

While the RF method is based on the induction of decision trees with conditions *'variable value x < split value B'* that define regions with axis-parallel borders, SVM allow to generate classifiers with discriminating borders that are not restricted to be parallel to the axes. Using SVM it was searched for individual variables as well as pairs and triplets of variables that provide the highest prediction accuracy $Q$ as determined by leave-one-out cross-validation.

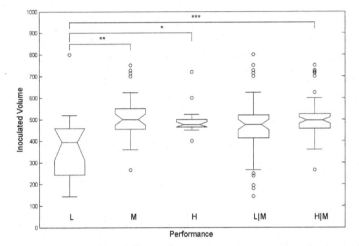

**Fig. 1.** Box plot of the values of the variable 'Inoculated_Volume'; each box shows the median, the lower and upper quartiles, the whiskers (length: 1.5-fold interquartile range) and the outliers (°, the rest of the data lies outside the whiskers); significant differences determined by the t-test and Wilcoxon's rank sum test are indicated by asterisks (*: $p=0.016$ and 0.0046, **: $p=0.0013$ and 0.0009, ***: $p=0.0003$ and 0.0005, respectively)

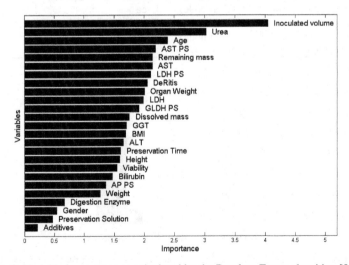

**Fig. 2.** Variable importance as calculated by the Random Forest algorithm [8, 9]

**Table 5.** Individual variables most frequently found as $1^{st}$ or $2^{nd}$ level split variable in the set of 5000 trees generated by the Random Forest algorithm; $N1$, $N2$ – number of trees in which the variable represented the $1^{st}$ level split variable or one of the two $2^{nd}$ level split variables

| $1^{st}$ Level Split Variable | $N1$ | $2^{nd}$ Level Split Variable | $N2$ |
|---|---|---|---|
| Inoculated_Volume | 508 | - (Leaf node) | 1054 |
| Urea | 415 | Inoculated_Volume | 671 |
| AST | 319 | AST_PS | 516 |
| Age | 302 | Age | 486 |
| AST_PS | 298 | Urea | 485 |

**Table 6.** Pairs of variables most frequently found as $1^{st}$ and $2^{nd}$ level split variables in the set of 5000 trees generated by the Random Forest algorithm; $N12$ – number of trees in which the pair represented the $1^{st}$ level split variable and one of the two $2^{nd}$ level split variables

| $1^{st}$ Level Split Variable | $2^{nd}$ Level Split Variable | $N12$ |
|---|---|---|
| Inoculated_Volume | - (Leaf node) | 242 |
| AST | - (Leaf node) | 217 |
| LDH | - (Leaf node) | 86 |
| Dissolved_Mass | - (Leaf node) | 83 |
| Urea | Inoculated_Volume | 67 |
| Inoculated_Volume | AST | 64 |
| Urea | DeRitis | 63 |
| Urea | AST_PS | 55 |
| AST | Inoculated_Volume | 52 |
| Remaining_Mass | Inoculated_Volume | 51 |

**Table 7.** Triplets of variables most frequently found as $1^{st}$ and $2^{nd}$ level split variables in the set of 5000 trees generated by the Random Forest algorithm; $N122$ – number of trees in which the triplet represented the $1^{st}$ level split variable and both $2^{nd}$ level split variables

| $1^{st}$ Level Split Variable | $2^{nd}$ Level Split Variables | | $N122$ |
|---|---|---|---|
| AST | - (Leaf node), | Inoculated_Volume | 31 |
| AST | - (Leaf node), | Urea | 24 |
| Inoculated_Volume | - (Leaf node), | Dissolved_Mass | 17 |
| AST | - (Leaf node), | Height | 17 |
| Inoculated_Volume | - (Leaf node), | AST_PS | 16 |

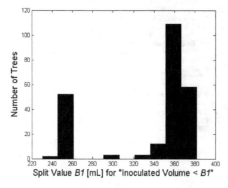

**Fig. 3.** Histogram of the split values $B1$ in the condition of decision rule (1) as obtained by the Random Forest algorithm [8, 9]; (number of trees out of the 5000 generated versus split value)

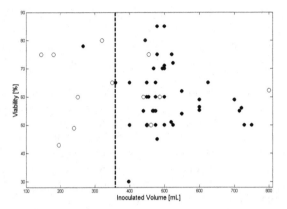

**Fig. 4.** 'Low' (o) versus 'high or medium' (•) bioreactor performance can be satisfactorily predicted by the variable 'Inoculated_Volume' alone; using the variable 'Viability' in addition does not improve the prediction accuracy considerably as was shown using SVM

Using SVM to distinguish between the performances 'low' and 'high or medium' yielded the highest values $Q$ for the individual variable 'Inoculated_Volume' ($Q$=0.836), for the variable pair 'Inoculated_Volume'/'Viability' ($Q$=0.844; Fig. 4) and for the variable triplet 'Inoculated_Volume'/'Viability'/'DeRitis' ($Q$=0.900). The value $Q$ was compared to the accuracy $Q_0$ of a dummy prediction 'all values are high or medium'. With a total number of 70 bioreactor runs and 18 low performance ones, $Q_0$ equals 0.74 (=1-18/70). However, neglecting runs with missing values, $Q_0$ varies: 0.74 for 'Inoculated_Volume', 0.78 for the pair 'Inoculated_Volume'/'Viability' and 0.77 for the triplet 'Inoculated_Volume'/'Viability'/'DeRitis'. Judged by the ratio $Q/Q_0$, the improvement in prediction accuracy by combining the 'Inoculated_Volume' as most important variable with other variables is rather small.

## 4  Conclusion

The variability of the performance (as single output) of 70 human liver cell bioreactor runs was studied based on 25 variables (as multiple inputs) that characterize donor organ properties, organ preservation, cell isolation and inoculation prior to bioreactor operation. The input-output relation was analyzed by various methods, in particular statistical tests (t-test, Wilcoxon's rank sum test, exact Fisher's test), Random Forests (RF, [8, 9]) and Support Vector Machines (SVM, [11, 12]) with a linear kernel as described in this paper. In addition, further methods were applied that yielded quite similar results which are not presented here: multivariate analysis by Principal Component Analysis (PCA), Independent Component Analysis (ICA) and Correspondence Analysis (CA), generation of classifiers by Induction of Decision Trees (See5, [13], with boosting and leave-one-out cross-validation), Support Vector Machines with polynomial and radial kernels, Recursive Partitioning, k-Nearest-Neighbor Classifiers, Naive Bayes Classifiers as well as cluster based rule generation [14] after clustering of the runs for the metric variables by the fuzzy c-means algorithm and minimum variance analysis with two clusters or an optimized cluster number (using 12 criteria).

The 'Inoculated_Volume' was found by all the applied methods as most important variable for the prediction of the liver cell bioreactor performance, in particular to predict 'low' performance. It is often followed next by 'Urea' (as shown in Table 3 for the t-test and Wilcoxon's rank sum test as well as in Fig. 2 and Tables 4 to 7 for RF).

No robust results were obtained combining two or three variables. These more complex classifiers depend on the chosen method for classifier construction as well as on the selected data subset. In particular, only a maximum of 242 (5 %) out of the 5000 trees generated by Breiman's RF algorithm were found to have the first level split variable, i.e. the 'Inoculated_Volume', and a leaf node (instead of one of the two second level split variables) in common (non-genuine variable pair; genuine pairs are even far less frequent; see Table 6). The split value $B1$ for the first level split variable 'Inoculated_Volume' has a bimodal distribution between 230.5 and 381.5 mL (Fig. 3). In addition, just 31 or less (0.6 % or less) out of the 5000 trees generated by the RF algorithm have the first level split variable, one of the two second level split variables and a leaf node (instead of the other second level split variable) in common (non-genuine variable triplet; genuine triplets are almost not existent; see Table 7).

This unsatisfactory robustness of classifiers combining two or three of the variables also showed when SVM were employed. The 'Inoculated_Volume' on its own again proved to be the most discriminating variable to predict bioreactor performance, here with an accuracy of 84%. The inclusion of other variables, however, does not improve prediction accuracy considerably (see Fig. 4).

The positive correlation of the inoculated volume of cells with the bioreactor performance shown by different methods indicates that an improved performance can be achieved by increasing the cell volume that is inoculated into the bioreactor. Based on this theoretical result, it has to be established experimentally which maximum inoculation volume is practically feasible and which bioreactor performance can actually be achieved by this under real operating conditions.

Due to the widely consistent picture that evolved from the study using different methods (with the 'Inoculated_Volume' as the by far most important single variable for bioreactor performance prediction prior to operation) it may be concluded with respect to the problem investigated that only limited further information can be gained from the 25 variables analyzed here.

Further work should therefore be directed towards the analysis of derived variables that for instance relate to more cell-specific than solely bioreactor-specific characteristics. Also, other variables not included so far for various reasons should be part of an extended analysis. This for instance applies to additional donor organ properties, such as the liver injury (e.g. steatosis, cirrhosis, fibrosis and others) that led to the exclusion of the organ from transplantation and to the use of its cells in the bioreactor culture. Due to the diversity of these injuries and still low case numbers for several of them, the variable was not included in the present study.

## Acknowledgement

This work was supported by the German Federal Ministry for Education and Research BMBF within the Programme 'Systems of Life – Systems Biology' (FKZ 0313079B, FKZ 0313079A).

# References

1. Gerlach, J.C., Botsch, M., Kardassis, D., Lemmens, P., Schon, M., Janke, J., Puhl, G., Unger, J., Kraemer, M., Busse, B., Bohmer, C., Belal, R., Ingenlath, M., Kosan, M., Kosan, B., Sultmann, J., Patzold, A., Tietze, S., Rossaint, R., Mueller, C., Monch, E., Sauer, I.M., Neuhaus, P.: Experimental evaluation of a cell module for hybrid liver support. Int. J. Artif. Organs 24 (2001) 793-98
2. Zeilinger, K., Holland, G., Sauer, I.M., Efimova, E., Kardassis, D., Obermayer, N., Liu, M., Neuhaus, P., Gerlach, J.C.: Time course of primary liver cell reorganization in three-dimensional high-density bioreactors for extracorporeal liver support: an immunohisto-chemical and ultrastructural study. Tissue Eng. 10 (2004) 1113-24
3. Gerlach, J.C., Mutig, K., Sauer, I.M., Schrade, P., Efimova, E., Mieder, T., Naumann, G., Grunwald, A.., Pless, G., Mas, A.., Bachmann, S., Neuhaus, P., Zeilinger, K.: Use of primary human liver cells originating from discarded grafts in a bioreactor for liver support therapy and the prospects of culturing adult liver stem cells in bioreactors: a morphologic study. Transplantation 76 (2003) 781-86
4. Gerlach, J.C., Brombacher, J., Kloeppel, K., Smith, M., Schnoy, N., Neuhaus, P.: Comparison of four methods for mass hepatocyte isolation from pig and human livers. Transplantation 57 (1994) 1318-22
5. Pfaff, M., Toepfer, S., Woetzel, D., Driesch, D., Zeilinger, K., Pless, G., Neuhaus, P., Gerlach, J.C., Schmidt-Heck, W., Guthke, R.: Fuzzy cluster and rule based analysis of the system dynamics of a bioartificial 3D human liver cell bioreactor for liver support therapy. In: Dounias, G., Magoulas, G., Linkens, D. (eds.): Intelligent Technologies in Bioinformatics and Medicine. Special Session. Proceedings of the EUNITE 2004 Symposium. A Publication of the University of the Aegean (2004) 57
6. Schmidt-Heck, W., Zeilinger, K., Pfaff, M., Toepfer, S., Driesch, D., Pless, G., Neuhaus, P., Gerlach, J.C., Guthke, R.: Network analysis of the kinetics of amino acid metabolism in a liver cell bioreactor. Lect. Notes Comput. Sc. 3337 (2004) 427-38
7. http://www.stat.unibe.ch/~duembgen/software/
8. Breiman, L.: Random forests. Technical Report, Stat. Dept. UCB (2001)
9. Breiman, L.: Random forests. Mach. Learn. 45 (2001) 5-32
10. http://www.r-project.org/
11. Vapnik, V.: Statistical Learning Theory. New York Wiley, 1998
12. http://www.eleceng.ohio-state.edu/~maj/osu_svm
13. http://www.rulequest.com/
14. Guthke, R., Schmidt-Heck, W., Pfaff, M.: Knowledge acquisition and knowledge based control in bio-process engineering. J. Biotechnol. 65 (1998) 37-46

# A Bioinformatic Approach to Epigenetic Susceptibility in Non-disjunctional Diseases

Ismael Ejarque[1,2], Guillermo López-Campos[2], Michel Herranz[3],
Francisco-Javier Vicente[2], and Fernando Martín-Sánchez[2]

[1] Ambulatorio di Genetica Medica, Ospedale Galliera
Via Volta 6 / rosso, - 16128 Genova, Italy
ismael.ejarque@galliera.it
[2] Medical Bioinformatics Department. Institute of Health Carlos III
Ctra. Majadahonda a Pozuelo Km 2, E – 28220 Majadahonda. Madrid, Spain
[3] Cancer Epigenetics Laboratory, Molecular Pathology Program
Spanish National Cancer Centre,
Calle Melchor Fernández Almagro, 3. E – 28029 Madrid, Spain

**Abstract.** The aim of this work is to present a fully "in silico" approach for the identification of genes that might be involved in the susceptibility for non disjunction diseases and their regulation by methylation processes. We have carried out a strategy based on the use of online available bioinformatics databases and programs for the retrieval and identification of interesting genes. As result we have obtained 29 putative susceptibility genes regulated by methylation processes. We were neither on the need of developing new software nor carry out clinical laboratory experiments for the identification of these genes. We consider that this "in silico" methodology is robust enough to provide candidate genes that must be checked "in vivo" due to the clinical relevance of non disjunction diseases with the aim of providing new tools and criteria for their diagnostics.

## 1 Introduction

The risk to conceive foetuses with Down's syndrome (and other constitutional aneuploidies) depends especially on the mother's age. Furthermore, different epidemiologic studies support the idea that there is a certain degree of genetic predisposition in young women which have conceived an aneuploid foetus. Their risk of recurrence for trisomies is very high if we compare it with their risk due only to the mother's age. Approximately ten times their risk associated to age [1, 2].

Nowadays there are several prenatal screening tests directed to identify those pregnancies with a more high risk to conceive foetuses with such pathologies. These techniques, double test and triple test, which are based on biochemical parameters on the mother's peripheral blood, together with foetal biophysical parameters, provide a probabilistic result, with no great certainty. Due to this uncertainty, these results are often difficult to be managed by the future parents.

The specificity and sensitivity of double test and triple test are away to be the very best. Foetuses which carry constitutional aneuploidies (T-21, T-13, and T-18) are still being born with a non-high risk result in spite of these screening techniques. These – apparently, trivial and methodological- facts are a great problem in Public Health because of the great emotional impact that having a baby with Down's syndrome has for a couple.

J.L. Oliveira et al. (Eds.): ISBMDA 2005, LNBI 3745, pp. 120–129, 2005.

Amniocentesis and CVS (Chorionic Villus Sampling) are the only diagnostic procedures which provide a certainty result for constitutional aneuploidies. But because of their aggressiveness, there is always a possibility of foetal loss due to the technique.

Nowadays there is not a pre-conceptional test that could provide data about the possibility to conceive foetuses with this kind of problems. The availability of such tests could be a great help in order to a deeper planning and supervision of the pregnancy.

The biological basis of the non-disjunction (the failure of homologous chromosomes or chromatids to segregate during mitosis or meiosis with the result that one daughter cell has both of a pair of parental chromosomes or chromatids and the other has none) has been not still cleared, despite the fact that a number of publications relate it to several polymorphisms in the folate pathway [3, 4, 5] and the fact that the new discoveries only give a very partial answer to this problem [6].

In the other hand, there are a series of oncology disorders which their clinical progression depends on specific phenomena related to non-disjunctional events but which nature has not been elucidated yet. These non-disjunctional phenomena are related to the survival and prognosis of the affected persons by several kind of tumours [7].

It is known that specific methylation patterns exist for each tumour. And it is feasible to discover the tumour type by diagnosing the methylation pattern in a determined cell line [8, 9, 10]. These data invite to argue that they could be correlated to common epigenetic basis both for constitutional aneuploidies and the somatic aneuploidies taking in account the methylation pattern of the promoters of the involved genes in meiosis and mitosis, that in many times are shared.

In vertebrate DNA, the sequence CpG is a signal for methylation by a specific cytosine DNA methyltransferase, which adds a methyl group to the 5' carbon of the cytosine. A CpG island is a short stretch of DNA, often <1 kb, containing high frequency of CpG dinucleotides. CpG islands tend to mark the 5' ends of genes in the promoter region, though they can localize inside the coding sequence of a gene [11]. Such islands of normal CpG density (CpG islands) are comparatively GC-rich (typically over 50% GC) and extend over hundreds of nucleotides. In the case of genes showing widespread expression, associated CpG islands are almost always found at the 5' ends of genes, usually in the promoter region and often extending into the first exon.

About 56% of human genes and 47% of mouse genes are associated with CpG islands [12]. Often CpG islands overlap the promoter and extend about 1000 base pairs downstream into the transcription unit. Identification of potential CpG islands during sequence analysis helps to define the extreme 5' ends of genes, something that is notoriously difficult with cDNA based approaches.

Hypermethylation of CpG islands located in the promoter regions of tumour suppressor genes is now firmly established as an important mechanism for gene inactivation. Besides, it's thought that many genes could be controlled differentially in the time (embryogenesis) or in different tissues through methylation gene expression regulation.

CpG islands could be a key target for the regulation of gene expression in a great number of genes of the human genome. The alteration in the epigenetic control of the gene expression in specific genes could be a reason to explain non-disjunctional processes leaving to constitutional and somatic aneuploidies.

Our hypothesis of work consists in that there must be an epigenetic susceptibility to conceive children with Down's syndrome (and other constitutional aneuploidies) and that could be due to different genes that intervene in the process of meiosis and where the control of their gene expression could be carried out through the methylation of their promoters. This is consistent with the fact that several polymorphisms of the folate pathway have an influence in this phenomenon given that this pathway is related with the generation of methyl radicals. The mentioned radicals would be those ones that, by means of the DNMTs (DNA methyl-transferases), would hypermethylate the promoters of such genes of susceptibility for non-disjunction.

In this work we present an approach for the identification of genes involved in the physical and biological process of meiosis and how their promoters are suitable to be regulated by means of methylation events as epigenetic phenomena.

## 2  Material and Methods

The interest of this work was in the identification and analysis of putative regulation by methylation of genes associated with meiosis and the non-disjunction associated diseases in humans.

The availability of large data sources is enabling the use of "in silico" analysis of sequences and biological processes. We use several bioinformatics resources to analyse our hypothesis of work. The selection of the genetic and biomedical information databases is crucial for the purpose of the project. NAR (Nucleic Acid Research) publishes yearly the "The Molecular Biology Database Collection" [13], a list of the most important databases hierarchically classified by areas in our field of interest. In the 2005 update, there were 719 databases registered in the journal, 171 more than the previous year, a number, although appreciably lower than the one obtained in Google, still unmanageable by a primary care physician. Each database has its own domain, its own architecture, its own interface and its own way to carry out a query, namely these resources are distributed and heterogeneous.

To select the appropriate biomedical and genetic resources, we choose those supported by a prestigious organization, frequently updated, with free and public access by internet. On this premises, the databases selected are HGNC, OMIM and PUBMED, close to GO (Gene Ontology), a consortium that produces a dynamic controlled vocabulary that can be applied to all organisms to describe molecular function, cellular location and biological process of genes and proteins. HGNC (Genew) [14] is the Human Gene Nomenclature Database, the only resource that provides data for all human genes, which have approved symbols. OMIM (Online Mendelian Inheritance in Man) [15] is a catalogue of human genetic and genomic disorders, authored and edited by Dr. Victor A. McKusick, that correlates pathology with genes and links with the rest of the databases of NCBI, including PubMed, a service of the National Library of Medicine that includes over 14 million citations for biomedical articles back to the 1950's.

A workflow was established for the development of this study in order to retrieve and analyse those genes of interest in the non-disjunction processes (Figure 1).

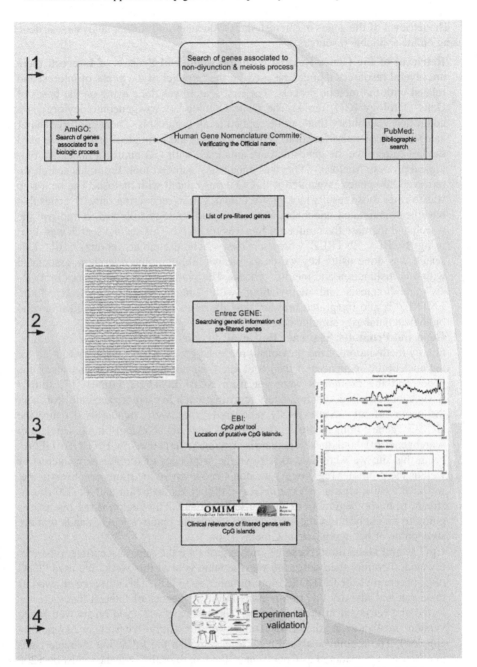

**Fig. 1.** Workflow followed during this work. Steps of the method. 1. Retrieval of the genes related to the biological process of interest. 2. Retrieval of gene and promoter sequences. 3. CpG islands detection on the promoter sequences. 4. Biological experimental validation and detection of methylation in the genes of interest (this step is not done yet)

The retrieval of the genes of interest, their sequences and further analysis was done using online available resources.

1. **Retrieval of the Genes Related with the Biological Process of Interest:** there are several resources that can be used for the retrieval of the genes of interest and related with the meiotic process. The first search was done using on the bases of Gene Ontology (GO) terms. The Gene Ontology (www.geneontology.org) is a controlled vocabulary that can be applied to all organisms. There are three major categories: molecular function, cellular location and biological process. We used the last to retrieve the genes that are annotated with GO numbers under the biological process "meiosis". This was done using AmiGO tool. From this search we retrieved 890 genes associated with GO terms related with meiosis. The next step was to filter those results looking for human genes, recovering only 21 genes that have been annotated under this term. A second strategy followed to improve the search and increase the number of hits was followed. This second search was done using NCBI's ENTREZ browser (www.ncbi.nlm.nih.gov/entrez) [16]. This search was done using key words to retrieve bibliographic references associated with non-disjunction and meiotic process. The search terms were "genes in meiosis" and "meiosis and methylation" with "human species" as search limit. ENTREZ is and integrated browser that allows us to do a search against several biological databases simultaneously.

2. **Gene and Promoter Sequence Retrieval:** In order to analyse the possible regulation by methylation of the previously retrieved genes it was necessary to look for the nucleotidic sequences of the genes and their promoters. The sequences of the genes and the promoters were got from NCBI ENTREZ GENE [17] (http://www.ncbi.nlm.nih.gov/entrez/query.fcgi?db=gene). For a more accurate search all the names of all the previously recovered genes were standardised using the HUGO gene nomenclature (http://www.gene.ucl.ac.uk/cgi-bin/nomenclature/searchgenes.pl), and these standardised names were used in ENTREZ GENE. From this site we were able to retrieve the sequences of the genes of interest as well as the promoter regions. As we don't know where the promoters exactly are, we have used a region that comprises -2.000 bases upstream and +1.000 downstream from the beginning of the transcription site. This region is the one that is going to be analysed in the next step to locate the putative CpG islands that are susceptible of being regulated by methylation.

3. **CpG Island Detection:** These GC rich regions are the ones susceptible to be methylated, therefore their detection was the main goal of this work. We used "CpG plot" program of the EMBOSS bioinformatics suite [18]. This software allowed us to search and identify GC rich regions in the sequences of interest that were selected and retrieved in previous steps. The software was used on its web based version available at the EBI Tools website (http://www.ebi.ac.uk/emboss/cpgplot/). The regions analysed have to accomplish the following characteristics in order to be considered as CpG islands, the length of the sequence must be >200 with >50% of GC content and Observed / Expected ratio is > 0.6. The results of this analysis are three plots of the three considered parameters along the sequence and also the starting and finishing bases of the putative CpG islands in the case of detecting any. Figure 2

4. **CpG Island Scoring:** The resulting putative CpG Islands were ranked into five degrees using as qualifying characteristics their length and their position referred to the transcription starting point (TSP). Table 1.

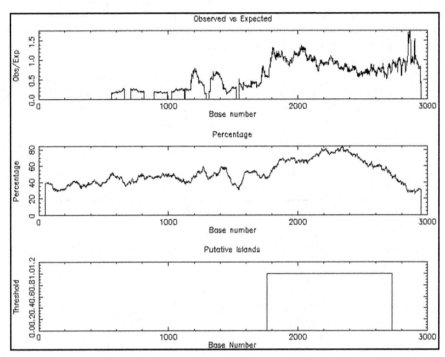

**Fig. 2.** Example of CpG island isolated using "CpG Plot program". In this case, the promoter region of KIF23 gene is the sequence comprised between -2.000 and + 1.000 nucleotides, taking as 0 (position 2000 in the lower plot) the transcription starting point (TSP). The upper plot represents the Obs/Exp ratio, that should be >0.6 to be considered as a possible CpG Island. In the middle plot the CG% is represented, the lowest limit to consider a CpG island is 50%. In the lowest plot it's possible to see the identified CpG Island, it's also possible to see from the 2 other graphs that all the conditions for detecting a CpG Island were satisfied as well as the highest quality rank proposed

**Table 1.** CpG Islands ranking criteria. The detected CpG islands were classified depending on their characteristics for further analysis in an attempt of measuring their reliability

| Rank | Classification Criteria |
|------|-------------------------|
| 1* | the island does not cover the start of transcription, so it's further than 50 nucleotides upstream or downstream of the TSP |
| 2* | the island covers the TSP but only for less than 50 nucleotides |
| 3* | the island covers the TSP, only for less than 50 nucleotides and the island has less than 350 bp |
| 4* | the island covers the TSP, more than 50 nucleotides and the isle has less than 350 bp |
| 5* | the island covers the TSP, more than 50 nucleotides and the isle has more than 350 bp |

# 3 Results

We have finally analyzed 63 genes that are directly related with the biological process of meiosis and therefore can be involved in the non-disjunction processes. These 63 genes come from the searches done in PubMed and in AmiGo using GO terms.

From the AmiGO search we retrieve up to 890 genes with go terms associated with meiosis. After filtering those genes by selecting only those ones annotated in humans in the different databases, we recover only 21 genes that will be studied in the following steps of our work. The search done in PubMed and ENTREZ using the criteria previously described in the methodology section, gives us 43 genes. It is interesting to remark that we have found only one gene that was retrieved in both searches. The complete list with the 63 retrieved genes is available in Table 2.

**Table 2.** List of the 63 retrieved through the search in AmiGO or/and PubMed tools. These selected genes suffered the undergoing process to analyse whether the might be regulated by methylation or not genes

| Gene Name | Source | Gene Name | Source |
|-----------|--------|-----------|--------|
| SYCP1 | AmiGO | FMN2 | PubMed |
| TUBG1 | AmiGO | PFN1 | PubMed |
| NPM2 | AmiGO | PFN2 | PubMed |
| CCNA1 | AmiGO | CFL1 | PubMed |
| HSPA1B | AmiGO | CFL2 | PubMed |
| PIM2 | AmiGO | ANLN | PubMed |
| SUV39H2 | AmiGO | RHOA | PubMed |
| TAF1L | AmiGO | ECT2 | PubMed |
| C8ORF1 | AmiGO | ROCK1 | PubMed |
| DMC1 | AmiGO | ROCK2 | PubMed |
| DMWD | AmiGO | PPP1R12A | PubMed |
| DUSP13 | AmiGO | PPP1R12B | PubMed |
| MSH5 | AmiGO | CIT | PubMed |
| RAD51 | AmiGO | SPIN | PubMed |
| ATRX | AmiGO | SPIN2 | PubMed |
| SGOL2 | PubMed | SPIN3 | PubMed |
| SMC1L1 | AmiGO | CDCA8 | PubMed |
| CSPG6 | AmiGO | SEC8L1 | PubMed |
| TOP3A | AmiGO | RAD21 | PubMed |
| XRCC2 | AmiGO | STAG2 | PubMed |
| ZW10 | AmiGO | STAG3 | PubMed |
| KIF23 | PubMed | SMC1L1 | PubMed |
| KRT4 | PubMed | SMC1L2 | PubMed |
| PRC1 | PubMed | ESPL1 | PubMed |
| AURKB | PubMed | PTTG1 | PubMed |
| INCENP | PubMed | SPO11 | PubMed |
| BIRC5 | PubMed | PLK1 | PubMed |
| KIF4A | PubMed | AURKA | PubMed |
| KIF4B | PubMed | AURKC | PubMed |
| FMN1 | PubMed | CDC14A | PubMed |
| SGOL1 | PubMed | CDC14B | PubMed |
| REC8L1 | AmiGO + PubMed | | |

Sequences from the selected genes are retrieved in FASTA format in all cases but in two genes (KIF4B and AURKA) due to the lack of annotation for them. The search for the CpG Islands is then carried out applying the methodology described in the previous section. From " CpG plot " output it is possible to identify 29 genes Table 3. which contains a putative CpG Islands. The quality of each of these detected Islands is then assessed giving a rank related with their reliability.

The analysis of the DNA sequences of the selected genes results in the detection of putative CpG islands in 29 of them. The presence of these Islands could explain an epigenetic regulation by methylation of their promoter sequences.

**Table 3.** List of the genes containing putative CpG Islands after the bioinformatic analysis. In the table it's shown the length of the Islands that can be detected as well as the results of the quality assessment for each gene

| Gene name (HUGO) | Data Source | Length (each island) | Quality of CpG island |
|---|---|---|---|
| AURKC | PubMed | 532 | 5 |
| CDC14A | PubMed | 267, 334, 359, 262, 257 | 5 |
| CDCA8 | PubMed | 244, 206, 700 | 5 |
| KIF23 | PubMed | 961 | 5 |
| PLK1 | PubMed | 910 | 5 |
| PPP1R12B | PubMed | 623 | 5 |
| RAD51 | AmiGO | 989 | 5 |
| RHOA | PubMed | 616, 378 | 5 |
| SUV39H2 | AmiGO | 1206 | 5 |
| BUB1B | PubMed | 388, 214 | 4 |
| MSH5 | AmiGO | 566 | 4 |
| REC8L1 | AmiGo + PubMed | 273, 423, 310 | 4 |
| SGOL2 | PubMed | 269, 569 | 4 |
| CCNA1 | AmiGO | 1203 | 3 |
| STAG3 | PubMed | 394, 294 | 3 |
| ANLN | PubMed | 376 | 2 |
| BIRC5 | PubMed | 341 | 2 |
| C8ORF1 | AmiGO | 431, 887 | 2 |
| CSPG6 | AmiGO | 236, 303 | 2 |
| HSPA1B | AmiGO | 591, 344 | 2 |
| PIM2 | AmiGO | 835 | 2 |
| PTTG1 | PubMed | 396 | 2 |
| SPO11 | PubMed | 270 | 2 |
| SYCP1 | AmiGO | 393 | 2 |
| TUBG1 | AmiGO | 580 | 2 |
| ESPL1 | PubMed | 275, 633 | 1 |
| KIF4A | PubMed | 368 | 1 |
| SEC8L1 | PubMed | 259, 254 | 1 |
| STAG2 | PubMed | 503 | 1 |

# 4  Conclusions

The availability of bioinformatics tools online as well as the huge amounts of data available nowadays in public databases is enabling the possibility of doing "in silico" experiments without the need of developing new tools nor carrying out laboratory analysis. These kind of theoretical approaches needs to be validated biologically afterwards by the corresponding experiments.

In our work we have focused on the detection of genes that could be related with the susceptibility for non-disjunction diseases and their epigenetic control by methylation. From our results we have been able to detect a set of 29 genes that are good candidates to be involved in these processes. The analysis has been performed using a totally "in silico" approach. All the study was done using the information available in public databases as well as public bioinformatics tools.

This approach might also open new scopes to the development of new predictive and pre-conceptional testing for non-disjunction, and for the identification of new candidate genes for other biological problem with the same "in silico" procedure.

The application of the proposed approach may help in the identification of therapy molecular targets for therapies. Nowadays some de-methylating agents such as decitabine (5-aza-2-deoxy-cytidine) are undergoing clinical trials for cancer therapies as well as for some congenital anomalies with a chromosomal character. The identification of such targets would enable the possibility of avoiding invasive techniques and improving planning and management of pregnancy at pre-conceptional stages. Epigenetic techniques are very robust and efficient needing a small amount of biological samples, the implementation of new tests based in epigenetics and the selection of targets by "in silico" analysis could substitute and improve the current screening techniques.

We think that it could be very interesting to carry out the experimental validation of these "in silico" results, so we propose to accomplish MSP (Methylation Specific PCR) tests, studying the methylation patterns of all those filtered genes in cell lines with trisomy 21, as well as in Down's syndrome patients, their parents and siblings to try to verify a relationship between their methylation pattern and the susceptibility for a non-disjunction disorder.

# References

1. Warburton D, Dallaire L, Thangavelu M, Ross L, Levin B, Kline J. Trisomy recurrence: a reconsideration based on North American data. Am J Hum Genet. (2004);75(3):376-85.
2. Gardner RMJ, Sutherland GR. Down syndrome, other full aneuploidies and polyploidy. In: Chromosome Abnormalities and Genetic Counselling (Oxford Monographs on Medical Genetics, No. 46). 3rd edn. Oxford University Press. (2003) 243-258.
3. James SJ, Pogribna M, Pogribny IP, Melnyk S, Hing RJ, Gibson JB, Yi P, Tafoya DL, Swenson DH, Wilson VL, Gaylor DW. Abnormal folate metabolism and mutation in the methylenetetrahydrofolate reductase gene may be maternal risk factors for Down syndrome. Am J Clin Nutr. (1990) 70. 495-501.
4. Hassold T, Hunt P. To err (meiotically) is human: the genesis of human aneuploidy. Nature Rev Genet. (2002) 2. 280-29.
5. Lucock M, Yates Z. Folic acid – vitamin and panacea or genetic time bomb? Nature Rev Genet. (2005) 6. 235–240.
6. Hanks S, Coleman K, Reid S, Plaja A, Firth H, Fitzpatrick D, Kidd A, Mehes K, Nash R, Robin N, Shannon N, Tolmie J, Swansbury J, Irrthum A, Douglas J, Rahman N. Constitutional aneuploidy and cancer predisposition caused by biallelic mutations in BUB1B. Nat Genet. (2004);36(11):1159-1161.

7. Heim S, Mitelman F.: Nonrandom chromosome abnormalities in cancer. In: Cancer Cytogenetics. 2nd edn. Ed Wiley-Liss. (1995), 19-32.

8. Esteller M, Corn PG, Baylin SB, Herman JG. A gene hypermethilation profile of human cancer. Cancer Res. (2001);61(8):3225-9.

9. Paz MF, Fraga MF, Avila S, Guo M, Pollan M, Herman JG, Esteller M. A systematic profile of DNA methylation in human cancer cell lines. Cancer Res. (2003);63(5):1114-21.

10. Paz MF, Avila S, Fraga MF, Pollan M, Capella G, Peinado MA, Sanchez-Cespedes M, Herman JG, Esteller M. Germ-line variants in methyl-group metabolism genes and susceptibility to DNA methylation in normal tissues and human primary tumours. Cancer Res. (2002);62(15):4519-24.

11. Esteller M. CpG island hypermethylation and tumor suppressor genes: a booming present, a brighter future. Oncogene. (2002);21(35):5427-40.

12. Antequera F, Bird A. Number of CpG islands and genes in human and mouse. Proc Natl Acad Sci U S A. (1993);90(24):11995-9.

13. Galperin MY. The Molecular Biology Database Collection: 2005 update. Nucleic Acids Research, (2005),33, Database issue D5–D24.

14. Wain HM, Lush M, Ducluzeau F, Povey S. Genew: the human gene nomenclature database. Nucleic Acids Res. (2002);30(1):169-71.

15. Hamosh A, Scott AF, Amberger JS, Bocchini CA, McKusick VA. Online Mendelian Inheritance in Man (OMIM), a knowledgebase of human genes and genetic disorders. Nucleic Acids Res. (2005);33(Database issue):D514-7Wheeler DL, Church DM, Edgar R, Federhen S, Helmberg W, Madden TL, Pontius JU, Schuler GD, Schriml LM, Sequeira E, Suzek TO, Tatusova TA, Wagner L. Database resources of the National Center for Biotechnology Information: update. Nucleic Acids Res. (2004)1;32(Database issue):D35-40.

17. Maglott D, Ostell J, Pruitt KD, Tatusova T. Entrez Gene: gene-centered information at NCBI. Nucleic Acids Res. (2005);33(Database issue):D54-8.

18. Rice P, Longden I, Bleasby A: EMBOSS: the European Molecular Biology Open Software Suite. Trends Genet. 2000 Jun;16(6):276-7.

# Foreseeing Promising Bio-medical Findings
# for Effective Applications of Data Mining

Stefano Bonacina, Marco Masseroli, and Francesco Pinciroli

Department of Bioengineering, Politecnico of Milan,
Piazza Leonardo da Vinci 32, 20133 Milan, Italy
{stefano.bonacina,marco.masseroli,francesco.pinciroli}@polimi.it
http://www.medinfopoli.polimi.it/

**Abstract.** The increasing availability of automated data collection tools, database technologies and Information and Communication Technologies in biomedicine and health care have led to huge amounts of biomedical and health-care data accumulated in several repositories. Unfortunately, the process of analysis of such data represents a complex task also because data volumes grow exponentially so manual analysis and interpretation become impractical. Fortunately, knowledge discovery in databases (KDD) and data mining (DM) are powerful tools available to medical and research people for help them in explore data and discover useful knowledge. To assess the spread of DM and KDD in biomedicine and health care, we designed and performed a search database of biomedical and health-care scientific literature, for the year interval 1997-2004, and analyzed the obtained results. There has been an increase of application of DM methods in literature of bio-medical informatics research most of which in bioinformatics and genomic area.

## 1 Introduction

In the medical arena, the increasing availability of automated data collection tools, database technologies and Information and Communication Technologies have led to huge amounts of biomedical and health-care data accumulated in several repositories. Extracting useful knowledge from these volumes of data is a complex task, not only because manual data analysis is impractical. Furthermore, medical data have a wide range of characteristics: for example, they are heterogeneous (alphanumeric data, signals, bio-images, audio, movies, genomic sequence), and are collected on human subjects (so the abuse and misuse of such data have to be prevented) [1]. Knowledge discovery in databases (KDD) and data mining (DM) are powerful tools available to medical and research people for help them in explore data and discover useful knowledge. For example, Lucas [2] outlined general trends in data mining and applications of Bayesian analysis in health care and illustrated recent results obtained in critical care. Wilson et al. [3] reviewed the current use of data mining in pharmacovigilance. and provide an overview of the data mining process. They found that KDD will not replace traditional methods of pharmacovigilance, but if used in conjunction may reduce the time required for detection of adverse drug events identification. As explained by Lee and Abbott [4] the large collections of valuable nursing/ healthcare data can be mined using Bayesian networks. For example, structural learning of Bayesian networks can assist researchers in identifying the contributing factors relating to a specific patient outcome [4]. Liu and Wong [5] described a methodology for analyzing genomic sequence data in order to recognized biologically meaningful

J.L. Oliveira et al. (Eds.): ISBMDA 2005, LNBI 3745, pp. 130–136, 2005.

functional site. To assess the spread of DM and KDD in biomedicine and health care, we designed and performed a search in PubMed database of biomedical and health-care scientific literature, for the years 1997-2004, and analyzed the obtained results.

## 2   Basic Concepts

The problem of knowledge extraction from large databases involves many steps, ranging from data manipulation and retrieval to fundamental mathematical and statistical inference, search, and reasoning [6]. KDD focuses on the overall process of knowledge discovery from data, including how the data is stored and accessed, how algorithms can be scaled to massive datasets and still run efficiently, how results can be interpreted and visualized, and how the overall human-machine interaction can be modeled and supported [6]. Data Mining is a step in KDD process consisting of applying computational techniques that, under acceptable computational efficiency limitations, produces a particular enumeration of patterns (or models) over the data [7]. A Data Mining technique may have two possible objective: prediction and description. Predictive Data Mining: that is the analysis of a set of data with the aim of predicting activities, classes or deduce other characteristics [8]. Descriptive Data Mining: that is the analysis of a set of data with the aim of describing the characteristics of one or more subsets.

A Data Mining strategy indicates how learning from data: kind of learning are supervised and unsupervised. Supervised learning methods are deployed when variables (inputs) are used to make prediction about another variable (considered as output) with known output. Unsupervised learning methods are deployed on data for which an output variable exists, but its value for input cases is unknown.

### 2.1   Data Mining Methods

Data Mining is not only the application of a single computational technique. It combines a group of techniques, getting results that the single one would not be able to obtain. Any technique that helps extract more out of your data is useful [9]. The following DM methods we considered: 1) decision trees, 2) neural networks, 3) clustering, 4) associations rules, 5) genetic algorithms, 6) visualization techniques. The considered techniques are here summarized.

Decision trees: Decision Trees are used to classify a set of data by the analysis of attributes and the determination of discriminating values. The result is a representation (tree) of an automatic classification procedure. A training set is used to create the tree and a testing set is used to verify it on new data.

Neural networks: the human brain consists of a very large number of neurons. Although neurons could be described as the simple building blocks of the brain, the human brain can  handle very complex tasks despite this relative simplicity. This analogy offers an interesting model for the creation of more complex learning machines, and has led to the creation of the artificial neural networks.

Clustering: a cluster is a collection of data objects that are similar to one another within the same cluster but dissimilar to the objects in other clusters. Cluster analysis consists in Grouping a set of data objects into clusters. Clustering is unsupervised classification, no predefined classes are defined.

Association rules: finding frequent patterns, associations, correlations, or causal structures among sets of items or objects in transaction databases, relational databases, and other information repositories. A frequent pattern is a set of items, sequence that occurs frequently in a database. An association rule is an implication like X => Y. X and Y are attribute values that occur simultaneously.

Genetic algorithms: nature is often a source of inspiration for technical breakthroughs. Evolutionary computing consist in problem solving by the application of mechanisms inspired to Darwin's theory of evolution.

Visualization techniques: visual data exploration aims at integrating the human in the data exploration process, applying its perceptual abilities to the large data sets available in today's computer systems.

## 3   Methods

On January 17, 2005, we performed biomedical literature searches using PubMed, the US National Library of Medicine publication database, and Entrez, the text-based search and retrieval system of the National Center for Biotechnology Information (NCBI) [10]. As main search string, we designed a query composed by "data mining" quoted phrase combined with an expression containing the words knowledge, discovery, database as shown in Table 1.

**Table 1.** The main query (Q1) submitted to PubMed to retrieve citations involving Data Mining and Knowledge Discovery and the number of obtained citations (Q.ty =Quantity)

| Q | Search string | Query submitted to PubMed (Main query) | Q.ty |
|---|---|---|---|
| 1 | Data mining OR Knowledge discovery | ("data mining"[TIAB] OR "data-mining"[TIAB] OR "datamining"[TIAB]) OR ("Knowledge discovery"[All Fields] AND ((("databases"[TIAB] NOT Medline[SB]) OR "databases"[MeSH Terms] OR database[Text Word]) OR ("databases"[MeSH Terms] OR databases[Text Word]))) AND (Journal Article[pt] NOT review[pt] NOT Congresses[pt] AND english [la])AND("1997/01/01"[PDAT]:"2004/12/31"[PDAT]) | 668 |

We established a temporal limit of retrieval: from January 1st, 1997 to December 31th, 2004. Considered publication types were Journal Article excludes Letter, Editorial, News, Review, Congresses, published in English language. Second, we formulated search strings describing biomedical fields - such as a) medicine OR medical OR medic, b) genome OR genomic, c) protein OR proteins, d) clinical OR clinic, e) pharmaceutical preparations OR drug, f) medical record OR patient record OR health record, g) delivery of health care, h) tumor OR cancer OR neoplasm in singular or plural form - and performed queries against PubMed. Then, the listed search strings were combined with the main search string and new queries are performed against PubMed. The detailed queries are shown in Table 2. We considered the union of the obtained result sets, i.e. (Q2 OR Q3 OR ... OR Q9), in order to count the number of publication and compare with the number of publication obtained from Q1 query. For each publication contained in the union, we established the year of publication, the journal in which the publication appeared, and the value of the Impact Factor according to the Science Citation Index [11], even tough not all journal have an impact factor. For each year, we counted the publications that were published or not in a journal with impact factor, we calculated for each year a total score by adding the single impact factors of publications in that year. Then, we considered the publications appeared into main query (Q1) results, but not into the union query results. In order to obtain information about DM methods applied in biomedical research publications, we formulated query involving names of DM methods - such as 1) neural networks, 2) clustering, 3) decision trees, 4) regression, 5) association rules, 6) genetic algorithms, 7) visualization techniques - and performed queries against PubMed. Then, the listed search strings were combined with the main search string and new queries are performed against PubMed. The detailed queries are shown in Table 3.

## 4 Results

As shown in Table 1, using query Q1 we retrieved a total of 668 publications indexed in PubMed. Of these 668 publications, 570 (85,33%) appeared in at least one of the bio-medical categories we defined (see Table 2). The Fig. 1 reports, for each year from 1997 to 2004, and for each bio-medical category defined in Table 2, the values of obtained publications. Of the 570 publications 266 (46,67%) were classified belonging to the "genome" category.

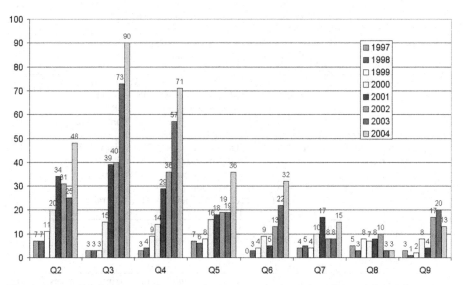

**Fig. 1.** Yearly distribution of all selected and categorized articles (570) published from January 1997 to December 2004. For explanation of Q1, Q2,...,Q9, see the captions of Tables 1 and 2.Vertical axis shows the amount of articles

Furthermore, of the 570 publications 337 (65,44%) were published on a journal with an impact factor. Details about results of number of Journals, number of Articles, and Impact Factor are shown in Table 4.

Of the 668 publications, 185 (27,69%) contained information about one or more DM methods involved in the studies (Table 3). Applications of neural networks are most diffused (67 studies on 185), while genetic algorithm and visualization techniques are the less considered.

**Table 2.** Search strings (Categories) from Biomedical fields and relative queries (Q2 to Q9) submitted to PubMed and the number of obtained citations for each of them (Q.ty=Quantity)

| Q | Search strings (Categories) | Queries submitted to PubMed: (Main query) AND ... | Q.ty |
|---|---|---|---|
| 2 | medicine OR medical OR medic | ("medicine" [MeSH Terms] OR medicine [Text Word]) OR medical [All Fields] OR medic [All Fields] | 183 |
| 3 | genome OR genomic | (("genome"[MeSH Terms] OR genome[Text Word]) OR genomic[All Fields] OR ("Genomics"[MeSH Terms] OR genomics[Text Word]) OR ("genes"[MeSH Terms] OR gene[Text Word])) OR ((microarray[All Fields] OR microarrays[All Fields]) OR "DNA chip"[All Fields] OR "DNA chips"[All Fields] OR ("oligonucleotide chips"[All Fields] AND "genechips"[All Fields])) | 266 |
| 4 | protein OR proteins | (proteins [MeSH Terms] OR protein[Text Word] OR proteins [Text Word]) | 233 |
| 5 | clinical OR clinic | clinical [All Fields] OR clinic [All Fields] OR clinics [All Fields] OR clinicals[All Fields] | 129 |
| 6 | pharmaceutical preparations OR drug | ("pharmaceutical preparations" [MeSH Terms] OR drug [Text Word]) OR ("pharmaceutical preparations" [MeSH Terms] OR drugs [Text Word]) OR pharmacovigilance [All Fields] | 88 |
| 7 | delivery of health care | ("delivery of health care" [MeSH Terms] OR healthcare [Text Word]) OR ("delivery of health care"[MeSH Terms] OR health care[Text Word]) | 71 |
| 8 | medical record OR patient record OR health record | ("medical records systems, computerized" [MeSH Terms]) OR ("medi-cal records" [MeSH Terms] OR medical record [Text Word]) OR patient records [Text Word] OR ("electronics" [MeSH Terms] OR Electronic [Text Word]) AND ("medical records" [MeSH Terms] OR Medical Re-cord[Text Word] OR patient records [Text Word]) OR ("health" [MeSH Terms] OR health [Text Word]) AND ("records" [MeSH Terms] OR records [Text Word]) OR ((medical [All Fields] OR medic [All Fields]) AND reports [All Fields]) OR (medical [All Fields] AND ("referral and consultation" [MeSH Terms] OR referral [Text Word])) | 47 |
| 9 | tumor OR cancer OR neoplasm | tumors [Text Word] OR tumours [Text Word] OR tumor[Text Word] OR tumour [Text Word] OR "neoplasms" [MeSH Terms] OR neoplasms [Text Word] OR cancer [Text Word] OR cancers [Text Word] | 68 |
| | | Union of the obtained result sets | 570 |

## 5   Discussion

In this study we designed and performed searches in PubMed database of biomedical and health-care scientific literature, regarding DM and KDD studies in biomedicine and healthcare, for the years 1997-2004.

**Table 3.** Search strings regarding DM Methods and relative queries (Q10 to Q16) submitted to PubMed and the number of obtained citations for each of them (Q.ty=Quantity)

| Q | Search strings (Methods) | Queries submitted to PubMed: (Main query) AND ... | Q.ty |
|---|---|---|---|
| 10 | neural networks | ("neural network"[All Fields] OR "neural networks"[All Fields]) | 67 |
| 11 | clustering | (partition[All Fields] OR ("cluster analysis"[MeSH Terms] OR clustering [Text Word])) | 53 |
| 12 | decision trees | ("decision tree"[All Fields] OR "decision trees"[All Fields]) | 44 |
| 13 | regression | (regression[Text Word]) | 31 |
| 14 | association rules | ("association rule"[All Fields] OR "association rules"[All Fields]) | 22 |
| 15 | genetic algorithms | ("genetic algorithm"[All Fields] OR "genetic algorithms"[All Fields]) | 12 |
| 16 | visualization techniques | ("visualization technique"[All Fields] OR "visualization techniques" [All Fields]) | 3 |
| | | Union of the obtained result sets | 185 |

**Table 4.** Yearly distribution of the results about number of Journals (#J), Journals with an Impact Factor (#J IF), Articles (#A), Articles with an Impact Factor (#A IF), and Total Impact Factor

| Year | #J | #J IF | #A | #A IF | Total IF |
|---|---|---|---|---|---|
| 1997 | 10 | 3 | 14 | 3 | 3,71 |
| 1998 | 11 | 4 | 15 | 5 | 6,84 |
| 1999 | 18 | 9 | 26 | 12 | 26,33 |
| 2000 | 36 | 25 | 49 | 28 | 119,17 |
| 2001 | 50 | 37 | 86 | 57 | 150,77 |
| 2002 | 63 | 49 | 95 | 73 | 210,59 |
| 2003 | 68 | 49 | 117 | 84 | 180,70 |
| 2004 | 99 | 72 | 168 | 111 | 278,04 |

Because of "Data mining" is not a MeSH Term we had to established a main query containing different forms of the term (such as "data mining", "data-mining"). The use of a quoted phrase is fundamental in order to avoid retrieving the studies on miner's health care.

The data in this paper suggest that there has been an augmented interest in DM and KDD from 1997 to 2004. The range of journals that had published articles in DM is greatly increased (from 10 to 99), also for journals with impact factor (72% in 2004).

Values of Total Impact Factor (IF) are increased from 3,71 in 1997 to 278,04 in 2004, this is a remarkable result for DM in Biomedical field. The value of Total IF in 2002 is determined by the publication of DM related articles in Journal having a great IF value: one study appeared on *Nature Genetics* (IF 2002 = 26,71), and another study appeared on *Nature Neuroscience* (IF 2002 = 14,85).

# 6  Conclusions

Database technology was born in the '60s with the first electronic data collections. At present, data base management systems (DBMS) are subject to a continuous evolution that causes the so known "Data Explosion Problem". Automated data collection tools and database technology lead to tremendous amounts of data accumulated in databases, data warehouses, and other information repositories. We are drowning in data, but starving for knowledge! The solution is the application of the Knowledge Discovery in Databases Process (KDD) and in particular the development and use of Data Mining techniques.

Today KDD applications are largely diffused. Also in biomedical contexts, there is a growing interest and an always more frequent use of various data mining techniques, that are adapted to specific goals with the development of new specific application fields.

The aim of our work, summarized in this paper, is giving a view of the situation, identifying the most significant biomedical application areas in which DM and KDD are used.

The most important application fields are the medical and genomic research. In particular, in medical field applications concern decision and diagnosis support systems, automatic text analysis, and risk factors analysis. In genomic research, applications concern genomic information retrieval and gene identification.

# References

1. Cios, K.J., Moore, G.W.: Uniqueness of medical data mining. Artif. Intell. Med. 26(1-2) (2002) 1-24
2. Lucas, P.: Bayesian analysis, pattern analysis, and data mining in health care. Curr. Opin. Crit. Care. 10(5) (2004) 399-403
3. Wilson, A.M., Thabane, L., Holbrook, A.: Application of data mining techniques in pharmacovigilance. Br. J. Clin. Pharmacol. 57(2) (2004) 127-34
4. Lee, S.M., Abbott, P.A.: Bayesian networks for knowledge discovery in large datasets: basics for nurse researchers. J. Biomed. Inform. 36(4-5) (2003) 389-99
5. Liu, H., Wong, L.: Data mining tools for biological sequences. J. Bioinform. Comput. Biol. 1(1) (2003) 139-67
6. Fayyad, U., Piatetsky-Shapiro, G., Smyth, P.: The KDD process for extracting useful knowledge from volumes of data. Communication of ACM. 39(11) (1996) 27-34
7. Fayyad, U.M., Piatetsky-Shapiro, G., Smyth, P.: From data mining to knowledge discovery: An overview. AI Mag. 17(3) (1996) 37-54
8. Weiss, S.M., Indurkhya, N.: Predictive data mining: a practical guide. Morgan Kaufmann Publishers, San Francisco (1997)
9. Han, J., Kamber, M.: Data mining: concepts and techniques. Morgan Kaufmann Publishers, San Francisco (2000)
10. National Library of Medicine. Entrez PubMed. Available at: http://www.ncbi.nlm.nih.gov/entrez/query.fcgi. Last access: June 1st, 2005
11. ISINET. The Journal Citation Report. Available at: http://www.isinet.com/isi/products/citation/jcr Last access: June 1st, 2005

# Hybridizing Sparse Component Analysis with Genetic Algorithms for Blind Source Separation

Kurt Stadlthanner[1], Fabian J. Theis[1], Carlos G. Puntonet[2], Juan M. Górriz[2], Ana Maria Tomé[3], and Elmar W. Lang[1]

[1] Institute of Biophysics, University of Regensburg, 93040 Regensburg, Germany
`kusta@web.de`
[2] Dept. Arquitectura y Tecnología de Computadores, Universidad de Granada, E-18071 Granada, Spain
[3] Dept. de Electrónica e Telecomunicações / IEETA, Universidade de Aveiro, 3810-Aveiro, Portugal

**Abstract.** Nonnegative Matrix Factorization (NMF) has proven to be a useful tool for the analysis of nonnegative multivariate data. However, it is known not to lead to unique results when applied to nonnegative Blind Source Separation (BSS) problems. In this paper we present first results of an extension to the NMF algorithm which solves the BSS problem when the underlying sources are sufficiently sparse. As the proposed target function has many local minima, we use a genetic algorithm for its minimization.

## 1 Matrix Factorization and Blind Source Separation

In the field of modern data analysis mathematical transforms of the observed data are often used to unveil hidden principles. Especially in situations where different observations of the same process are available matrix factorization techniques have been used very successfully in recent years. Thereby, the $m \times T$ observation matrix $\mathbf{X}$ is decomposed into a $m \times n$ matrix $\mathbf{W}$ and a $n \times T$ matrix $\mathbf{H}$

$$\mathbf{X} = \mathbf{WH}. \tag{1}$$

Here, it is assumed that $m$ observations, consisting of $T$ samples, constitute the rows of $\mathbf{X}$ and that $m \leq n$.

One application of matrix factorization is linear blind source separation (BSS), where the observations $\mathbf{X}$ are known to be weighted sums of $n$ underlying sources. If the sources form the rows of the $n \times T$ matrix $\mathbf{S}$, and the element $a_{ij}$ of the so-called mixing matrix $\mathbf{A}$ is the weight with which the $j$-th source contributes to the $i$-th observation, then $\mathbf{X}$ can be decomposed as

$$\mathbf{X} = \mathbf{AS}. \tag{2}$$

In BSS now, given only the matrix $\mathbf{X}$, a matrix factorization as in (1) is sought such that $\mathbf{A}$ and $\mathbf{S}$ are essentially equal to $\mathbf{W}$ and $\mathbf{H}$, i.e. they are identical up

J.L. Oliveira et al. (Eds.): ISBMDA 2005, LNBI 3745, pp. 137–148, 2005.

to some scaling and permutation indeterminacies. Obviously, the BSS problem is highly underdetermined such that it can only be solved uniquely if additional assumptions on the sources or the mixing matrix are made.

Note, that in the sequel we will confine ourselves to the quadratic BSS problem where the number of sources to be recovered equals the number of available observations, i.e. $m = n$.

## 2   Sparse Nonnegative Blind Source Separation

An often used variant of matrix factorization is nonnegative matrix factorization (NMF) where the source matrix $\mathbf{S}$, the mixing matrix $\mathbf{A}$ as well as the observation matrix $\mathbf{X}$ are assumed to be strictly nonnegative. Albeit NMF has been used successfully in the field of image and text analysis [1], it cannot solve the BSS problem uniquely up to scaling and permutation indeterminacies, which means, that additional constraints are needed.

In literature, the assumption has often been made that the sources are sparsely represented, i.e. have many zero entries. Such sparseness constraints have already been exploited successfully in NMF based image analysis methods [2] as well as in other BSS algorithms. Therefore, we adopt this idea and require in our approach that the sources are sparsely represented.

The basic idea of our approach is to estimate the original source matrix $\mathbf{A}$ and mixing matrix $\mathbf{S}$, respectively, by determining two nonnegative matrices $\hat{\mathbf{A}}$ and $\hat{\mathbf{S}}$ such that

1. the rows of the matrix $\hat{\mathbf{S}}$ are as sparse as possible,
2. the reconstruction error of the mixtures $||\mathbf{X} - \hat{\mathbf{A}}\hat{\mathbf{S}}||^2$ is as small as possible.

Our approach to solve this problem algorithmically is to find two nonnegative matrices $\hat{\mathbf{A}}$ and $\hat{\mathbf{S}}$ which minimize the following target function $E(\tilde{\mathbf{A}}, \tilde{\mathbf{S}})$

$$E(\tilde{\mathbf{A}}, \tilde{\mathbf{S}}) = ||\mathbf{X} - \tilde{\mathbf{A}}\tilde{\mathbf{S}}||^2 + \lambda \sum_{i=1}^{N} \sigma(\tilde{\mathbf{S}}_i), \qquad (3)$$

where $\sigma$ is an appropriate sparseness measure, $\lambda$ is a weighting factor, and $\tilde{\mathbf{S}}_i$ denotes the $i$-th row of the matrix $\tilde{\mathbf{S}}$.

Thereby, the matrix $\tilde{\mathbf{S}}$ is obtained from the matrix $\tilde{\mathbf{S}}_- = \tilde{\mathbf{A}}^{-1}\mathbf{X}$ by setting the negative elements of $\tilde{\mathbf{S}}_-$ to zero. Note, that given the matrix $\tilde{\mathbf{A}}$ the matrix $\tilde{\mathbf{S}}$ is already defined which means that the above optimization problem depends only on the matrix $\tilde{\mathbf{A}}$.

## 3   Sparseness Measure

For the sparse nonnegative BSS problem at hand, we define the sparseness $\sigma$ of a vector $\mathbf{s}$ as the fraction of its zero to its nonzero elements. However, as in real life experiments measurements are always corrupted by noise, also small nonzero entries of a vector should be treated as zero elements. Hence, we use a nonnegative threshold $\tau$ which defines the maximum value an entry of $\mathbf{s}$ may have in order to be regarded as a zero element. This leads to the following sparseness measure $\sigma$:

$$\sigma(\mathbf{s}) = \frac{\text{number of elements of } \mathbf{s} < \tau}{\text{number of elements of } \mathbf{s}}, \tag{4}$$

It may be noted that in literature another sparseness measure has already been proposed in the context of NMF which defines the spareness of a vector by the ratio of its $l_1$ norm to its $l_2$ norm. Even if such a sparseness measure is computationally less demanding it is inappropriate for the BSS task as will be shown in the simulations section.

## 4    Genetic Algorithm Based Optimization

As the target function defined in Eq. 3 has many local minima we use a Genetic Algorithm (GA) for its minimization.

GAs are stochastic global search and optimization methods inspired by natural biological evolution. The core of a GA is a population of possible solutions, called individuals, to a given optimization problem as well as a set of operators borrowed from natural genetics. At each generation of a GA, a new set of approximations is created by the process of selecting individuals according to their level of fitness in the problem domain and reproducing them using the genetically motivated operators. This process leads to the evolution of populations of individuals that better solve the optimization problem than the individuals from which they were created. Finally, this process should lead to the optimal solution of the optimization problem even if many suboptimal solutions exist, i.e. if the target function to be optimized has many local minima.

For the minimization of the target function in Eq. 3 the $m^2$ elements of the solution matrix $\hat{\mathbf{A}}$ have to be determined. Taking advantage of the scaling indeterminacy inherent in the linear mixture model (2) we may assume that the columns of the original mixing matrix $\mathbf{A}$ are normalized such that its diagonal elements are ones. Hence, only the $m^2 - m$ off elements of the matrix $\hat{\mathbf{A}}$ have to be determined by the GA. Accordingly, each of the $N_{ind}$ individuals of the GA algorithm consists of $m^2 - m$ parameters which are usually referred to as genes. As the original mixing matrix is known to have only nonnegative entries it seems self-evident to confine the genes to be nonnegative, too. However, we allow the genes to be negative throughout the optimization procedure as we have observed in our experiments that otherwise the GA often fails to find the global minimum of the target function.

In every generation of the GA, the fitness of each individual for the optimization task has to be computed in order to determine the number of offsprings it will be allowed to produce. For this purpose, the target function (3) is evaluated for all individuals. These function values are not used directly as fitness values as otherwise the fittest individuals often produce too many offsprings such that the needed diversity in the population is destroyed and the algorithm converges prematurely to a suboptimal solution. Hence, we use a linear scaling procedure to transform target function values to fitness values.

In order to compute the target function values, for every individual a matrix $\tilde{\mathbf{A}}_-$ is generated which off elements consist of the genes as stored in the individual and which diagonal elements are set to one. As in the next step the matrix $\tilde{\mathbf{A}}_-$ has to be inverted, care must be taken that it is not singular. Therefore, we replace matrices $\tilde{\mathbf{A}}_-$ with a conditional number with respect to inversion which is higher than a user defined threshold $\tau_{sing}$ by a random matrix (also with ones on its diagonal) with a conditional number lower than $\tau_{sing}$. Accordingly, the genes of the corresponding individual are adjusted.

Next, the matrices $\tilde{\mathbf{A}}$ and $\tilde{\mathbf{S}}$ are needed in order to evaluate the target function (3). For this purpose, the inverse $\tilde{\mathbf{W}}_-$ of $\tilde{\mathbf{A}}_-$ is computed and the matrices $\tilde{\mathbf{S}}$ and $\tilde{\mathbf{A}}$ are then obtained by setting the negative elements of the matrices $\tilde{\mathbf{S}}_- = \tilde{\mathbf{W}}_- \mathbf{X}$ and $\tilde{\mathbf{A}}_-$, respectively, to zero.

After inserting the matrices $\tilde{\mathbf{S}}$ and $\tilde{\mathbf{A}}$ into (3) the resulting target function value is assigned to the corresponding individual. The individuals are then arranged in ascending order according to their target function values and their fitness values $F(p^{(i)})$, $i = 1, \ldots, N_{ind}$, are determined by

$$F(p^{(i)}) = 2 - \mu + 2(\mu - 1)\frac{p^{(i)} - 1}{N_{ind} - 1}, \tag{5}$$

where $p^{(i)}$ is the position of individual $i$ in the ordered population. The scalar parameter $\mu$, which is usually chosen to be between 1.1 and 2.0, denotes the selective pressure towards the fittest individuals.

We have used Stochastic Universal Sampling (SUS) to determine the absolute number of offsprings an individual may produce. Thereby, an arc $R_i$ of length $F(p^{(i)})$ is assigned to the $i$-th individual, $i = 1, \ldots, N_{ind}$, on a circle of circumference $C = \sum_{i=1}^{N_{ind}} F(x^{(i)})$. Starting from a randomly selected position, $2N_{off}$ marker points are allocated on the circle, whereas the distance between two consecutive marker points is $C/2N_{off}$ and $N_{off}$ is the total number of offsprings to be created. The $i$-th individual may then produce as many offsprings as there are marker points in its corresponding arc $R_i$ on the circle.

The offsprings are created in a two step procedure. In the first step, two individuals, which are eligible for reproduction according to the SUS criterion, are chosen at random and are used to create a new individual. Thereby, the genes of the new individual are generated by uniform crossover, i.e. each gene of the new individual is created by copying, each time with a probability of 50 %, the corresponding gene of the first or the second parent individuum.

In the second step, called mutation, the actual offsprings are obtained by altering a certain fraction $r_{mut}$ of the genes of the new individuals. These genes are chosen at random and are increased or decreased by a random number in the range of $[0, m_{max}]$. The role of mutation is often seen as providing a guarantee that the probability of searching any given parameter set will never be zero and acting as a safety net to recover good genetic material that may be lost through the action of selection and crossover.

The last action occurring during each generation of a GA is the replacement of the parent individuals by their offsprings. We use an elitist reinsertion scheme meaning that a certain fraction $r_{elit}$ of the fittest individuals is deterministically allowed to propagate through successive generations. Hence, only the $(1-r_{elit})N_{ind}$ less fittest parent individuals are replaced by their fittest offsprings which ensures that the best solution found so far remains in the population.

In order to keep the algorithm from converging permaturely we make use of the concept of multiple populations. Thereby, a number $N_{pop}$ of populations, each consisting of $N_{ind}$ individuals, are propagating independently in parallel and are only allowed to exchange their fittest individuals after every $T_{ex}$-th generation. Hence, as long as not all populations have converged to the same solution they will regain some diversity after every $T_{ex}$-th iteration step. We use the complete net structure scheme for the exchange of individuals which means that every population is exchanging a fraction $r_{mig}$ of its fittest individuals with all other populations.

Finally, it must be noted that we have used the functions provided by the *Genetic Algorithm Toolbox* [4] for all GA procedures apart from the mutation operator which was implemented by ourselves.

## 5   Algorithm Repetitions

Despite the use of the mutation operator and multiple populations, the algorithm failed in many experiments to recover the source and mixing matrix after its first run. In order to keep the computational cost of the algorithm reasonable, this problem could not be overcome by simply increasing the number $N_{ind}$ of individuals and $N_{pop}$ of populations to arbitrarily large values.

We still managed to achieve satisfying results by applying the algorithm repeatedly. As usually, the algorithm is provided with the observation matrix $\mathbf{X}$ in its first run which is then decomposed into a first estimate of the source matrix $\tilde{\mathbf{S}}^{(1)}$ and the first estimate of the mixing matrix $\tilde{\mathbf{A}}^{(1)}$, i.e. $\mathbf{X} = \tilde{\mathbf{A}}^{(1)}\tilde{\mathbf{S}}^{(1)}$. In order to make use of the suboptimal results already achieved in the first run, the matrix $\tilde{\mathbf{S}}^{(1)}$ is provided to the algorithm instead of the matrix $\mathbf{X}$ in the second run. The matrix $\tilde{\mathbf{S}}^{(1)}$ is then factorized into the matrices $\tilde{\mathbf{A}}^{(2)}$ and $\tilde{\mathbf{S}}^{(2)}$, which means that the matrix $\mathbf{X}$ can now be factorized as $\mathbf{X} = \tilde{\mathbf{A}}^{(1)}\tilde{\mathbf{A}}^{(2)}\tilde{\mathbf{S}}^{(2)}$. This procedure is repeated $K$ times until the newly determined mixing matrix $\mathbf{A}^{(K)}$ differs only marginally from the identity matrix. With this procedure the final estimates of the mixing matrix $\hat{\mathbf{A}}$ and of the source matrix $\hat{\mathbf{S}}$ are determined as

$$\hat{\mathbf{A}} = \prod_{j=1}^{K} \tilde{\mathbf{A}}^{(j)} \tag{6}$$

and

$$\hat{\mathbf{S}} = \tilde{\mathbf{S}}^{(K)}, \tag{7}$$

respectively, as the matrix $\mathbf{X}$ can be factorized as $\mathbf{X} = \prod_{j=1}^{K} \tilde{\mathbf{A}}^{(j)}\tilde{\mathbf{S}}^{(K)}$.

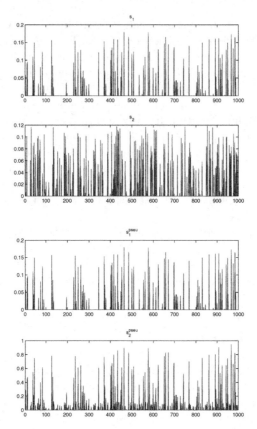

**Fig. 1.** Top: The original sources $\mathbf{s}_1$ and $\mathbf{s}_2$. Bottom: The pseudo sources $\mathbf{s}_1^{pseu}$ and $\mathbf{s}_2^{pseu}$. Even if the number of zero elements in $\mathbf{s}_2^{pseu}$ is lower than in the original source $\mathbf{s}$ the sparseness measure $sp$ assigns to it a higher value than to the original source

# 6   Simulations

## 6.1   Choice of Sparseness Measure

We want to point out that the often used sparseness measure $sp(\mathbf{s})$ of a $T$-dimensional vector $\mathbf{s}$

$$sp(\mathbf{s}) = \frac{\sqrt{n} - \sum_{i=1}^{T} |s_i| / \sqrt{\sum_{i=1}^{T} s_i^2}}{\sqrt{n} - 1}, \tag{8}$$

where $s_i$ is the $i$-th component of $\mathbf{s}$, cannot be used for the BSS task even if it measures reasonably the sparseness of vectors with only one nonzero entry ($sp(\mathbf{s}) = 1$) and the sparseness of vectors where all elements are equal ($sp(\mathbf{s}) = 0$).

To give an example for the ineligibility of the sparseness measure $sp$, we have generated two nonnegative random sources $\mathbf{s}_1$ and $\mathbf{s}_2$, each consisting of 1000 data points, and have randomly set 90% of the elements of the first source and 80% of the elements of the second source to zero. These two sources were normalized and then used to constitute the rows of the source matrix $\mathbf{S}$ (cf. Fig. 1). The matrix of observations $\mathbf{X}$ was obtained by mixing the sources with the following mixing matrix

$$\mathbf{A} = \begin{bmatrix} 5 & 1 \\ 6 & 1 \end{bmatrix} \tag{9}$$

according to Eq. 2.

Note, that as the first source is dominating in both of the mixtures the following alternative factorization of the observation matrix $\mathbf{X}$ is feasible. First, the original mixing matrix $\mathbf{A}$ can be replaced by the pseudo mixing matrix

$$\mathbf{A}^{pseu} = \begin{bmatrix} 0 & 1 \\ 1 & 1 \end{bmatrix}. \tag{10}$$

Correspondingly, the original source matrix $\mathbf{S}$ has then to be replaced by the matrix $\mathbf{S}^{pseu}$, which rows are constituted by the following pseudo sources (see Fig. 1)

$$\mathbf{s}_1^{pseu} = \mathbf{s}_1, \tag{11}$$
$$\mathbf{s}_2^{pseu} = 5\mathbf{s}_1 + \mathbf{s}_2. \tag{12}$$

Obviously, these matrices also factorize $\mathbf{X}$, i.e. $\mathbf{X} = \mathbf{A}^{pseu}\mathbf{S}^{pseu}$ still holds, but the number of zero elements in the second pseudo source $\mathbf{s}_2^{pseu}$ is about 8% lower than that of the original source $\mathbf{s}_2$ (cf. Tab. 1). Hence, a BSS algorithm based on the sparseness measure $\sigma$ as defined in Eq. 4 would correctly favor the original source and mixing matrix $\mathbf{S}$ and $\mathbf{A}$, respectively, over their pseudo variants $\mathbf{S}^{pseudo}$ and $\mathbf{A}^{pseudo}$.

In contrast, the sparseness measure $sp$ as defined in Eq. 8 assigns a higher sparseness value to the second pseudo source $\mathbf{s}_2^{pseudo}$ than to the original source $\mathbf{s}_2$ (cf. Tab. 1). Accordingly, a sparse BSS algorithm based on this sparseness measure would fail to recover the original source and mixing matrix $\mathbf{A}$ and $\mathbf{S}$, respectively.

## 6.2   Reliability of the Proposed Algorithm

We have generated 25 different observation matrices $\mathbf{X}^{(j)}$, $j = 1,\ldots,25$, in order to evaluate the reliability of the proposed algorithm. For the generation of each

**Table 1.** The sparsenesses $sp$ as defined in Eq. 8 and $\sigma$ as defined in Eq. 4 ($\tau = 0$) of the original and the pseudo sources. Note, that contradicting the fact that the number of zero elements of the pseudo source $\mathbf{s}_2^{pseu}$ is lower than that of the original source $\mathbf{s}_2$, the sparseness measure $sp$ reaches a higher value for the second pseudo source than for the second original source

|  | $sp(\mathbf{s})$ | $\sigma(\mathbf{s})$ |
|---|---|---|
| $\mathbf{s}_1$ | 0.76 | 0.90 |
| $\mathbf{s}_2$ | 0.62 | 0.80 |
| $\mathbf{s}_1^{pseu}$ | 0.76 | 0.90 |
| $\mathbf{s}_2^{pseu}$ | 0.69 | 0.72 |

of the $j$-th observations, three nonnegative sources $s_i^{(j)}$, $i = 1, \ldots, 3$, have been created, whereas each source consisted of 1000 nonnegative random elements uniformly distributed in the interval $(0, 1)$. We have set 900 randomly selected elements of the first source, 800 of the second source and 700 of the third source, respectively, to zero before adding some low level random noise with a maximum amplitude of 0.001. Accordingly, elements smaller than $\tau = 0.001$ (cf. Eq. 4) were treated as zero elements leading to sparseness values of $\sigma_1 = 0.9$, $\sigma_2 = 0.8$ and $\sigma_3 = 0.7$, respectively, of the sources. These sources were used to constitute the rows of 25 different source matrices $\mathbf{S}^{(j)}$. Next, 25 random nonnegative $3 \times 3$ mixing matrices $\mathbf{A}^{(j)}$ have been generated and the observation matrices $\mathbf{X}^{(j)}$ were computed as $\mathbf{X}^{(j)} = \mathbf{A}^{(j)}\mathbf{S}^{(j)}$. Based on these observation matrices only, the presented algorithm was used to recover the source and the mixing matrices, respectively.

It turned out that multiple populations are indispensable for the success of the algorithm as it otherwise converged prematurely to suboptimal solutions. Hence, we used $N_{pop} = 8$ populations, each consisting of $N_{ind} = 50$ individuals, and allowed them every $T_{ex} = 100$ generations to exchange $r_{mig} = 20\%$ of their fittest individuals. Thereby, each individual consisted of 6 genes corresponding to the 6 off elements of the mixing matrix $\mathbf{A}$.

As individuals corresponding to singular mixing matrices lead to problems during the optimization procedure, we have replaced matrices with a conditional number larger than $\tau_{sing} = 100$ by random matrices with a lower conditional number.

For the computation of the target function values for every individual, the weight factor $\lambda$ in (3) was set to 0.01 and kept fixed troughout the experiments.

The selective pressure $\mu$ used for the fitness assignment was set to 1.5 while $r_{mut} = 10\%$ of the genes of each individual were increased or decreased by maximally $m_{max} = 0.1$ during the mutation step.

Furthermore, we used an elitist reinsertion scheme where 98% of the individuals were replaced by their offsprings, i.e. only the best individual of the parent generation was passed to the new generation.

We noticed, that the algorithm seemed to converge after about 1500 iterations and finally stopped it after 2000 iterations. Finally between $K = 2$ and $K = 5$ repetitions of the algorithm were necessary in order to obtain reasonable estimates $\hat{\mathbf{A}}$ and $\hat{\mathbf{S}}$ of the mixing and source matrix according to (6) and (7), respectively.

The results of the algorithm were evaluated by computing the correlation coefficients between the original and the estimated sources on the one hand and the cross-talking error (CTE) between the original and the estimated mixing matrix on the other hand. As can be seen in Fig. 2 the algorithm lead in 76% of its runs to correlation coefficients higher than 0.99. Likewise, also the the cross-talking error between the estimated and the original mixing matrix was below 1 in 76% of the runs. As expected, problems have arisen when the conditional number of the original mixing matrix was high as we replaced possible solution matrices with a conditional number larger than $\tau_{sing} = 100$ during the optimization phase of the algorithm. However, even if we use higher $\tau_{sing}$'s the algorithm still fails to

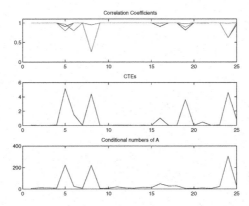

**Fig. 2.** Top: the correlation coefficients between the estimated and the original sources. Middle: crosstalking error between the estimated and the original mixing matrix. Bottom: the condition numbers of the original mixing matrices. If the condition numbers of the original matrices become too high the algorithm fails to recover the source and the mixing matrix

recover the source and mixing matrix sufficiently well. The reason is that in the case of poorly conditioned mixing matrices the global minimum is very narrow and therefore hard to find during the optimization process. On the other hand, we have noticed that choosing too high values for $\tau_{sing}$ leads to worse results when the conditional number of the original mixing matrix is low. This happens because a low $\tau_{sing}$ narrows the search space and the global minimum is found easier. Hence, our algorithm is especially eligible for problems where the mixing matrix is not extremely poorly conditioned.

### 6.3   Recovery of Correlated Sources

In this section we show that the presented method is capable of solving the BSS problem even if the underlying sources are correlated. This case is especially interesting as a very popular alternative BSS technique, called Independent Component Analysis (ICA), fails in such a situation as it is based on the assumption that the underlying sources are statistically independent.

For the simulation, we have generated three sources $s_i$, $i = 1, \ldots, 3$, as follows. The first and the second source were generated as nonnegative random vectors where 90% and 80%, respectively, of the elements were randomly set to zero. The

**Table 2.** Results obtained with the presented method (abbreviated as sparse NN BSS) and the fastICA algorithm. Displayed are the correlation coefficients $c_i$ between the $i$-th original source and its corresponding estimate as well as the cross-talking error (CTE) between the estimated and the original mixing matrix

|  | $c_1$ | $c_2$ | $c_3$ | $CTE$ |
|---|---|---|---|---|
| sparse NN BSS | 1.00 | 1.00 | 1.00 | 0.39 |
| fastICA | 1.00 | 1.00 | 0.75 | 5.19 |

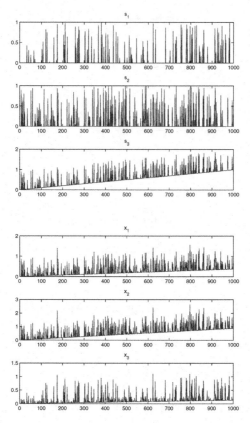

**Fig. 3.** Top: The original sources $\mathbf{s}_i$. Note, that $\mathbf{s}_3$ was obtained from $\mathbf{s}_2$ by adding a linear function. Bottom: The rows $\mathbf{x}_i$ of the mixture matrix $\mathbf{X}$ as provided to the algorithms

third source was generated from the second source by adding a linear function, i.e.

$$\mathbf{s}_3(n) = \mathbf{s}_2(n) + 0.001(n - 1), \tag{13}$$

where $\mathbf{s}_i(n)$ denotes the $n$-th element of the source $\mathbf{s}_i$ and $n = 1, \ldots, 1000$ (cf. Fig. 3).

This procedure lead to a non-vanishing correlation coefficient of $c = 0.65$ between the second and the third source, while the sources $\mathbf{s}_1$ and $\mathbf{s}_2$ as well as $\mathbf{s}_1$ and $\mathbf{s}_3$ were uncorrelated. As before, these sources were used to constitute the source matrix $\mathbf{S}$.

The observation matrix $\mathbf{X}$ (cf. Fig. 3) was generated as in Eq. 2 by multiplying the source matrix $\mathbf{S}$ with the following well conditioned (condition number about 5.5) mixing matrix

$$\mathbf{A} = \begin{bmatrix} 0.4554 & 0.5833 & 0.3739 \\ 0.8916 & 0.3988 & 0.8736 \\ 0.9042 & 0.0604 & 0.1326 \end{bmatrix}. \tag{14}$$

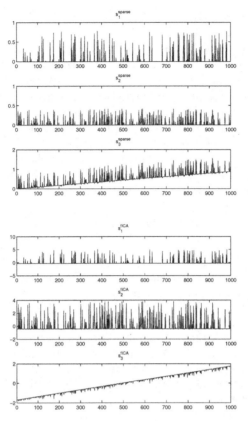

**Fig. 4.** Top: The estimates $\mathbf{s}_i^{sparse}$ of the sources as obtained by the nonnegative sparse BSS algorithm. Bottom: The estimates $\mathbf{s}_i^{fICA}$ of the sources as obtained by the fastICA algorithm. Note, that fastICA fails to recover the third source

When we used the presented nonnegative sparse BSS algorithm with the parameters set as in the last section, we could recover the sources as well as the mixing matrix almost perfectly as can be seen in Tab. 2 and Fig. 4. Thereby, only $K = 2$ successive runs of the algorithm were needed.

In contrast, such a perfect recovery seems to be impossible by ICA based BSS. To show this, we have used the famous fastICA algorithm [3] in order to recover the sources $\mathbf{s}_i$, $i = 1, \ldots, 3$, and the mixing matrix $\mathbf{A}$. This algorithm also succeeded in recovering the sources $\mathbf{s}_1$ and $\mathbf{s}_2$ almost perfectly, but it failed to recover the third source $\mathbf{s}_3$ (cf. Fig. 4). Accordingly, the cross-talking error between the estimated and the original mixing matrix is more than five times higher than the CTE achieved with the sparse nonnegative BSS approach (cf. Tab. 2). Surely, these poor results are not surprising as we have violated the independence assumption by creating correlated sources.

Hence, the presented algorithm seems to be capable of solving BSS problems where other well established BSS algorithms, like fastICA, principally fail.

# 7   Conclusions

In this paper we have presented a new BSS algorithm which is appropriate for problems where the observations, the underlying sources as well as the mixing matrix are nonnegative and where the sources are sparse. As the used target function has many local minima we have used a GA for its minimization. Furthermore, we have discussed which sparseness measure is eligible for our approach. As shown by our simulations the proposed algorithm solves well the BSS problem as long as the mixing matrix is not close to singular. Furthermore, we have shown that our approach is capable of solving the BSS problem even if the underlying sources are not uncorrelated or statistically independent. Future research will focus on alternative optimization methods for the target function like simulated annealing.

# References

1. D.D. Lee and H.S. Seung. Learning the parts of objects by non-negative matrix factorization. Nature, 40:788-791, 1999.
2. P.O. Hoyer. Non-negative matrix factorization with sparseness constraints. Journal of Machine Learning Research, 5:1457-1469, 2004
3. A. Hyvärinen, Fast and robust fixed-point algorithms for independent component analysis, IEEE Transactions on Neuronal Networks, 10(3), 626-634, 1999
4. A. Chipperfield, P. Fleming, H. Pohlheim, C. Fonseca, Genetic Algorithm Toolbox, Evolutionary Computation Research Group, University fo Sheffield, www.shef.ac.uk/acse/research/ecrg/

# Hardware Approach
# to the Artificial Hand Control Algorithm Realization

Andrzej R. Wolczowski[1], Przemyslaw M. Szecówka[2],
Krzysztof Krysztoforski[3], and Mateusz Kowalski[2]

[1] Institute of Engineering Cybernetics
andrzej.wolczowski@pwr.wroc.pl
[2] Faculty of Microsystem Electronics and Photonics
[3] Institute of Machine Design and Operations, Wroclaw University of Technology,
Wybrzeze Wyspianskiego 27, 50-370 Wroclaw, Poland

**Abstract.** The concept of the bioprosthesis control system implementation in the dedicated hardware is presented. The complete control algorithm was analysed and the decomposition revealing the parts which could be calculated concurrently was made. Specialized digital circuits providing the wavelet transform and the neural network calculations were designed and successfully verified. The experiment results show that the proposed solution provides the desired dexterity and agility of the artificial hand.

## 1 Introduction

The hand is the most universal part of the human motion system. Its loss has tremendous impact on the whole human life. That's why any devices regaining the manipulating-grasping functions (the lost ones) improve the live of the handicapped persons and help them to regain independence of the care of others [13,14].

The construction issues of the artificial hand, which could reliably copy the functionality of the living original, have been extensively investigated in various research centres [1,2,7,9,14]. The natural way of supervising and control of the contemporary prostheses utilizes the signals from the human organism, normally associated with the movements of the original limb. The electrical potentials accompanying the activity of the skeleton muscles, measured on the surface of the skin, seem the best solution here, due to the non-invasive character of its acquisition. They are called the electromyography signals (EMG) [1].

The myoelectric control consists in the recognition of the users intention, based on the analysis of the features of the signals generated by the muscles of the amputated limb. These signals are forced by the intended movement of the fingers. The recognized intentions of movements (classes) may be perceived as the decisions for the lower level control units performing the follow-up regulation of the prosthesis joints.

Taking into account various inferences accompanying the EMG signals measurements, the recognition of the decision is a hard problem [1,4,5,9,13], requiring considerable computational effort. It may be simplified by limiting the operation of the hand to the fingers grip opening-closing and possible rotation of the wrist. Such solution is commonly applied in the commercial prostheses and is forced by cost limitations.

The prosthesis providing the comfort closer to the original hand must be dexterous (adapting to the features of a grabbed object, like shape, size, elasticity, surface, etc.).

J.L. Oliveira et al. (Eds.): ISBMDA 2005, LNBI 3745, pp. 149–160, 2005.

This dexterity requires independent control of many degrees of freedom [2,12], which induces a huge number of recognized classes in the EMG signals, additionally increasing the already mentioned computational effort. The prosthesis should also be agile, i.e. fast enough to react accordingly to the dynamics of the environment (e.g. to grab an object which is falling down). This requires sufficiently fast control.

Eventually the contradicting requirements appear here – high complexity of calculations (to ensure dexterity), performed in a very short time (to ensure agility) and implemented in a portable, low-power device. The solution may rely on the decomposition of the control algorithm to the operations which may be performed concurrently, especially when implemented in the dedicated digital circuit, providing the optimal utilization of the resources.

Section 2 presents the problem of the bioprosthesis control and the possibilities of its decomposition to the elements processed concurrently, leading to a model with parallel structure. A brief analysis of the software and hardware implementations of the artificial hand control algorithms, followed by the proposed architectures of the dedicated digital circuits is presented in Section 3. In Section 4 the time dependencies analysis of the elaborated model and the proposed hardware implementation are discussed. The conclusions are given in Section 5.

## 2   Bioprosthesis Control Model

The hand (palm) prosthesis is a kinematical structure of numerous degrees of freedom, copying the skeletal system of a palm. The state of this structure is described by a vector of discreet angular positions of its joints:

$$q =< \alpha_1, \alpha_2,..., \alpha_r > \quad \alpha \in N_1^\xi, \quad q \in Q \tag{1}$$

where: $\alpha_i$ – discreet angular position of joint $i$, $Q$ – set of possible prosthesis states.

The activity of an artificial palm is described by the trajectory $q(t)$ of its joints' motion. Prosthesis control consists in the recognition of patient's intention expressed by the EMG signals (from the stump muscles of the prosthesised limb) and determination of the desired motion trajectory $q*(t)$ on this basis. Disregarding the dynamic properties of the prosthesis, trajectory $q*(t)$ constitutes a vector of the set signals for the follow-up regulators of the particular prosthesis joints:

$$EMG \xrightarrow{recognition} q*(t) =< \alpha_1^*(t), \alpha_2^*(t),..., \alpha_k^*(t) > \tag{2}$$

Practical realization of the recognition process requires a defined time period, thus the vector of set values is available periodically at discreet moments of time:

$$q*(t) = q*(t_1), q*(t_2), ...,q*(t_k) \tag{3}$$

where: $t_1, t_2, ..., t_k$ – discreet moments of time.

Thus the recognition of a movement intention is a multistage decision process and the values of vector $q*$, recognised at subsequent discreet moments, are *stage decisions*. The execution of these decisions is a task of the execution control level. Prosthesis movements, performed as a result of the subsequent decisions tracking, are called the elementary movements (and marked by $e(k)$ symbol).

The succeeding stages of the decision process are performed in the discrete time periods, marked by the successive natural numbers: 1, 2, ..., k, ... . The *basic model* assumes that each stage lasts one discreet time unit (**1** d.t.u.). In this time interval, the prosthesis control system subsequently performs the following operations (see Fig.1.):

- EMG signal measurement – which results in a series of the independent sequences of samples, one for each measurement channel, marked as $<s_j>$, where $j$ is the channel index, $j = 1..J$, and $J$ is the number of channels,
- feature vector extraction – which provides $J$ vectors of features $<v_j>$,
- concatenation of $J$ independent sequences of features, into the resultant vector $v^{\cdot}$,
- movement intent recognition and stage control decision $q^*(k)=<a_r^*>$ calculation,
- measurement of $r$-joint position $<a_r(t_i)>$ of prosthesis, and
- calculation of $r$ signals $<u_r(t_i)>$ of driver control.

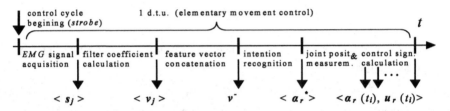

**Fig. 1.** An overview of the component processes of the single elementary movement control. (The symbols are described in the text.)

The listed component processes, excluding the recognition and decision $q^*$ determination, are multidimensional ones (i.e. they encompass several concurrent subprocesses of the same kind). For example, a measurement of the EMG signal consists in the measurement of 6 independent signals (arising from the 6 channels) and its result is $J = 6$ independent sequences of the discrete values. The extraction of the features encompasses the processing of these 6 signals. Only at the stages of the features vectors linking, intention recognition and multi-degree control decision determination, the process is a 1-dimensional one. The measurement of the prosthesis joints positions and the determination of the driver control, again involves $r$ processes associated with $r$ joints of the prosthesis. These component sub-processes may be executed independently, what leads to the parallel structure of the control system. All the listed component processes are described below.

*EMG Signals Measurement Block*
The purpose of the block is acquisition of myosignals from the selected limb points and delivering them to the prosthesis control system in a form convenient for the processing purposes. The basic element of the block is a multi-channel A/D converter which samples pre-amplified signals and converts them into digital values sequences (sample sequences) [5]. These operations can be realised concurrently for the particular signals.

*Movement Intention Recognition Block*
The block encompasses two consecutive processes: the extraction of particular signal features and, on the basis of these features, the recognition of classes corresponding to the specific hand movements. A block diagram of the recognition process is presented in Fig. 2.

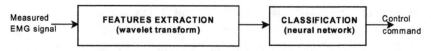

**Fig. 2.** Movement intention recognition block diagram

*Features extraction model* is based on a numerical wavelet analysis algorithm [4,8]. The wavelet analysis of signal f(t) $\epsilon L^2(R)$ consists in the determination of wavelet transform coefficient:

$$CWT(a,b) = \int_{-\infty}^{+\infty} f(t)\Psi^*_{ab}(t),  \tag{4}$$

where: $\Psi^*_{ab}(t)$ – dual function of the wavelet family: $\Psi_{ab}(t) = \dfrac{1}{\sqrt{a}}\Psi\!\left(\dfrac{t-b}{a}\right), a\in R,^+ b\in R$ .

For the discretesized parameters $a=2^m$ and $b=n2^m$; $m,n\in Z^2$, the wavelet family $\Psi_{ab}(t)$ takes a form:

$$\Psi_{mn} = 2^{-m/2}\Psi(2^{-m}t-n).  \tag{5}$$

This is a dyadic orthonormal wavelet basis of space $L^2$. It means that the change of *m* by 1 changes the scale twice (hence the name dyadic scaling).

A discrete wavelet transformation is associated with the multi-distributive analysis of the signals which uses the orthogonal basic functions, which span the signal in the appropriate sub-spaces of details and signal approximations. The signal approximation *f(t)* with $2^m$ resolution is defined as an orthogonal projection *f(t)* on the (approximation) subspace $V_m$. When moving from $V_m$ space to $V_{m+1}$ space (i.e. the less-detailed one) some information is lost. The rejected details may be defined in the $W_{m+1}$ space which makes an orthogonal completion of $V_{m+1}$ to $V_m$ space:

$$V_m = V_{m+1} \oplus W_{m+1},  \tag{6}$$

Thus the decomposition to the defined approximation level $m_0$ may be described as:

$$V_m = V_{m_0} \cup \bigoplus_{j=m+1}^{m_0} W_j, \quad m_0 > m, \quad m_0 \in Z.  \tag{7}$$

This is the basic equation of the wavelet decomposition. $W_j$ spaces are called the *detail spaces* and $V_{m0}$ makes a signal *approximation space* of the final decomposition level.

The numerical realization of the discrete wavelet transformation algorithm does not use the $\Psi(t)$ wavelet and scaling function directly. It uses only the filters associated with them. Such approach was proposed by Mallat [8] where a recurrent dependency between wavelet decomposition coefficients from two consecutive resolution levels was introduced using a multi-distributive analysis (7).

The realization of the digital filtration process consists in a convolution operation in time domain. The appropriate equations are presented below:

$$c_m(n) = \sum_{\kappa} l^*_{2n-\kappa}\, c_{m-1}(\kappa)$$
$$d_m(n) = \sum_{\kappa} h^*_{2n-\kappa}\, c_{m-1}(\kappa)  \tag{8}$$

This algorithm can be presented in a graphic form of a tree (Fig. 2).

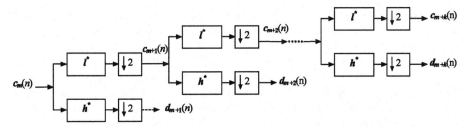

**Fig. 3.** The diagram of the wavelet decomposition coefficients calculation based on the fast Mallat algorithm. The $l^*$ and $h^*$ symbols denote the low-pass and high-pass filters respectively

In the decomposition (analysis), the scaling function is associated with a low-pass filter with $l^*$ coefficients which gives the signal approximation, whilst the wavelet is associated with a high-pass filter with $h^*$ coefficients which provides the signal details. The algorithm makes the $f(n) = c_0(n)$ signal filtration (where $n \in Z$) using this set of filters, simultaneously decreasing the sampling frequency twice. In practice the *Finite Impulse Response-Quadrature Mirror Filters* (FIR-QMF) are used. On the basis of the scaling function associated with the $l^*$ low-pass filter, the filter's impulse response is determined. It forms the basis for the remaining filters, according to the rule:

$$h_n^* = (-1)^n l_{L-1-n}^*,$$
$$l_n = l_{L-1-n}^* \tag{9}$$
$$h_n = (-1)^{n+1} l_n^*$$

where: $n = 0, 1,.., L\text{-}1$; $L$ – number of coefficients in a filter (level of the filter).

The module of the features extraction from a single signal, is based on the Mallat's fast wavelet algorithm (the number of samples is a power of 2), with $db2$ Daubechies wavelet utilised [8] (Fig. 3). In each cycle (a single stage of the control system) the algorithm operates on the input signal segment, represented by an $N$-element set of samples ($N$=256, for 1 kHz sampling and EMG spectrum). On the consecutive levels of the wavelet decomposition, the signal frequency resolution decreases twice (together with the reduction of the number of signal samples). The 7 levels of decomposition (numbered 0 to 6) per cycle were applied. It was forced by the analysed frequency spectrum of the EMG, which shall be 4Hz÷500Hz, to preserve the valuable information. Eventually, after processing in the 7 pairs of filters, the number of samples is reduced to two. The coefficients obtained at the particular levels (as a result of the high-pass filtration) create the sets of features. To reduce the total number of obtained features, each set is replaced by a single feature through the calculation of the energy included in these coefficients, calculated as a mean absolute value of the subsequent high-pass filter responses:

$$\phi_m = \sum_{i=0}^{N-1/2m} |d_m[i]| \tag{10}$$

where: $m$ is the Mallat's algorithm level number.

Hence the resultant features vector $v_j$ for a single EMG signal contains 7 values. The prosthesis control algorithm uses 6 independently measured EMG signals – $s_j$, (6 channels). This means that the module of wavelet decomposition should be replicated 6 times. Finally the total process of features extraction gives a 42-element vector of features $v^-$ in each cycle.

*Recognition module* is realised by the *Learning Vector Quantization (LVQ)* type neural network [3]. The *LVQ* network consists of a single layer containing $\rho$ neurons. Neuron number $i$ is connected with all elements of the input vector $v^-$ by the weight vector $w_i$. Each feature vector and each neuron are connected with a particular $q^*$class. The recognition mechanism is similar to the *K-Nearest Neighbour*, with $K = 1$, [6]. Every input vector $v^-$ is qualified to the class, associated with the winning neuron $\omega$ (i.e. the one with $w_i$ nearest to the $v^-$, in the sense of the applied metric):

$$\omega = \min_{1 \le i \le \rho} d\left( m_i, v^- \right). \tag{11}$$

To ensure the appropriate functioning of the prosthesis, a network with 42 inputs and 90 neurons was developed, (10 recognised classes and 9 neurons for each of the class). The implemented weight values, i.e. the *codebook*, were determined in the training process based on the data set obtained by the measurements of the EMG signals on the surface of the moving human forearm.

*Follow-Up Regulation Block*
This block contains $r$ modules of arithmetic units which implement the equation of the follow-up regulators:

$$u_r(t_i) = f_r(\alpha_r^*(t_k) - \alpha_r(t_i)) \tag{12}$$

where: $\alpha_r^*(t_k)$ – the set value for the $r$ joint angular position of the prosthesis, at the $k$ stage of the decision process; $\alpha_r(t_i)$, $u_r(t_i)$ – the measured value of the angular position and the $r$ joint control signal at $i$ moment of time, respectively; $f$ - regulation function of a PID type. (Within a single movement $e(k)$, in 1 d.t.u. there are numerous follow-up regulation cycles at consecutive moments $t_i$).

Follow-up control has a different (much shorter) time cycle than the decision process, hence the set value $\alpha_r^*(t_k)$ is treated as a constant whilst the $\alpha_r(t_i)$ and $u_r(t_i)$ are temporary values.

*Prosthesis State Measurement Block*
The task of this block consists in collecting $r$ signals from the angular positions converters of the prosthesis joints and delivering them to the prosthesis control system in the form acceptable for the digital processing.

# 3   Digital Circuit Architecture

The main advantage of the dedicated digital circuit, when compared with the software implementation of a typical computational algorithm is the higher processing speed obtained for similar or smaller complexity of the circuit. In the microprocessor, equipped with a single processing unit, the arithmetic operations are executed one after another in the consecutive clock cycles. Eventually the time needed for processing complex algorithms may be estimated as the number of arithmetic operations multiplied by the number of clock cycles needed for each of them. Moreover when the variables needed for the processing are stored in a memory, accessible via the classic interface with data, address and read-write lines, only one variable may be collected in the current clock cycle. This is another bottleneck for the algorithms requiring simultaneous access to the huge amount of data. Eventually the processing speed available for the classic microprocessor architecture is quite slow, especially if to take into account the complexity of the structure and the resources available.

Digital hardware, dedicated for the computational algorithm may be organized in three ways, depending on the features of the algorithm and the actual need for speed:

- If the problem may be decomposed to a series of independent calculations, which may be done in parallel, the dedicated piece of hardware may be designed and allotted to each of them. This approach provides the ultimate speed of processing, generally not depending on the computational complexity.

- When the operations are not independent, i.e. some of them must be made one after another, this kind of concurrency is not possible. However the appropriate set of the processing units may be designed again and connected serially, to provide the pipelining operation mode. For the massive stream of input data sets, the time of processing of a single set of variables may be estimated as a single cycle again. Together with the simultaneous access to all data it makes this kind of circuits work much faster than a microprocessor. Eventually the throughput available for these two hardware solutions may be similar, exhaustively outperforming the software approach. The only disadvantage is the cost of the implementation, growing with the complexity of calculations.

- Sometimes the high performance is not affordable and/or not necessary, encouraging the selection of some low-cost variant. If more or less the same operation shall be performed on all the input variables, a single processing unit may be used for all of them. This kind of solution is somewhat similar to already described "primitive" microprocessor approach. But the processing unit, optimised for the required calculations, together with a simultaneous access to data stored in "unlimited" number of registers make such low-cost solution outperform the software approach again.

The three kinds of a hardware approach presented above shall be perceived as simplified model solutions. The real world applications are usually a combination of all of them. In the case of this project the computational algorithm is primarily decomposed to the wavelet transform block and the neural network block, operating concurrently in a pipelining mode. Simultaneously each of them is constructed of the elements operating in various modes.

## 3.1  Wavelet Transform Block

The architecture of the wavelet transform module is presented in Fig. 4. The low-pass and high-pass filters, described in the previous section, are implemented in the separate blocks, providing a parallel mode of calculation. The results of the low-pass filter calculation $c_i$ are stored in the registers and fed-back to the input of the arithmetic units, which are this way reused for the calculation of the following wavelet decomposition factors, one after another. The specialized control unit provides a series of switching signals, controlled by the counters, connecting the appropriate sets of data registers to the high-pass and low-pass blocks in the appropriate phases of calculation.

The new samples are delivered to the circuit via the serial-in/parallel-out buffer presented in Fig. 5. Another control unit enables the introduction of the new set of samples, and latching the accumulated sum in the appropriate register, depending on which $d_i$ factor is currently being calculated. All the synchronization is based on the strobe signal (Fig. 1) which determines the moment when the calculation process shall start (i.e., the new samples are delivered). In-house developed modules were used for the floating-point arithmetic operations performed on the data represented directly by the vectors of bits. For the considerations in the next section it is assumed that this module is replicated 6 times to provide parallel processing of the EMG signals arising from each of the 6 channels. Shall be stated, however, that the single module working on all the channels one after another could also considered as a lower-cost/lower-performance solution, as well as some intermediate approach with e.g. 2 or 3 modules replicated. For any parallel solution, a little redesign may be performed, leading to the single control unit switching the data in all the arithmetic blocks.

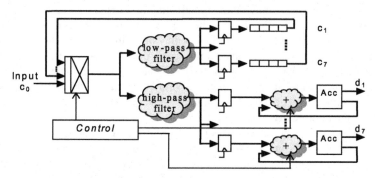

**Fig. 4.** Wavelet transform module architecture overview

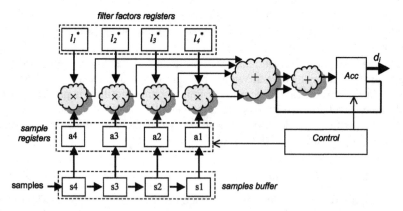

**Fig. 5.** Wavelet transform high-pass filter architecture with the samples delivery buffer

## 3.2  Neural Network Block

For the purpose of this project the new concept of the neural network hardware implementation was developed, balanced between the low-cost serial and high-performance parallel approaches, partially based on the previous experience [11]. The concept is presented in Fig. 6. It is assumed that a single arithmetic unit is shared by all the neurons from a single layer. The neuron inputs and weights are stored in the shift and rotating registers respectively, and this way delivered to the arithmetic units, in the appropriate phase of calculation. For this kind of neural networks the arithmetic processing in the neuron consists in the subtraction and the accumulation leading to the calculation of the distance between the weights and input vector (the *Manhattan* metrics was applied, due to the relatively low computational complexity). Further processing involves a series of comparisons, targeting in finding the neuron with the lowest response, indicating the appropriate class. The in-house developed floating point summation/subtraction modules, operating directly on bit vectors, were applied again.

### 3.3 Implementation

Both the wavelet and neural network blocks were designed using VHDL - the hardware description language. Most of the signals were represented by the binary std_logic type (IEEE standard No 1164). The VHDL code may be processed by simulation tools, providing the verification of the design functionality, and the synthesis tools leading to the real hardware – the *Application Specific Integrated Circuit* (ASIC) to be fabricated or the *Field Programmable Gate Array* (FPGA) to be programmed.

The verification process involved the computer simulation of the circuitry operation stimulated by the varying input signals. This way the operation of the two modules was found correct, especially in the context of the calculation results, which were compared with the ones performed for the wavelet and neural algorithms originally implemented in a PC [9].

Finally the FPGA synthesis was performed, separately for the wavelet and neural network modules, using the Quartus II tools delivered by ALTERA. The wavelet module was successfully fitted to the *Stratix II* series device *EP2S15F484C* utilizing 64% of its resources [10]. The neural network was fitted to the *EP2S60F1020C3* of the same series, with 19% of the resources utilized. This way the possibility of the hardware implementation of the prosthesis control algorithm was proved. Taking into account the size of the devices involved, a single-chip implementation in the currently available programmable logic circuits is possible. The speed estimations, based on the Static Timing Analysis results are presented in the next section.

## 4  Time Analysis in Prosthesis Control Model

The critical question of the prosthesis operation is the time of the single control cycle (i.e. the time of the single elementary movement $e(k)$, denoted **1** d.t.u.).

The considered premises are as follows:

a) the delay of the prosthesis reaction for the *control intention* (expressed by the muscle excitation) $T_r$ must not exceed 0.3 s (to preserve the dexterity);
b) the band of the significant harmonics of EMG signal is 4 Hz – 500 Hz;
c) the number of the elementary movements $\geq 10$;
d) the joint move range $\leq 90°$;
e) the joint positioning precision $\leq 10°$.

Point (b) induces:

the sampling frequency $f_s \geq 1000$ samples per second (i.e. 2*500 Hz);

the sampling window $T_w \geq 0.25$ s ($1/T_w \leq 4$ Hz), which makes 250 samples ($T_w f_s = 0.25 \cdot 1000$);

Point (a) induces that the time available for the all necessary operations associated with the elementary movement realisation, must not exceed 300 ms.

The full cycle of the elementary movement execution $e(k)$, involves the acquisition of the sequence of the EMG signal samples $<s_j>$, the recognition of the intention, the determining the control decision, and finally the realisation of the movement made by the prosthesis actuators (the control sequence $<u_r(t_i)>$).

Eventually the time of the single movement execution ($T_{mov}$) is a sum of the supervision cycle ($T_{spv}$) and the regulation cycle ($T_{reg}$).

$$T_{mov} = T_{spv} + T_{reg} \tag{13}$$

These time dependences are illustrated in Fig. 6.

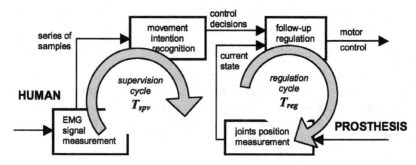

**Fig. 6.** The diagram of the prosthesis control process

The calculation speed of the proposed implementation of the prosthesis control algorithm may be estimated on the basis of the *Static Timing Analysis* results obtained for the synthesized digital circuits - wavelet and neural. For the wavelet circuit the maximum operation frequency was estimated to 21 MHz, giving the period $T_{c1} = 47$ ns. Significant part of the calculations may be performed before the last sample is delivered. Eventually after the complete sequence of samples $s_j$ is available, only 40 clock cycles are needed to finish the calculation of the wavelet coefficients. The time of the wavelet processing, (assuming that the hardware is replicated 6 times to maintain the 6 channels) may be estimated to $T_{wave} = 40 * 47$ ns $\approx 0.002$ ms.

For the neural network circuit the maximum clock frequency was estimated to 20 MHz, giving the period $T_{c2} = 50$ ns. In this case the number of clock cycles needed to perform the processing, is approximately equal to the number of the hidden neurons multiplied by the number of inputs, which for the assumed structure is $42*90 = 3780$. This determines the time needed for the neural processing to $T_{NN} = 3780*50$ ns $\approx 0.2$ ms.

The supervision cycle equals:

$$T_{spv} = T_w + T_{wave} + T_{NN} \tag{14}$$

Both estimated times - $T_{wave}$ and $T_{NN}$, are practically insignificant when compared with the sampling window size $T_w$= 256 ms (for the sampling period $T_s$ = 1/fs= 1 ms). Eventually the time of prosthesis reaction onto the control intention amounts:

$$(T_{wave} + T_{NN} \ll T_w) \rightarrow T_{spv} \approx T_w , \tag{15}$$

whereas the total time of elementary movement (**1** d.t.u.), is:

$$T_{mov} = T_w + T_{reg} . \tag{16}$$

## 5 Conclusions

The complex algorithm providing the smart control of the artificial hand was proposed. The main calculation effort consists in the wavelet transform and the supervised Kohonen neural network. The previous experiments with the software implementation of the algorithm have shown that the simple microcontroller cannot handle the required calculations in the acceptable time [13]. Such calculation speed is hardly possible for the high performance processors, if we consider only the ones meeting the portability requirements.

The combination of the high-performance and the mobility requirements for the considered device encouraged the hardware approach, well fitted to the computational algorithms involved. The wavelet transform and neural network algorithms were implemented in the dedicated digital circuits with the reasonably balanced speed/size trade-off. For this purpose the algorithm was analysed and decomposed to the parts which may be performed simultaneously. Despite the relatively low-cost approach selected at the current stage of the investigations, the performance of the designed modules was found substantially higher than the timing requirements imaginable for the basic model of the prosthesis control system. Having a huge reserve of the computational power, the more advanced solutions may be considered, e.g., the same hardware and overlapping of the sample windows, which shall provide much better dexterity and agility, outperforming the microprocessor based approaches [7,9]. It should be noted that the computational power of the neural network module, being currently the bottleneck of the control algorithm, may be significantly increased by applying another architecture, providing more parallelism.

The obtained simulation results are very promising, encouraging to the experiments with the physically implemented prototype device.

## References

1. De Luca, C.J.: The use of Surface Electromyography in Biomechanics, Journal of Applied Biomechanics, Vol. 13, No. 2, May 1997.
2. Hudgins, B., Parker, P., Scott, R.N.: A New Strategy for Multifunction Myoelectric Control, IEEE BME 40, 1, 1993, 82 - 94.
3. Kohonen, T.K.: Self-Organizing Maps, Springer, Berlin, 1995.
4. Krysztoforski, K., Wolczowski, A., Bedzinski, R., Helt, K.: Recognition of Palm Finger Movements on the Basis of EMG Signals with Application of Wavelets, TASK Quarterly, vol. 8, No 2. (2004), 269 – 280.
5. Krysztoforski, K., Wolczowski, A.: Laboratory Test-Bed for EMG Signal Measurement (in Polish), Postepy Robotyki, WKiL, Warszawa 2004.
6. Kurzynski, M.: Pattern Recognition. Statistical Methods, (in Polish), Oficyna Wydawnicza PWr, Wroclaw, 1997.

7. Light, C.M., Chappell, P.H., Hudgins, B., Englehart, K.: Intelligent multifunction myoelectric control of hand prostheses, Journal of Medical Engineering & Technology, Volume 26, Number 4, (July/August 2002), 139– 146.
8. Mallat, S.: A wavelet tour of signal processing, Academic Press, San Diego, 1998.
9. Nishikawa, D.: Studies on Electromyogram to Motion Classifier, Ph.D. dissertation, Graduate school of engineering, Hokkaido University, Sapporo, 2001.
10. Stratix II Device Handbook. Altera Corporation, San Jose, 2005.
11. Szecówka, P.M, Charytoniuk, A., Najbert, R.: Automated implementation of feedforward neural network in digital integrated circuit, Proc. 10th MIXDES, Lodz, 2003, 164-167.
12. Wolczowski, A.: Smart Hand: The Concept of Sensor based Control, Proc. of 7th IEEE Int. Symposium on Methods and Models in Automation and Robotics, Miedzyzdroje, 2001.
13. Wolczowski A., Krysztoforski, K., Bedzinski, R.: Control issues of Myoelectric Hand, XII National Biocybernetic and Biomedical Eng. Conference, Warszawa, 2001, 419-423.
14. www.ottobock.us.com; Patient information about Sensor Hand.

# Improving the Therapeutic Performance of a Medical Bayesian Network Using Noisy Threshold Models

Stefan Visscher[1], Peter Lucas[2], Marc Bonten[1], and Karin Schurink[1]

[1] University Medical Center Utrecht,
Dept. of Internal Medicine and Infectious Diseases,
HP F.02.126, Heidelberglaan 100, 3584 CX Utrecht, The Netherlands
{S.Visscher,K.Schurink,M.J.M.Bonten}@umcutrecht.nl
[2] Radboud University Nijmegen, Institute for Computing and Information Sciences,
Toernooiveld 1, 6525 ED Nijmegen, The Netherlands
peterl@cs.ru.nl

**Abstract.** Treatment management in critically ill patients needs to be efficient, as delay in treatment may give rise to deterioration in the patient's condition. Ventilator-associated pneumonia (VAP) occurs in patients who are mechanically ventilated in intensive care units. As it is quite difficult to diagnose and treat VAP, some form of computer-based decision support might be helpful. As diagnosing and treating disorders in medicine involves reasoning with uncertainty, we have used a Bayesian network as our primary tool for building a decision-support system for the clinical management of VAP. The effects of antibiotics on colonisation with various pathogens and subsequent antibiotic choices in case of VAP were modelled in the Bayesian network using the notion of causal independence. In particular, the conditional probability distribution of the random variable that represents the overall coverage of pathogens by antibiotics was modelled in terms of the conjunctive effect of the seven different pathogens, usually referred to as the *noisy-AND gate*. In this paper, we investigate generalisations of the noisy-AND, called *noisy threshold models*. It is shown that they offer a means for further improvement to the performance of the Bayesian network.

## 1 Introduction

Establishing an accurate diagnosis and choosing appropriate treatment are desirable especially when it concerns critically ill patients. In the intensive care unit (ICU), patients are often severely ill. Patients who depend on respiratory support in the ICU are even more vulnerable than other patients, and are at risk of developing *ventilator-associated pneumonia*, or VAP for short. Thus, it is important to start antimicrobial treatment against VAP as soon as possible in these patients. However, unnecessary antimicrobial treatment will enhance selection of antibiotic-resistant pathogens, which may subsequently cause difficulty in treating future infections adequately. Since only time-consuming and

J.L. Oliveira et al. (Eds.): ISBMDA 2005, LNBI 3745, pp. 161–172, 2005.

patient-unfriendly diagnostic tests are available for diagnosing VAP, some form of computer-based decision support could be helpful in the process of early diagnosis and treatment of VAP.

Previously, we have developed a computer-based decision-support system (DSS) that is aimed at assisting physicians in the diagnosis and treatment of VAP. The model underlying the DSS consists of a Bayesian network with an associated decision-theoretic part. The structure as well as the conditional probabilities and utilities were elicitated with the help of two infectious disease specialists. The resulting decision-theoretic model, or influence diagram, was translated into a Bayesian network, and this is the model currently used (Cf. Ref. [1] for details concerning the model and the translation process). The probability of VAP is computed using the diagnostic part of the Bayesian network; the best possible combination of antibiotics can be determined using the therapeutic part of the network.

When prescribing antimicrobial treatment a physician wishes to cover all microorganisms causing the infection, with a spectrum of antibiotics as narrow as possible. This policy aims at preventing the creation of antibiotic resistance and at saving financial costs [2]. This was already taken into account when constructing the DSS, described in more detail in Ref. [1]. To cover as many of the pathogens as possible by the antibiotic treatment advised by the DSS, a noisy-AND gate was used in the Bayesian network for the modelling of the probabilistic interactions of the effects of the prescribed antibiotics on the pathogens. However, it was found that this way of modelling often yields an antibiotic spectrum which is too broad. In the research reported in this paper, noisy threshold functions replace the noisy-AND gate used previously. We investigate whether the therapeutic performance of the Bayesian network for VAP improves in this way. Thus, the aim of the research was to refine the Bayesian network so that it prescribes antibiotics with a spectrum that is less broad.

The paper is organised as follows. In the next section, our earlier work on the development of a Bayesian network that is able to assist physicians in the diagnosis and treatment of VAP is briefly reviewed. In Section 3, the mathematical principles of causal independence models are discussed and noisy threshold functions are introduced. In Section 4, the data and methods used in evaluating the Bayesian networks incorporating the noisy threshold functions are described. The results achieved are commented on in Section 5. The paper is rounded off by some conclusions in Section 6.

## 2    A Bayesian Network for the Management of VAP

Bayesian networks, or BNs for short, have been introduced in the 1980s as a formalism to compactly represent and reason efficiently with joint probability distributions. Bayesian networks are in particular well suited for the representation of causal relations within a specific domain of expertise.

Formally, a Bayesian network $\mathcal{B} = (G, \mathrm{Pr})$ is a directed acyclic graph $G = (\mathbf{V}(G), \mathbf{A}(G))$ with set of vertices $\mathbf{V}(G) = \{V_1, \ldots, V_n\}$, corresponding to stochastic variables, here denoted by the same indexed letters, and a set of arcs

$\mathbf{A}(G) \subseteq \mathbf{V}(G) \times \mathbf{V}(G)$, representing statistical dependences and independences among the variables. On the set of stochastic variables, a joint probability distribution $\Pr(V_1, \ldots, V_n)$ is defined that is factorised respecting the independences represented in the graph:

$$\Pr(V_1, \ldots, V_n) = \prod_{i=1}^{n} \Pr(V_i \mid \pi(V_i)),$$

where $\pi(V_i)$ stands for the variables corresponding to the parents of vertex $V_i$.

The formalism of BNs supports the kind of the reasoning under uncertainty that is typical for medicine when dealing with diagnosis, treatment selection, planning, and prediction of prognosis. Our medical domain is restricted to patients who are mechanically ventilated and are at risk of developing ventilator-associated pneumonia. Entities that play an important role in the development of VAP and that belong to the diagnostic part of the Bayesian network for VAP include: the duration of *mechanical ventilation*, the amount of *sputum*, *radiological signs*, i.e., whether the chest radiograph shows signs of an infection, body *temperature* of the patient and the number of *leukocytes* (white blood cells) [3]. The structure of the Bayesian network for VAP is shown in Fig. 1. Mechanically ventilated ICU patients become colonised by bacteria. When colonisation of the lower respiratory tract occurs within 2–4 days after intubation, this is usually caused by antibiotic-sensitive bacteria, whereas after one week of intubation often antibiotic-resistant bacteria are involved in colonisation and infection. Such infections are more difficult to treat and immediate start of appropriate treatment is, therefore, important. Duration of hospital stay and severity of illness are associated with an increased risk of colonisation and infection with Gram-negative bacteria. We modelled seven groups of microorganisms, each as one vertex in the Bayesian network. Also, for each modelled microorganism, the pathogenicity, i.e., the influence of that particular microorganism on the development of VAP, was included in the model. The presence of certain bacteria is influenced by antimicrobial therapy. Each microorganism is susceptible to some particular antibiotics. Susceptibility, in this case, is stated as the sensitivity to or degree to which a microorganism is affected by treatment with a specific antibiotic. The susceptibility of each microorganism was taken into account while constructing the model. The infectious-disease experts assigned utilities, by definition quantitative measures of the strength of the preference for an outcome [4], to each combination of microorganism(s) and antimicrobial drug(s) using a decision-theoretic model. These variables are included in the therapeutic part of the Bayesian network for VAP.

## 3   Causal Independence Modelling

Causal independence is a popular means to facility the specification of conditional probability distributions $\Pr(V_i \mid \pi(V_i))$ involving many parent variables $\pi(V_i)$. Its basic principles and some special forms are briefly discussed below.

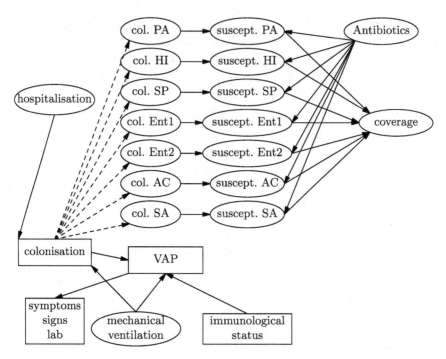

**Fig. 1.** Abstract model of the Bayesian network for the management of VAP. Colonisation and pneumonia play a central role in this model. The duration of hospitalisation and mechanical ventilation have influence on colonisation (col.) of the patient. PA: Pseudomonas aeruginosa; HI: Haemophilus influenzae; SP: Streptococcus pneumoniae; Ent{1,2}: Enterobacteriaceae{1,2}; SA: Staphylococcus Aureus; AC: Acinetobacter. Each pathogen is susceptible (suscept.) to particular antibiotics and an optimal coverage of the pathogens is what the model tries to achieve. The duration of mechanical ventilation, immunological status and colonisation have influence on the development of VAP. When a patient is diagnosed with VAP, the patient often has symptoms like for example an increased body temperature. Boxes denote entities or processes which are observed; processes that change or can be changed are denoted by ellipses

### 3.1 Basic Principles

Consider the conditional probability distribution $\Pr(E \mid C_1, \ldots, C_n)$, where the variable $E$ stands for an *effect*, e.g., coverage, and the variables $C_j$, $j = 1, \ldots, n$, denote *causes*, e.g., colonisation by pathogens. By taking a number of assumptions into account, which are summarised in Fig. 2, it is possible to simplify the specification of $\Pr(E \mid C_1, \ldots, C_n)$. These assumptions are: (1) the causes $C_j$ are assumed to be mutually independent, and (2) the variable $E$ is conditionally independent of any cause variable $C_j$ given the intermediate variables $I_1, \ldots, I_n$. In our domain the intermediate variable $I_j$ stands for susceptibility of pathogen$_j$ to a specific antibiotic. The, using basic probability theory, it follows that:

$$\Pr(e \mid C_1, \ldots, C_n) = \sum_{I_1, \ldots, I_n} \Pr(e \mid I_1, \ldots, I_n) \prod_{j=1}^{n} \Pr(I_j \mid C_j). \qquad (1)$$

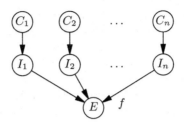

**Fig. 2.** Causal independence model

Now, if we assume that the probability distribution $\Pr(E \mid I_1, \ldots, I_n)$ that is specified for variable $E$ expresses some deterministic function $f : I_1 \times \cdots \times I_n \to E$, called an *interaction function*, an alternative formalisation is possible. Using the interaction function $f$ and the causal parameters $\Pr(I_j \mid C_j)$, it follows that [5–7]:

$$\Pr(e \mid C_1, \ldots, C_n) = \sum_{f(I_1, \ldots, I_n) = e} \prod_{j=1}^{n} \Pr(I_j \mid C_j). \qquad (2)$$

The result is called a *causal independence model* [5, 6, 8]. In this paper we assume that the function $f$ in Equation (2) is a Boolean function. Systematic analyses of the global probabilistic patterns in causal independence models based on restricted Boolean functions were presented in Ref. [6] and Ref. [9]. However, there are $2^{2^n}$ different $n$-ary Boolean functions [10, 11]; thus, the potential number of causal interaction models is huge. However, if we assume that the order of the cause variables does not matter, the Boolean functions become symmetric; formally, an interaction function $f$ is called *symmetric* if

$$f(I_1, \ldots, I_n) = f(I_{j_1}, \ldots, I_{j_n})$$

for any index function $j : \{1, \ldots, n\} \to \{1, \ldots, n\}$ [11]. The number of different Boolean function reduces then to $2^{n+1}$. Examples of symmetric binary Boolean functions include the logical OR, AND, exclusive OR and bi-implication. An example of a general symmetric Boolean functions is the *exact* Boolean function $e_k$, which is defined as:

$$e_k(I_1, \ldots, I_n) = \begin{cases} \top \text{ if } \sum_{j=1}^{n} \nu(I_j) = k \\ \bot \text{ otherwise} \end{cases} \qquad (3)$$

with $k \in \mathbb{N}$, and

$$\nu(I) = \begin{cases} 1 \text{ if } I = \top \\ 0 \text{ otherwise} \end{cases}$$

where $\top$ stands for 'true', and $\bot$ for 'false'. The interaction among variables modelled by the susceptibility, or coverage variables, as shown in Fig. 1, was modelled by assuming $f$ to be a logical AND. The resulting probabilistic model $\Pr(E \mid C_1, \ldots, C_n)$ is sometimes called the *noisy-AND* or *noisy-AND gate*. The probability distribution of the variable that represents the overall susceptibility (coverage in Fig. 1), models the conjunctive effect of the seven different pathogens.

This principle is modelled by a probability distribution $\Pr(E \mid C_1, \ldots, C_n)$ that is defined as in Equation (1) by the noisy-AND, yielding the following equation:

$$\Pr(\text{coverage} \mid \text{Col}_1, \ldots, \text{Col}_n, \text{Antibiotics}) =$$

$$\prod_{j=1}^{n} \Pr(\text{susceptibility-pathogen}_j \mid \text{Col}_j, \text{Antibiotics}).$$

By adopting this modelling approach, the network attempts to cover all pathogens in choosing appropriate antimicrobial treatment.

Evidence has shown that a patient can be colonised by at most 3 pathogens. Therefore, covering the 7 possible groups of pathogens is simply too much and results most of time in choosing antimicrobial treatment with a spectrum that is too broad. This casts doubts on the appropriateness of the noisy-AND for the modelling of interactions concerning coverage of bacteria by antibiotics.

### 3.2   Threshold Functions

As argued before, clinicians need to be careful in the prescription of antibiotics as they have a tendency to prescribe antibiotics with a spectrum that is too broad. A symmetric Boolean function that is useful in designing a generalised version of the noisy-AND is the *threshold function* $\tau_k$, which simply checks whether there are at least $k$ trues among its arguments, i.e., $\tau_k(I_1, \ldots, I_n) = \top$ (i.e., true), if $\sum_{j=1}^{n} \nu(I_j) \geq k$ with $\nu(I_j)$ equals 1, if $I_j$ equals $\top$ (true) and 0 otherwise [11]. Note that the noisy-AND gate corresponds to the threshold function $\tau_k$ with $k = n$. Hence, the noisy-AND can be taken as one extreme of a spectrum of Boolean functions based on the threshold function.

Using the threshold function $\tau_k$ with $k \neq 1, n$, may result in a better model. More intuitively, using the noisy threshold functions the network would only cover for 1 ($k = 1$), i.e. noisy-OR, 2 ($k = 2$), 3 ($k = 3$), 4 ($k = 4$), 5 ($k = 5$) or 6 ($k = 6$) pathogens compared to the noisy-AND gate, where all pathogens, i.e. $k = 7$, are taken into account. In the following we therefore investigate properties of the threshold function, and subsequently study its use in improving the Bayesian network model shown in Fig. 1.

### 3.3   The Noisy Threshold Model

Symmetric Boolean functions can be decomposed in terms of the exact functions $e_k$ as follows [11]:

$$f(I_1, \ldots, I_n) = \bigvee_{k=0}^{n} e_k(I_1, \ldots, I_n) \wedge \gamma_k \tag{4}$$

where $\gamma_k$ are Boolean constants only dependent of the function $f$. Using this result, the conditional probability of the occurrence of the effect $E$ given the causes $C_1, \ldots, C_n$ can be decomposed in terms of probabilities that exactly $l$ amongst the intermediate variables $I_1, \ldots, I_n$ are true, as follows:

$$\Pr(e \mid C_1, \ldots, C_n) = \sum_{\substack{0 \leq l \leq n \\ \gamma_l}} \sum_{e_l(I_1,\ldots,I_n)} \prod_{j=1}^{n} \Pr(I_j \mid C_j). \tag{5}$$

Thus, Equation (5) yields a general formula to compute the probability of the effect in terms of exact functions in any causal independence model where an interaction function $f$ is a symmetric Boolean function.

Let us denote a conditional probability of the effect $E$ given causes $C_1, \ldots, C_n$ in a noisy threshold model with interaction function $\tau_k$ as $\Pr_{\tau_k}(e \mid C_1, \ldots, C_n)$. Then, from Equation (5) it follows that:

$$\Pr_{\tau_k}(e \mid C_1, \ldots, C_n) = \sum_{k \leq l \leq n} \sum_{e_l(I_1,\ldots,I_n)} \prod_{j=1}^{n} \Pr(I_j \mid C_j). \tag{6}$$

## 4    Data and Methods

In our attempt to improve the performance of therapeutic advice provided by the Bayesian network for VAP, the following data and methods were used.

### 4.1    Data

We used a temporal database with 17710 records, each record representing a period of 24 hours of a mechanically ventilated patient in the intensive care unit. The database contains information of 2233 distinct patients, admitted to the ICU of the University Medical Center Utrecht between 1999 and 2002. For 157 of these 17710 episodes, a VAP was diagnosed according to the judgement of two infectious-disease specialists (IDS). We considered the period from admission to the ICU until discharge from the ICU of the patient as a *time series* $\langle X_t \rangle$, $t = 0, \ldots, n_p$, where $t = n_p$ is the time of discharge of patient $p$. The time-point at which VAP was diagnosed was denoted by $t_p^{VAP}$, $t = 0 \leq t_p^{VAP} \leq t = n_p$.

For each patient day, we collected the output of the Bayesian network, i.e., the best possible antimicrobial treatment. As reasoning with the network is time-consuming, certainly when varying therapy advice, we (randomly) selected 6 patients, with a total of 40 patient days, who were diagnosed with VAP.

### 4.2    Therapy Advice

During the period of seven days from the time-point of diagnosis, the patient is treated with antibiotics. Table 1 shows information for the 6 patients using the original Bayesian network. When the number of days following the day of the diagnosis of VAP is less than 7, we assume that this patient recovered, or died. We furthermore assume that when a patient is colonised by one or more microorganisms on a given day $t_c$, that after three days, i.e. $t_c + 1$, $t_c + 2$, $t_c + 3$, this patient is still colonised.

**Table 1.** This table shows the therapeutic advises by the ICU physician (column 5) and the Bayesian network (column 6), as well as the clinical culture data (column 4): acinetb = Acinetobacter; entbct{1,2} = Enterobacteriaceae{1,2}; hinflu = H. influenzae; paeru = P. aeruginosa; spneumon = S. pneumoniae; negative = 'no microorganisms found'

| Patient | VAP | day | Colonised by | Antibiotics selected by Physician | Antibiotics selected by BN |
|---------|-----|-----|--------------|-----------------------------------|----------------------------|
| 1 | 1 | 6 | entbct1 | augmentin (n) | meropenem (b) |
| 1 | 0 | 7 | | cefpirom (i) | meropenem (b) |
| 1 | 0 | 8 | | cefpirom (i) | meropenem (b) |
| 2 | 1 | 7 | paeru, hinflu, entbct2 | cefpirom (i) | clindam.+ciprox. (b) |
| 2 | 0 | 8 | | cefpirom (i) | clindam.+ciprox. (b) |
| 2 | 0 | 9 | | cefpirom (i) | clindam.+ciprox. (b) |
| 2 | 0 | 10 | paeru | cefpirom (i) | ceftazidime (i) |
| 2 | 0 | 11 | | ciproxin (b) | ceftazidime (i) |
| 2 | 0 | 12 | | ciproxin (b) | ceftazidime (i) |
| 2 | 0 | 13 | | ciproxin (b) | ceftazidime (i) |
| 2 | 0 | 14 | | ciproxin (b) | none |
| 3 | 1 | 5 | | augm/erytro/gent (i) | meropenem (b) |
| 3 | 0 | 6 | entbct1 | augm/erytro (i) | meropenem (b) |
| 3 | 0 | 7 | | erytro/ceftriaxon (i) | meropenem (b) |
| 3 | 0 | 8 | entbct1 | erytro/ceftriaxon (i) | meropenem (b) |
| 3 | 0 | 9 | | ceftriaxon (i) | meropenem (b) |
| 3 | 0 | 10 | | ceftriaxon (i) | meropenem (b) |
| 3 | 0 | 11 | negative | ceftriaxon (i) | ceftriaxone (i) |
| 3 | 0 | 12 | | ceftriaxon (i) | ceftriaxone (i) |
| 4 | 1 | 30 | acinetb, entbct1 | cefpirom (i) | meropenem (b) |
| 4 | 0 | 31 | | cefpirom (i) | meropenem (b) |
| 4 | 0 | 32 | acinetb | cefpirom (i) | meropenem (b) |
| 4 | 0 | 33 | acinetb | cefpirom (i) | meropenem (b) |
| 4 | 0 | 34 | | cefpirom (i) | meropenem (b) |
| 4 | 0 | 35 | | cefpirom (i) | meropenem (b) |
| 4 | 0 | 36 | | cefpirom (i) | meropenem (b) |
| 4 | 0 | 37 | | cefpirom (i) | meropenem (b) |
| 5 | 1 | 5 | hinflu, spneumon | augmentin (n) | meropenem (b) |
| 5 | 0 | 6 | | augmentin (n) | meropenem (b) |
| 5 | 0 | 7 | | augmentin (n) | meropenem (b) |
| 5 | 0 | 8 | | augmentin (n) | meropenem (b) |
| 5 | 0 | 9 | | augmentin (n) | none |
| 6 | 1 | 11 | paeru, entbct1 | augm/pipcil (i) | clindam.+ciprox. (b) |
| 6 | 0 | 12 | | pipcil (i) | clindam.+ciprox.(b) |
| 6 | 0 | 13 | | pipcil (i) | clindam.+ciprox. (b) |
| 6 | 0 | 14 | | pipcil (i) | clindam.+ciprox. (b) |
| 6 | 0 | 15 | | pipcil (i) | clindam.+ciprox. (b) |
| 6 | 0 | 16 | | pipcil/ciprox.(b) | clindam.+ciprox. (b) |
| 6 | 0 | 17 | paeru, entbct1 | ciproxin (b) | clindam.+ciprox. (b) |
| 6 | 0 | 18 | | ciproxin (b) | clindam.+ciprox. (b) |

The columns in Table 1 have the following meaning:

1. the patient number;
2. the day at which VAP was diagnosed (indicated by prefix '1' instead of '0');
3. the number of days a patient has been mechanically ventilated;
4. the microorganism(s) found in the sputum culture; abbreviations:
   - acinetb = Acinetobacter;
   - entbct{1,2} = Enterobacteriaceae{1,2};
   - hinflu = H. influenzae;
   - paeru = P. aeruginosa;
   - spneumon = S. pneumoniae.
5. antibiotics, as mentioned above, can be divided in spectral groups. Used abbreviations are:
   - $v$ : very narrow;
   - $n$ : narrow;
   - $i$ : intermediate;
   - $b$ : broad.

Comparison of antimicrobial spectrum imposes some difficulty. There are, for example, several intermediate-spectrum antibiotics available, possibly produced by different vendors. Thus, it is possible that in the table two different antibiotics are mentioned, even through they have the same effect. For example, in the table we see that for patient 2 on day 10, the ICU physician prescribes *cefpirom*, whereas the Bayesian network advises to prescribe *ceftazidime*. The antimicrobial therapy prescribed by the ICU physician and corresponding spectrum indicated between parentheses, as well as the therapy advice given by the Bayesian Network with associated spectrum, are mentioned in the last two columns of the table. Note that 'none' means that the Bayesian network advises not to prescribe any antibiotics.

## 4.3   Methods

In order to improve the therapeutic performance of the Bayesian network, the network was inspected in detail. Points for possible improvement that were identified included the utilities, used in the selection of antibiotics, and the noisy-AND, used as a basis for the assessment of the conditional probability distribution

$$\Pr(\text{Coverage} \mid \text{Col}_1, \ldots, \text{Col}_7, \text{Antibiotics}).$$

Based on the inspection, the following actions were taken:

- It became clear that the preferences of antibiotics with a broad spectrum were overestimated by the experts while assigning utilities. By giving the broad-spectrum antibiotics a lower utility, it was expected that this might result in a more appropriate treatment advice. On a scale between 0 and 100, with 0 representing 'not preferred' and 100 'preferred' new utilities were assessed by infectious disease experts; both the old and new utilities are summarised in Table 2. In redefining the utilities, it was assumed that each patient had VAP and that we wished to cover all pathogens present.

**Table 2.** Old and new utilities

| Spectrum | Utilities | |
|---|---|---|
| | Old | New |
| none | 29 | 29 |
| very narrow | 96 | 96 |
| narrow | 89 | 89 |
| intermediate | 82 | 82 |
| broad | **71** | **60** |

- We studied the use of noisy threshold functions as explained in Section 3.2, in achieving a better therapeutic performance by the Bayesian network.

Two infectious-disease specialists were requested to prescribe antibiotics for each of the 6 patients. Their treatments were considered to be the *gold standard*. This allowed us to validate the outcomes of the study.

## 5   Results

When prescribing antibiotics, choice of the spectrum should be based on taking into account the susceptibilities of causative pathogens. When the number of different causative pathogens increases, the necessary coverage will become more broad and often more than one antibiotic will be prescribed. Preliminary results, shown in Table 1, indicate that the antimicrobial spectrum advised by the Bayesian network is often broad, even when only two causative pathogens are present. This effect is summarised in Table 3 in the column with header 'Noisy-AND old'. Note that the treatments given by the ICU doctors included in the table are not considered to be *gold standard*, as is it known that ICU doctors have a tendency to prescribe antibiotics with a spectrum that is often too broad.

Table 3 indicates in the column with header 'Noisy-AND new' that the redefinition of the utilities already resulted in a better therapeutic performance. Yet, the Bayesian network still advised antibiotics with a spectrum that was often too broad. Hence, there was a clear need for further refinement of the network, which was subsequently undertaken using noisy threshold models.

For each noisy threshold model, we collected the output in the same manner as for Table 1. We have summarised the resulting antibiotic spectra per patient in Table 3 for thresholds $k = 3$, $k = 4$, $k = 5$ and the two noisy-ANDs. The performance for the noisy-OR model (threshold function with $k = 1$), $k = 2$ (both not in the table), and $k = 3$ were rather poor: for all patients the resulting antimicrobial spectrum was too narrow. The networks with threshold function with $k = 6$ (not in the table) and the original network with the noisy-AND (threshold function with $k = 7$), had a poor therapeutic performance as well, but here the prescribed antibiotic spectrum was too broad in most of the cases. The Bayesian networks with threshold functions with $k = 4$ and $k = 5$, however, performed relatively well.

**Table 3.** Results of the prescription of antibiotics to 6 patients with VAP according to: ICU physician, the original network (noisy-AND old), network with new utilities, network with noisy threshold with $k = 3$, $k = 4$ and $k = 5$ in comparison to the Infectious Disease Specialists (IDS). Abbreviations of antibiotic spectrum: o: none; v: very narrow; n: narrow; i: intermediate; b: broad

| Patient <br> Model | 1 | 2 | 3 | 4 | 5 | 6 | total |
|---|---|---|---|---|---|---|---|
| **ICU Physician** | 1n 2i | 4i 4b | 8i | 8i | 5n | 5i 3b | 6n 27i 7b |
| **Noisy-AND old** ($k = 7$) | 3b | 1o 4i 3b | 2i 6b | 8b | 1o 4b | 8b | 2o 9i 29b |
| **Noisy-AND new** ($k = 7$) | 3i | 8i | 1n 4i 3b | 8b | 5i | 3i 5b | 1n 23i 16b |
| **Noisy threshold** ($k = 3$) | 3v | 8v | 8v | 8v | 5v | 8v | 40v |
| **Noisy threshold** ($k = 4$) | 3v | 4v 4n | 4v 4n | 2v 4n 2i | 1v 4n | 8n | 14v 24n 2i |
| **Noisy threshold** ($k = 5$) | 1n 2i | 4n 4i | 5v 1n 2i | 2n 6i | 3n 2i | 3n 5i | 5v 14n 21i |
| **IDS** *gold standard* | 3n | 8i | 8i | 8i | 5n | 8i | 8n 32i |

# 6    Conclusions and Discussion

In this paper, we have shown that by reconsidering the modelling of interactions between variables in a Bayesian network, it is possible to improve its performance. We used a Bayesian network for the diagnosis and treatment of ventilator-associated pneumonia as an example. Intensive use was made of the theory of causal independence, which not only facilitates the assessment of probability tables by allowing specifying a table in terms of a linear number of parameters $\Pr(I_j \mid C_j)$, but also allows taking into account domain characteristics [6]. This was clearly shown for our Bayesian network concerning VAP, where motivation was derived from the domain of infectious disease, indicating that a noisy threshold model might be appropriate for the modelling of the interaction between pathogens and antimicrobial susceptibility. It appeared that a noisy threshold function with $k = 5$ yielded the best results, according to the gold standard, i.e., the infectious disease experts. This also provides evidence that the noisy OR and noisy AND, which are very popular in Bayesian network modelling, might not be the best interaction functions for other application areas as well. In the near future, we intend to test our findings on a larger sample of our database.

To conclude, it was shown that the noisy threshold model is useful from a practical point of view by using it as a basis for the refinement of an existing real-world Bayesian network for the management of critically ill patients.

## Acknowledgements

This research, undertaken in the TimeBayes project, was funded by the Netherlands Science Foundation (NWO) under the ToKeN2000 programme. We would like to thank Rasa Jurgelenaite for her help with the theoretical work concerning noisy threshold models.

# References

1. Lucas PJF, Bruijn NC de, Schurink K, and Hoepelman A. A probabilistic and decision-theoretic approach to the management of infectious disease at the icu. *Artificial Intelligence in Medicine*, 19(3):251–279, 2000.
2. Bonten MJM. Prevention of infection in the intensive care unit. *Current Opinion in Critical Care*, 10(5), 2004.
3. Bonten MJM, Kollef MH, and Hall JB. Risk factors for ventilator-associated pneumonia: from epidemiology to patient management. *Clinical Infectious Diseases*, 38(8), 2004.
4. Sox HC, Blatt MA, Higgins MC, and Marton KI. *Medical Decision Making*. Butterworth-Heinemann, 1988.
5. Heckerman D. Causal independence for knowledge acquisition and inference. In *Proceedings of the Ninth Conference on Uncertainty in Artificial Intelligence*, 1993.
6. Lucas PJF. Bayesian network modelling through qualitative patterns. *Artificial Intelligence*, 163:233–263, 2005.
7. Zhang LH and Poole D. Exploiting causal independence in bayesian networks inference. *Journal of Artificial Intelligence Research*, 5:301–328, 1996.
8. Heckerman D and Breese JS. Causal independence for probabilistic assessment and inference using bayesian networks. *IEEE Transactions on Systems, Man and Cybernetics*, 26(6), 1996.
9. Jurgelenaite R and Lucas PJF. Exploiting causal independence in large bayesian networks. *Knowledge Based Systems Journal*, 18(4–5), 2005.
10. Enderton HB. *A Mathematical Introduction to Logic*. Academic Press, 1972.
11. Wegener I. *The Complexity of Boolean Functions*. John Wiley & Sons, 1987.

# SVM Detection of Premature Ectopic Excitations Based on Modified PCA

Stanisław Jankowski[1], Jacek J. Dusza[1], Mariusz Wierzbowski[3], and Artur Oręziak[2]

[1] Warsaw University of Technology, Institute of Electronic Systems
ul. Nowowiejska 15/19, 00-660 Warsaw, Poland
{sjank,jdusza}@ise.pw.edu.pl
[2] I[st] Department of Cardiology, Medical University of Warsaw,
ul. Banacha 1 A, 02-097 Warsaw, Poland
artur.oreziak@amwaw.edu.pl
[3] Military Academy of Technology, Institute of Electronic Fundamentals
ul.Kaliskiego 2, 00-908 Warsaw 49, Poland

**Abstract.** The paper presents a modified version of principal component analysis of 3-channel Holter recordings that enables to construct one SVM linear classifier for the selected group of patients with arrhythmias. Our classifier has perfect generalization properties. We studied the discrimination of premature ventricular excitation from normal ones. The high score of correct classification (95 %) is due to the orientation of the system of coordinates along the largest eigenvector of the normal heart action of every patient under study.

## 1 Introduction

The morphological analysis of ECG signal is one of basic non-invasive diagnostic methods [7]. It allows making an assessment of myocardium state. The examination of arrhythmia and sporadically episode detection requires the analysis of long sequences of heartbeats. It corresponds to approximately 100 000 of ECG cycles. Therefore, automatic morphological analysis of Holter ECG recordings can be considered as a useful diagnostic tool.

In [7] we reported the results of neural network and SVM classifiers that enabled fast and efficient detection of premature ventricular and supraventricular excitations with a high score of successful classification. Prior to automatic classification the ECG Holter recordings were preprocessed: filtered, segmented into separated heartbeats. Then we applied to each heartbeat segment the principal component analysis of its covariance matrix. Hence, the description of the ECG signal shape was reduced to two angles of the corresponding principal eigenvector. Consequently, the resulting classifiers were just linear and the number of support vectors were minimal. The advantage of this approach is the graphical presentation of classification on the plane and clear interpretation of results. However, this method of automatic shape recognition required to design a special classifier for each considered patient.

In this work automatic classifiers based on support vector machine (SVM) is presented. The statistical classifiers, as e.g. neural networks and SVMs, require large enough learning set of labelled examples The power of learning set, according to the Cover theorem [2], must be greater than $(2N + 1)$, where $N$ is the dimension of the input space Therefore it is reasonable to apply the principal component analysis[1, 3, 5] for the dimensionality reduction [4]. The computation speed of classification and

J.L. Oliveira et al. (Eds.): ISBMDA 2005, LNBI 3745, pp. 173–183, 2005.

its effectiveness is obtained due to signal compression and particular parameterisation method. For each heartbeat description only two parameters were used. Such a small number of descriptors allow us to apply a training set containing not too many patterns.

In this paper we introduce the modification of PCA approach to ECG morphological classification that enables to design one SVM classifier for the group of patients suffering from the same disease. We attempt to obtain one classifier for all group of selected patients that can perfectly discriminate pathological excitations from normal ones. The classifier efficiency is defined as a quotient of correctly classified patterns to total number of testing patterns. It reaches 95% for the classes of normal as well as pathological heartbeats. Our study is based on the Holter recordings from the I$^{st}$ Department of Cardiology, Medical University of Warsaw for a group of patients with arrhythmia caused by premature ventricular excitations.

## 2   PCA Parameterisation of ECG Holter Recordings

We studied the 3-channel 24-hours Holter ECG signals measured by magnetic type recorder in the I$^{st}$ Department of Cardiology of the Medical University of Warsaw. The signals are sampled by specialized hardware system at 128 Hz with 8-bit accuracy and preprocessed by the Oxford MEDILOG Excel 2 software package. The data are stored in the *fdb* format.

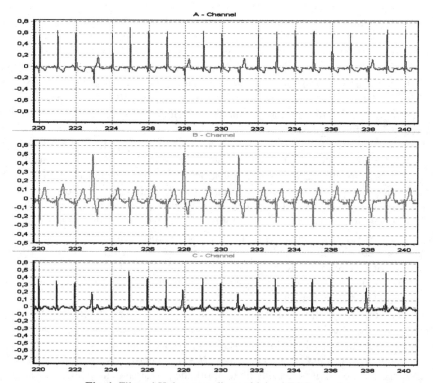

**Fig. 1.** Filtered Holter recordings of 3-lead ECG signal.

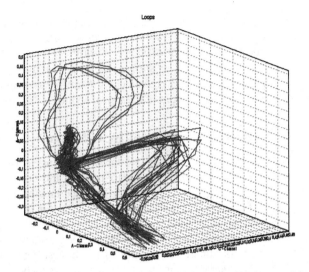

**Fig. 2.** Trajectory of ECG signal in 3-dimensional phase space reconstructed from signals x(t), y(t), z(t) shown in Fig. 1

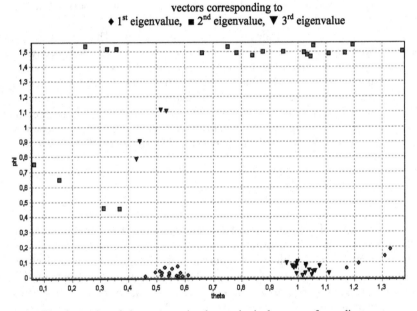

**Fig. 3.** Angles of eigenvectors in plane-spherical system of coordinates

The signals x(t), y(t) and z(t), shown in Fig. 1 can be treated as quasi-orthogonal components of 3-dimensional trajectory of a hypothetical dynamic system. The corresponding trajectory is reconstructed in phase space as shown in Fig. 2. As can be stated, the trajectory corresponding to a single heartbeat is set up of 3 loops.

The basic steps of our approach are as follows: ECG signal segmentation, calculation of covariance matrix of each signal segment and corresponding eigenvalues and eigenvectors. Thus each segment is represented by 9 numbers: 3 eigenvectors in 3-dimensional space. Taking into account only the direction of eigenvectors in spherical coordinate system reduces the number of shape descriptors to 6 angles in plane-spherical system of coordinates $(\theta, \varphi)$, as shown in Fig. 2. The method described in [7] used the orientation of the sum of 3 eigenvectors, therefore two angles in spherical coordinates were sufficient descriptors of the signal shape.

## 3  Modified PCA Parameterisation of ECG Signal

Our approach deals with the 3-channel Holter monitoring performing quasi-orthogonal system of coordinates that enables to register the electric potential in myocardium. The idea of presented modification of PCA parameterisation is to introduce a special system of coordinates that is oriented natural for the average normal heart action of every patient under study. The Ox axis of this orthogonal system is determined by the largest eigenvector of the average normal excitation. Hence, in plane-spherical system of coordinates the angles of the largest eigenvectors of normally excited beats are concentrated near the origin of the system coordinates (0, 0).

Suppose that each signal is a function of time x(t), y(t) and z(t). An example of these recordings is presented in Fig. 1. At any moment $t_0$ three coordinates of points in 3-dimensional space can be calculated. In this way we can reconstruct the trace of the electric vector in 3-dimensional space as a trajectory, as shown in Fig 2. For any single normal heartbeat the trajectory consists of 3 loops. The largest loop corresponds to QRS wave and two small loops represent P and T waves respectively. Single trajectory as a 3- dimensional object can be placed into rectangular prism. The lengths of the edges are proportional to eigenvalues and orientation of rectangular prism depends on eigenvectors of covariance matrix.

We attempt to reconstruct the trajectory of heart electric potential in three dimensional phase space by using signals from each channel as $(x(t), y(t), z(t))$ components. The 3-channel Holter ECG signal can be described by the matrix $\mathbf{S}$

$$\mathbf{S}_{3\times N} = \begin{bmatrix} x_1, x_2,..., x_N \\ y_1, y_2,..., y_N \\ z_1, z_2,..., z_N \end{bmatrix} \tag{1}$$

where: $N$ – number of sample points of a given segment of the signal, $x_i, y_i, z_i$ – values of sampled signals from 3 channels.

The evaluating person selects the interval that contains a reasonable number of subsequent normal heartbeats(in practice 20 to 30). This selected part of ECG recording is described by $F_{3\times k}$ matrix (where $k$ is the number of points of a given signal part). Its covariance matrix $\mathbf{K}$ is equal to:

$$\mathbf{K} = \mathbf{F} \cdot \mathbf{F}^T \tag{2}$$

The eigenvalues $\lambda_i$ and eigenvectors $\mathbf{w}_i$ of matrix $\mathbf{F}$ define a new matrix:

$$\mathbf{S}' = \mathbf{W}\mathbf{S} \tag{3}$$

where matrix $\mathbf{W}$ is set up of rows equal to eigenvectors of matrix $\mathbf{K}$.

This operation is equivalent to projection of a given trajectory into the coordinate system that is oriented along the largest eigenvector of the average normal excitation of the heart.

The orientation of the largest eigenvector (called principal component) is relevant to the shape of each beat. In order to improve the classification we calculate the orientation of each principal component with respect to the average orientation of the principal components corresponding to normal beats. The SVM classifier is used to discriminate the normal and abnormal beats upon the relative orientation of principal components.

The modification of the classical approach is aimed to define a new orthogonal system of coordinates for the average trajectory of normal beats so that the principal component (the largest eigenvector) is parallel to Ox axis or to (0, 0) point in plane-spherical system.

Thus we obtain compressed information about the signal energy (the elements on a diagonal are proportional to the square of RMS values of each channel) and correlation between the pairs of signal (the rest of elements of matrix $C$ are dot product of every pair of signals) is not lost. The elements of covariance matrix are averaged over time interval.

The projected components are shown as x'(t), y'(t) and z'(t) and the corresponding trajectory is shown in Figures 4 and 5. Fig. 6 presents the eigenvector in plane-spherical coordinate system.

The product describes the projection of the ECG signal trajectory onto axes determined by eigenvectors of the selected interval of normal heat action. Thus we obtain a new 3 orthogonal components x'(t), y'(t) and z'(t) of the signal S, shown in Fig. 1. These signals and the trace of transformed 3-dimensional trajectory are presented in Fig. 4 and in Fig.5.

The ECG Holter recording is subjected to segmentation into intervals corresponding to single heartbeats. Every cycle is related to R wave of ECG and contains 30% of $R_{n-1}R_n$ interval and 70% of $R_nR_{n+1}$ interval. The covariance matrix of $k$-th heartbeat is equal:

$$\mathbf{C}^k = \frac{1}{N}\mathbf{S}^{\prime k} \cdot \left(\mathbf{S}^{\prime k}\right)^T \tag{4}$$

where $N$ – number of sample points.

Then we calculate eigenvectors of these covariance matrices corresponding to subsequent heartbeats. The results are shown in plane-spherical coordinate system. As can be seen from Fig. 6, the angles of the largest eigenvector of normal heartbeats are concentrated in the vicinity of point (0, 0) while those corresponding to pathological ones are significantly distanced from the point (0, 0).

We applied the linear soft margin support vector machine classifier. The input of the classifier has the form of feature vector $\mathbf{x} = (x_1,...,x_n)$ (column vector) and its output is real-valued function $f\colon X \subseteq R^n \to R$. If $f(\mathbf{x}) \geq 0$ the input $\mathbf{x}$ is assigned to the positive class and otherwise to the negative class.

Linear classifier in general form can be expressed as

$$f(\mathbf{x}) = (\mathbf{w} \cdot \mathbf{x}) + b = \sum_{i=1}^{n} w_i x_i + b \tag{5}$$

where: $(\mathbf{w}, b) \in R^n \times R$ - parameters that control the function $f$. $\mathbf{w}$ - the weight vector, $b$ - the bias (threshold).

The learning paradigm says these parameters must be learned from the data. The decision rule of classification is

$$h(\mathbf{x}) = \text{sgn}(f(\mathbf{x})) \tag{6}$$

This classifier has the natural geometric interpretation. The input space X is divided into two parts by the hyperplane defined by the equation:

$$(\mathbf{w} \cdot \mathbf{x}) + b = 0 \tag{7}$$

which divides the space into two half spaces which correspond to the inputs of the two distinct classes. The vector $\mathbf{w}$ defines a direction perpendicular to the hyperplane, the value of $b$ is the distance of the hyperplane from the origin.

In order to separate a training set with a minimal number of errors we introduce some non-negative variables $\xi_i \geq 0$ (slack variables). The Lagrangian of the data set is equal

$$L(\mathbf{w}, b, \xi, \mathbf{\alpha}) = \frac{1}{2}(\mathbf{w} \cdot \mathbf{w}) + \frac{C}{2}\sum_{i=1}^{l}\xi_i^2 - \sum_{i=1}^{l}\alpha_i[y_i((\mathbf{w}_i \cdot \mathbf{x}_i) + b) - 1 + \xi_i]$$

$$\alpha_i \geq 0 \tag{8}$$

where: $\mathbf{w}$ – weight vector, b – bias, C –regularisation term, $\xi$ -slack variable, $\alpha$ - Lagrange multipliers, $l$ – number of examples

The Lagrangian L has to be minimised with respect to the primal variables $\mathbf{w}$ and $b$ and maximised with respect to the dual variables $\alpha_i$ - a saddle point has to be found.

Then the weight vector $\mathbf{w}^*$:

$$\mathbf{w}^* = \sum_{i=1}^{l} y_i \alpha_i^* \mathbf{x_i} \tag{9}$$

realises the maximal margin hyperplane with geometric margin $\gamma = 1/\|\mathbf{w}\|_2$.

In this expression only these points are involved that lie closest to the hyperplane because corresponding Lagrange multipliers are non-zero. These points are called support vectors.

Usually there are only few support vectors in the training set hence, the information compression property.

The fact that only a subset of the Lagrange multipliers is non-zero is referred to as sparseness and means that support vectors contain all the information necessary to construct the optimal separating hyperplane. The fewer number of support vectors the better generalisation can be expected. This property does not depend on the dimension of the feature space.

In our case the input space is just a plane and the linear classifier takes the form of an optimal separating line.

## 4 Experimental Results

Holter electrocardiography are clinical routine examinations that produce a large amount of data. Monitoring of the electrocardiogram during normal activity using Holter devices has become standard procedure for detection of cardiac arrhythmias.

Ambulatory electrocardiography was carried out using the Oxford Medilog MR 45 ECG recorder. The subjects with the history of myocardial infarction and heart failure post were encouraged and advised to undertake their usual daily activities except bathing. They were also advised to note the time and details of any symptoms perceived (the event diary).

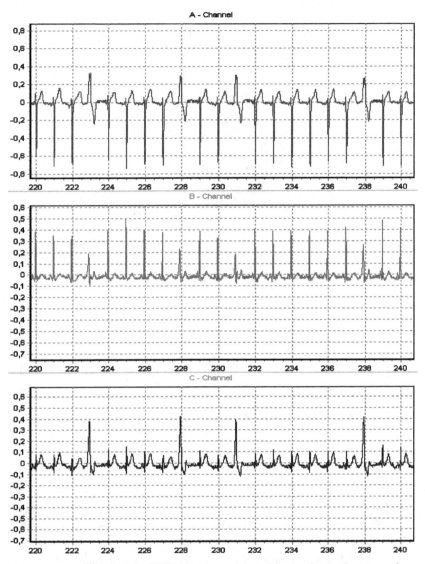

**Fig. 4.** Three components of the ECG trajectory projected into orthogonal system of coordinates

The 24-hour data was analysed by Medilog Excel 2 Holter Management System that divided the cardiac arrhythmias in supraventricular arrhythmias (supraventricular extasystoles, supraventricular couplets, supraventricular triplets, supraventricular bigeminy, supraventricular trigeminy, supraventricular tachycardia) and ventricular arrhythmias (ventricular extrasystoles, ventricular couplets, ventricular triplets, ventricular bigeminy, ventricular trigeminy, ventricular tachycardia and R on T phenomenon). The decisions of Medilog Excel 2 system were verified by the physician-cardiologist.

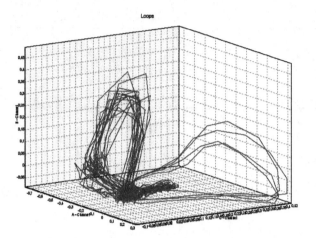

**Fig. 5.** ECG signal trajectory in orthogonal system of coordinates

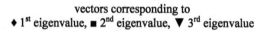

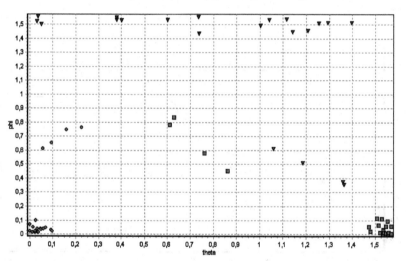

**Fig. 6.** Distribution of eigenvectors after transformation

The data from the Chair and Clinic of Cardiology, Medical University of Warsaw of 5 selected patients suffering from arrhythmia caused by premature ventricular excitations from various foci were examined. The signal files of 2000 heartbeats for each patient were analysed. The training set consisted of 200 cycles for each patient. The training set for SVM classifier consisted of 200 heartbeats for each patient, hence total number of examples was 1000 excitations. The linear SVM classifier trained for one patient is able to detect normal beats of another patient recording. The score of successful recognition of normal and pathological excitations is greater than 95%, in 4 considered cases, as listed in Table 1. The functionality of SVM classifier is illustrated in Figures 7 and 8.

**Table 1.** Results of support vector machine classification

| Patient | P003 | P011 | P018 | P101 | P103 | |
|---|---|---|---|---|---|---|
| Correct classi-fication | 1302 | 1344 | 1581 | 740 | 1775 | Normal ECG beats |
| Wrong classifi-cation | 15 | 56 | 0 | 2 | 7 | |
| Number of patterns | 1317 | 1400 | 1581 | 742 | 1782 | |
| Score | 0.987 | 0.960 | 1.000 | 0.997 | 0.996 | |
| Correct classi-fication | 384 | 325 | 417 | 329 | 202 | Pathological ECG beats |
| Wrong classifi-cation | 15 | 6 | 22 | 8 | 13 | |
| Number of patterns | 399 | 331 | 439 | 337 | 215 | |
| Score | 0.962 | 0.982 | 0.959 | 0.976 | 0.940 | |
| Total number of patterns | 1716 | 1731 | 2030 | 1079 | 1997 | |

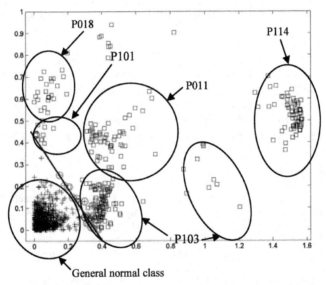

**Fig. 7.** Clusters of heartbeats representing the normal and the premature ventricular excitations for 5 patients (P011, P018, P101, P103, P114) represented by modified PCA descriptors and the linear SVM classifier

## 5 Conclusions

The modified PCA representation of single heartbeats obtained from 3-channel Holter monitoring enables efficient automatic classification of normal and pathological cases. Application of modified principal component analysis to data parameterisation allowed us to design simple and efficient classifier based on support vector machine. We emphasize that SVM trained on data set of one patient is able to classify the heart beats of other patients with approximately equal probability.

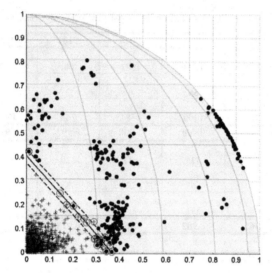

**Fig. 8.** Linear SVM classifier and margin of data from Fig. 5 presented in spherical coordinate system

Modified principal component analysis enables to perform one general classifier for all selected patients. The shape of clusters in plane-spherical system of coordinates can be used for unsupervised classification of large data sets.

The successful recognition is due to combination of PCA data suppression, as reported in [4] and to natural orientation of systems of coordinates for each patient along the largest eigenvector corresponding to individual normal heart action.

# References

1. Endo T., Yamaki M., Ikeda H., Kubota I., Tomoike H.: Relation of principal components of ECG maps to loci of wall abnormality in old myocardial infarction, *Am. J. Physiol.* 266 (*Heart Circ. Physiol.* 35), pp. H1604-H1609, 1994
2. Cover T. M.: Geometrical and Statistical Properties of Systems of Linear Inequalities with Application in Pattern Recognition, *IEEE Transaction on Electronic Computers*, June 1965, pp.326-334
3. Dubois, R.: Application des nouvelles méthodes d'apprentissage à la détection précoce d'anomalies en électrocardiographie, Ph.D. thesis, Université Paris 6, 2003
4. Cao, L. J., Chua, K. S., Chong, W. K., Lee, H. P., Gu, Q. M.: A comparison of PCA, KPCA and ICA for dimensionality reduction in support vector machine, *Neurocomputing* 55 (2003), pp. 321-336
5. Jankowski, S. Tijink, J., Vumbaca, G., Balsi, M., Karpiński, G.: Morphological analysis of ECG Holter recordings by support vector machines, Medical Data Analysis, Third International Symposium ISMDA 2002, Rome, Italy, October 2002, Lecture Notes in Computer Science LNCS 2526 (eds. A. Colosimo, A. Giuliani, P. Sirabella) Springer-Verlag, Berlin Heidelberg New York 2002, pp. 134-143
6. Jankowski, S., Oręziak, A., Skorupski, A., Kowalski, H., Szymański, Z., Piątkowska-Janko, E.: Computer-aided Morphological Analysis of Holter ECG Recordings Based on Support Vector Learning System, *Proc. IEEE International Conference Computers in Cardiology* 2003, Tessaloniki, pp. 597-600

7. Oxford Medilog Holter Management Systems - http://www.scanmed.co.uk/oxford.htm
8. Jankowski S., Dusza J. J., Wierzbowski M. and Oręziak A. "PCA representation as a useful tool for detection of premature ventricular beats in 3-channel Holter recording by neural network and support vector machine classifier", International Symposium on Biomedical Data Analysis ISBMDA 2004, Barcelona, Lecture Notes in Computer Sciences LNCS 3337, Springer Verlag, Berlin Heidelberg New York 2004, pp. 259-268.
9. Wagner, G. S.: Marriot's Practical Electrocardiography, Williams & Wilkins, 1994
10. Vapnik, N. V.: Statistical Learning Theory, John Wiley & Sons Inc., New York, 1998

# A Text Corpora-Based Estimation
# of the Familiarity of Health Terminology

Qing Zeng[1], Eunjung Kim[1], Jon Crowell[1], and Tony Tse[2]

[1] Decision Systems Group,
Harvard Medical School and Brigham & Women's Hospital, Boston, MA
{qzeng,ejkim,jcrowell}@dsg.harvard.edu
[2] Lister Hill National Center for Biomedical Communications,
National Library of Medicine, Bethesda, MD
tse@nlm.nih.gov

**Abstract.** In a pilot effort to improve health communication we created a method for measuring the familiarity of various medical terms. To obtain term familiarity data, we recruited 21 volunteers who agreed to take medical terminology quizzes containing 68 terms. We then created predictive models for familiarity based on term occurrence in text corpora and reader's demographics. Although the sample size was small, our preliminary results indicate that predicting the familiarity of medical terms based on an analysis of the frequency in text corpora is feasible. Further, individualized familiarity assessment is feasible when demographic features are included as predictors.

## 1 Introduction

Health literacy is "the degree to which individuals have the capacity to obtain, process, and understand basic health information and services needed to make appropriate health decisions" [1]. A large percentage of the US population has low health literacy, which has been linked to poor outcomes in health care [2,3,4,5]. Although health literacy includes reading, writing, speaking, listening, and numeric, cultural and conceptual knowledge components, reading comprehension of health terms and concepts has been the primary focus of health literacy tests.

Literacy is closely coupled to readability, a measure of the difficulty in reading text. Outside the health domain, grade level has been used to benchmark both a person's literacy level and the readability of text. Within the health domain, health literacy tests are modeled on functional literacy measures and grade-level scores [6,7]. For example, the two most frequently used tests – Test of Functional Health Literacy Assessment (TOFHLA) and Rapid Estimate of Adult Literacy in Medicine (REALM) – assign 3 and 4 levels of health literacy, respectively. However, because there is no comparable readability measure for health-related text, health care researchers rely on readability formulas developed since the 1940's for assessing general-domain text [8]. Key factors of these general readability formulas include vocabulary difficulty, sentence structure complexity, and text cohesion. Although existing formulas provide "rough" estimates of readability for health-related text [9], a more accurate approach is needed, since most computer-based health communication is textual.

Assessing term difficulty is clearly a shortcoming of applying general readability measures to health-related content. The usual techniques such as counting the number

J.L. Oliveira et al. (Eds.): ISBMDA 2005, LNBI 3745, pp. 184–192, 2005.

of letters or syllables in words, or appearance on "easy word lists," often do not apply to health-related text, which typically contains technical terms. For instance, common words such as *operation* have more letters and syllables than technical terms like *femur*. Further, although *diabetes* and *menopause* are commonly recognized terms due to their prevalence and media coverage, general readability formulas consider them "unfamiliar" along with truly difficult terms such as *atelectasis* and *alveoli*. We focus exclusively on health terminology because sentence complexity and cohesion are domain-independent factors.

Ideally, data on vocabulary comprehension of health terms in the population would be obtained for such a study. Since it is not feasible to test *all* health terms on every subpopulation, we conducted a preliminary study using a convenience sample to estimate familiarity with a set of 34 terms that frequently occurred in health-related text corpora.

## 2  Methods   .

### 2.1  Familiarity of Sample Terms

We had previously created an instrument for evaluating health vocabulary familiarity as part of the ongoing Consumer Health Vocabulary Initiative project. The questionnaire, modeled on the TOFHLA, contained 34 multiple-choice questions, each assessing a commonly used health-related concept [10].

Two synonyms were selected for each concept: a consumer-friendly but precise term or Consumer-Friendly Display (CFD) name (*kneecap*) and an ambiguous term or jargon (*patellae*). We created two versions of the questionnaire, each consisting of 17 statements containing CFD names and 17 with jargon. For example, statement 1 below contains jargon in version A (*geriatric*) and a corresponding CFD name in version B (*elderly*).

**Version A**

1. A geriatric person is one who is _____.
   A.  Very old
   B.  lanky and good looking
   C.  well groomed
   D.  aggressive and loud

2. You are in trouble when alcohol is detected _____.
   A.  in your skin while you're sun bathing
   B.  in your eye while you're reading
   C.  in your heart while you're exercising
   D.  in your blood while you're driving

3. If you have a cerebrovascular accident it means that _____.
   A.  you broke a bone
   B.  you were unable to make it to the bathroom on time
   C.  you had a heart attack
   D.  a blood vessel in your brain ruptured or clogged

**Version B**

1. An elderly person is one who is _____.
   A. Very old
   B. lanky and good looking
   C. well groomed
   D. aggressive and loud

2. You are in trouble when ethanol is detected _____.
   A. in your skin while you're sun bathing
   B. in your eye while you're reading
   C. in your heart while you're exercising
   D. in your blood while you're driving

3. If you have a stroke it means that _____.
   A. you broke a bone
   B. you were unable to make it to the bathroom on time
   C. you had a heart attack
   D. a blood vessel in your brain ruptured or clogged

We did not use TOFHLA for this preliminary study because it contains only a few difficult terms. This questionnaire, on the other hand, assesses a variety of technical terms.

The questionnaire was administered to a convenience sample of 21 people recruited from the Brigham and Women's Hospital in Boston and local churches. The inclusion criteria were: non-clinician, 18 years of age or older, and the ability to read and write in English. Each participant also provided demographic information: age, gender, race, ethnicity, first language, profession, and education level.

The authors used the completed questionnaires to calculate term familiarity scores. Each statement was given a score of 1 if completed with the correct term; otherwise, it was given a 0. We then estimated the familiarity score for each term across the population by averaging all scores for that term across the sample.

To obtain baseline data for term familiarity, we employed two methods commonly used by general readability formulas: (1) counting syllables per word and (2) consulting the Dale-Chall List [11]. Words that contain 3 syllables or more have been deemed "difficult" by some readability measurements. Words that do not appear on the Dale-Chall list, which contains about 3,000 words claimed to be understandable by 80% of fourth graders, have also been deemed "difficult." The familiarity score of difficult and easy words was calculated by syllable count as well as using the Dale-Chall list.

## 2.2 Health-Related, Text Corpora-Based Features

General readability formulas rely on term frequency counts from newspaper articles. (Term frequency is the number of occurrences of a term within a corpus). Because of the scarcity of health-related terms in such general text sources, we obtained three health-related corpora:

1. MEDLINE® abstracts: the National Library of Medicine (NLM) MEDLINE indexes publications from all disciplines in the health domain. While coverage of health-related terms is very broad, MEDLINE's content focuses more on biomedical research than clinical practice. Jargon usage is prevalent[1].
2. MedlinePlus®: MedlinePlus is a high quality consumer health information Web site developed by the NLM. Because it is tailored for a lay audience, MedlinePlus terminology consists of a mixture of lay terms and jargon[2].
3. MedlinePlus logs: Log data (i.e., user-submitted queries) are one of the best sources of consumer health language. A limitation is that the authors of consumer-generated text (e.g., newsgroup postings, email messages, or queries) tend to be more motivated and better educated than the general population.

**Table 1.** Text corpora used by the study

| Corpus | Size (no. of words) | Date | Author | Audience |
|---|---|---|---|---|
| MEDLINE | 45,924,958 | Jan 1987 - Dec 1991 | Professional | Professional |
| MedlinePlus | 3,717,365 | Sep. 2003 | Professional | Lay |
| MedlinePlus log | 28,797,199 | Oct. 2002 - Sep. 2003 | Lay | N/A |

### 2.3  Non-health Related Features

The three health-related corpora do not provide sufficient representation of the health term usage or exposure of lay people with lower literacy levels. Thus, we used the word list from the popular Dale-Chall readability formula as a supplement. The percentage of words in a term that belongs to the Dale list of easy words is treated as a feature.

Another non-health related characteristic is word length: difficult words tend to be longer than easy words. Even though there are many exceptions to the rule, word length is nonetheless a useful feature.

### 2.4  Familiarity Predication for the Sample Population

A support vector machine (SMV) for familiarity prediction was developed using the following feature variables: term frequency in the three health text corpora, percentage of easy words from the Dale-Chall list, and average word length. The mean familiarity score served as the outcome measure. For evaluation, 10-fold cross validation was performed.

### 2.5  Individualized Familiarity Predication

While predicating term feasibility for a population is useful, the variation in even a relatively small population can be large. For example, different ethnic groups in US may share characteristics (e.g., language), but have individual differences in other

---

[1] For information on MEDLINE, see http://www.nlm.nih.gov/pubs/factsheets/medline.html. A text corpus of MEDLINE abstracts is available from http://trec.nist.gov/data/t9_filtering.html
[2] For information on MedlinePlus, see http://www.nlm.nih.gov/medlineplus/faq/faq.html

ways (i.e., are heterogeneous) – such as age, education level, gender, profession, and other demographic factors. One approach is to acquire a large and diverse sample and treat demographic variables as features or predictor variables.

Despite the small sample size in this pilot study, the participants came from varying backgrounds, which allowed us to experiment with some demographic variables. Logistic regression model was used for familiarity prediction. The dependent variable familiar is a dichotomous variable coded "1" if the participant answer was right and "0" if wrong or missing. Five term variables (average length, query log frequency, Medline frequency, 4th grader level test, MedlinePlus frequency) and seven demographic variables (gender, native language, race, job, age, ethnicity, and education level) were used.

**Table 2.** Categorical Variables Coding

| Demographic Variable | Categories | Parameter Coding | |
|---|---|---|---|
| | | (1) | (2) |
| Education | High School | 1.000 | 0.000 |
| | College | 0.000 | 1.000 |
| | Graduate | 0.000 | 0.000 |
| Age | Middle | 1.000 | 0.000 |
| | Young | 0.000 | 1.000 |
| | Older | 0.000 | 0.000 |
| Language | English | 1.000 | |
| | Non-English | 0.000 | |
| Race | White | 1.000 | |
| | Non-White | 0.000 | |
| Job | Non-Professional | 1.000 | |
| | Professional | 0.000 | |
| Ethnicity | Non-Hispanic | 1.000 | |
| | Hispanic | 0.000 | |
| Gender | Female | 1.000 | |
| | Male | 0.000 | |

All term variables are continuous variables. Among the demographic variables, age was converted to categorical data with three levels. We coded age as "Young" when between 20 and 35 years; "Middle," between 36 and 50; and "Older" when over 51. Education levels also had three levels: "High school," "College," and "Graduate." Other variables were all dummy variables. Table 2 describes the categorical variables coding.

## 3   Results

### 3.1   Sample Data

We recruited 21 participants of varying socioeconomic background (see Table 3). All completed the questionnaire, although some questions were left blank, as we instructed participants not to guess the answers.

Among the 68 terms, 19 consisted of words on the Dale-Chall list and 31 consisted of 1- or 2-syllable words. The remainder were regarded as difficult words. The average familiarity score of the terms is 0.77, with more than half of the terms scoring 0.9

or higher. In other words, at least 50% of the terms were recognized by over 90% of the participants. The familiarity scores of the difficult and easy terms, as judged by the number of syllables and the Dale-Chall list alone, do not appear to be reliable indicators of term familiarity in the test population (see Figure 1).

**Table 3.** Demographics (n=21)

| Variables | Subgroup | Frequency |
|---|---|---|
| Gender | Female | 5 (23.8%) |
| | Male | 16 (76.2%) |
| Age | 20 ~ 35 | 9 (42.9%) |
| | 36 ~ 50 | 5 (23.8%) |
| | 51 over | 7 (33.3%) |
| Education | High School | 8 (38.1%) |
| | College | 5 (23.8%) |
| | Graduate | 8 (38.1%) |
| Native Language | English | 16 (76.2%) |
| | Non-English | 5 (23.8%) |
| Race | White | 13 (61.9%) |
| | Black | 4 (19.0%) |
| | Asian | 1 (4.8%) |
| | American Indian | 1 (4.8%) |
| | Other | 2 (9.5%) |
| Ethnicity | Hispanic | 1 (4.8%) |
| | Non-Hispanic | 17 (81%) |
| | N/A | 3 (14.2%) |
| Occupation | Non-Professional | 12 (57%) |
| | Professional | 9 (43%) |

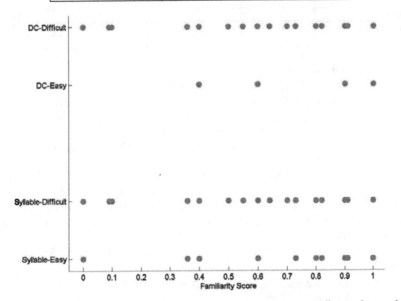

**Fig. 1.** Distribution of familiarity scores of difficult and easy terms according to the number of syllables and the Dale-Chall easy word list. Some easy terms scored low on the familiarity scale and some difficult terms scored high on the scale

## 3.2  Prediction of Average Familiarity

Using a support vector machine (SVM) to predict the average familiarity score of terms based on text corpora resulted in limited success. In 10-fold cross-validation (n=68), the average performance of the SVM is: Mean absolute error=0.196; Root mean squared error=0.293. When reviewing the evaluation results, we found that prediction of terms with the lowest familiarity scores was the least accurate. We suspect that this may be a result of lack of samples with low familiarity scores – only 9 terms scored less than 0.5.

## 3.3  Individualized Familiarity Predication

A logistic regression model was created for predicting whether a term is familiar to a person with particular background characteristics. As our analysis showed that MEDLINE frequency, MedlinePlus frequency, Gender, Age, and Ethnicity were not significant for predicting familiar terms for this sample, we removed these variables from the model.

**Table 4.** Logistic regression analysis result

| Variable | Coefficient | Standard Error | Significance (p) | Odds Ratio |
|----------|-------------|----------------|------------------|------------|
| AvgLength | 0.098 | 0.044 | 0.028 | 1.103 |
| QueryLog | 22.490 | 3.162 | 0.000 | 5.85E+09 |
| Dale-Chall | 1.370 | 0.353 | 0.000 | 3.937 |
| Language(1) | 0.790 | 0.309 | 0.010 | 2.204 |
| Race(1) | 0.624 | 0.241 | 0.010 | 1.866 |
| Job(1) | -0.647 | 0.245 | 0.008 | 0.523 |
| Education |  |  | 0.004 |  |
| Education(1) | -0.963 | 0.296 | 0.001 | 0.382 |
| Education(2) | -0.735 | 0.294 | 0.012 | 0.479 |
| Constant | -0.413 | 0.465 | 0.374 | 0.662 |

The final logistic regression model is as follows:

$$A = \text{Log(odds of finding the right answer)}$$
$$= -0.413 + (0.098 \times \text{average length}) + (22.490 \times \text{query log})$$
$$+ (1.37 \times \text{Dale-Chall}) + (0.790 \times \text{language}(1)) \quad (1)$$
$$+ (0.624 \times \text{race}) - (0.647 \times \text{job}(1))$$
$$- (0.963 \times \text{education}(1)) - (0.735 \times \text{education}(2))$$

$$\text{Probability that the term is familiar to the reader} = \frac{\exp(A)}{1 + \exp(A)} \quad (2)$$

Please note that the model and the variables reflect this study's participants' familiarity with a sample of terms. For a different population, the significant variables and the model will be likely to change.

Using 10-fold cross validation (n=714), the regression model performed moderately well:

- Correctly classified instances = 574 (80.4 %)
- Mean absolute error = 0.273
- Root mean squared error = 0.371
- Area under the ROC = 0.796

## 4  Discussion

This preliminary study showed that predicting text corpora-based term familiarity for health vocabulary is feasible. We measured 21 participants' familiarity with 68 terms through a TOFHLA-style questionnaire. Our attempt to predict average term familiarity from text corpra-based frequencies and other term characteristics resulted in moderate success (10-fold cross validation: mean absolute error = 0.196, root mean squared error = 0.293). Predicting term familiarity from the reader's demographics and the term characteristics generated reasonable results (10-fold cross validation: mean absolute error = 0.273, root mean squared error = 0.371, area under the ROC = 0.796).

Because of the considerable amount of health-related materials available and the large numbers and diverse consumers of these materials, there is a need to measure the readability of health content and make appropriate matches between consumers and content. Use of word length and the Dale-Chall easy word list to identify health terms that are difficult for consumers to comprehend can only provide a very rough estimate of term familiarity, as illustrated by Figure 1. For example, some short words might be incomprehensible and a single list does not reflect the health literacy level of all consumers. In contrast, our approach could provide a more refined and group-specific estimate of familiarity with health-related terms.

Because the number of terms and participants in this preliminary study is small, the predictive models are not likely to be applicable to the general population. In fact, it is not meaningful to estimate familiarity for the general population. Rather, models should be developed for the targeted audience populations. For instance, some public health campaigns may focus on low literacy group or minor groups, while other information might be intended for health-literate readers.

Another limitation is that we extracted term frequencies from the corpora without considering the morphological and lexical variations (e.g. number, tense) of the terms, which influences the calculation of term frequencies. For representation of term usage in media coverage and common English language, text corpora like the Reuters® collections will be better than the Dale-Chall list of easy words.

It may be argued that the underlying relationship between readability and the feature variables cannot be captured by support vector machines or logistic regression. On the other hand, applying some other methods including neural networks to this data set yielded almost identical or worse results – this may not be the case if more sample data are available or different features are used.

Another limitation of the reported approach is that only surface-level familiarity is measured, and not deeper knowledge of the concepts. For instance, although many consumers may recognize the term *heart attack*, few will know its precise definition or risk factors. That a participant answers a multiple choice question containing a term correctly does not indicate full comprehension of the underlying concept. In a related project (www.consumerhealthvocab.org), we manually reviewed concepts and assigned consumer-friendly display names to them. In the manual review process, reviewers not only consider whether a term is recognizable, but also its relationship with existing medical concept(s) found in the NLM Unified Medical Language System® (UMLS®). On the other hand, comprehensive, systematic manual review is limited by labor costs, while automated methods can be easily applied to large number of terms.

For future work, we would like to extend the questionnaire to include more terms and test their familiarity on a larger, more diverse sample population. It would also be interesting to evaluate the health literacy of participants and explore the relationship between health literacy level and term familiarity.

Research on readability and learning has indicated that providing material of an appropriate level is important to readers of all levels: providing materials that are either too difficult or too easy impairs a reader's ability to absorb new information. Suggesting materials at an appropriate readability level requires differentiating between health terms that consumers are likely to find familiar and unfamiliar (or "difficult") and knowing what a consumer or group is likely to comprehend. Thus, we believe our preliminary study on accurate, user-specific estimations of term familiarity is a necessary step towards improving health communication.

## Acknowledgement

We thank our collaborators at the Consumer Health Vocabulary Initiative: Guy Divita, Allen Browne, and Laura Roth. This research is funded in part by the NIH grant: R01 LM07222.

## References

1. Ratzan, S.C., and R.M. Parker. (2000). Introduction. In: National Library of Medicine Current Bibliographies in Medicine: Health Literacy. Selden, C.R., Zorn, M., Ratzan, S.C. and R.M. Parker (Eds). NLM Pub No. CBM 2000-1. Bethesda, MD: National Institutes of Health, U.S. Department of Health and Human Services.
2. Rudd, R., B. Moeykens, et al. (2000). Health and Literacy: A Review of Medical and Public Health Literature. Annual Review of Adult Learning and Literacy. J. Comings, B. Garner and C. Smith. San Francisco, CA, Jossey-Bass. 1: 158-199.
3. Osborne, H. (2004). Health Literacy From A To Z: Practical Ways To Communicate Your Health, Jones & Bartlett Pub.
4. (2004). "AHRQ, IOM weigh in on developing a health-literate America." Qual Lett Healthc Lead 16(5): 6-8.
5. McCray, A. T. (2005). "Promoting health literacy." J Am Med Inform Assoc 12(2): 152-63.
6. Davis, T. C., S. W. Long, et al. (1993). "Rapid estimate of adult literacy in medicine: a shortened screening instrument." Fam Med 25(6): 391-5.
7. Parker, R. M., D. W. Baker, et al. (1995). "The test of functional health literacy in adults: a new instrument for measuring patients' literacy skills." J Gen Intern Med 10(10): 537-41.
8. Zakaluk, B. L. and S. J. Samuels (1988). Readability: Its Past, Present, and Future, Intl Reading Assn.
9. Gemoets, D., Rosemblat, G., Tse, T., and R. Logan. (2004). Assessing readability of consumer health information: an exploratory study. Medinfo. 2004: 869-73.
10. Zeng, Q.T., Tse, T., Crowell, J., Divita, G., Roth, R., and Browne, A.C. (2005). Identifying consumer-friendly display (CFD) names for health concepts. Technical Report, DSG-TR-2005-003. Boston: Decision Systems Group (DSG), Brigham and Women's Hospital, Harvard Medical School.
11. Chall, J. S. and E. Dale (May 1, 1995). Readability Revisited: The New Dale-Chall Readability Formula, Brookline Books.

# On Sample Size and Classification Accuracy: A Performance Comparison

Margarita Sordo and Qing Zeng

Decision Systems Group, Harvard Medical School, Boston, MA, USA
{msordo,qzeng}@dsg.harvard.edu

**Abstract.** We investigate the dependency between sample size and classification accuracy of three classification techniques: Naïve Bayes, Support Vector Machines and Decision Trees over a set of ~8500 text excerpts extracted automatically from narrative reports from the Brigham & Women's Hospital, Boston, USA. Each excerpt refers to the smoking status of a patient as: current, past, never a smoker or, denies smoking. Our empirical results, consistent with [1], confirm that size of the training set and the classification rate are indeed correlated. Even though these algorithms perform reasonably well with small datasets, as the number of cases increases, both SMV and Decision Trees show a substantial improvement in performance, suggesting a more consistent learning process. Unlike the majority of evaluations, ours were carried out specifically in a medical domain where the limited amount of data is a common occurrence [13][14]. This study is part of the I2B2 project, Core 2[1].

## 1 Introduction

The performance of classification algorithms heavily depends on various parameters: sample size, number of features, etc, and, in some cases, it also depends on the particular nature and complexity of the problem under study [2].

It is well-known that classification of free-text is, on the one hand, a difficult task given the high-dimensionality of the feature space, and countless combinations of subtle, abstract relationships among features. On the other, this can be an easy task given the high level of redundancy that may exist in the bulk of available data. However, in order to succeed, it is necessary to extract important phrases, find meaningful, related words, and of course, gather enough examples, so patterns within the text can be identified and assigned correctly to one or more predefined classes.

Multiple statistical classification schemes and machine learning techniques have been applied to text classification, e.g. nearest neighbor classifier [3], Bayesian models [4], neural networks [5], symbolic learning [6], support vector machines [7] with very good results. However, they have focused mostly on learning speed, real-time classification speed, classification accuracy, the number of features, rather than in the sample size. Dumais et al [8] examined the performance of five different automatic learning algorithms. They explored the effect of sample size in a large training corpus as Reuters. They found that a SVM trained on 7147 cases had classification rates 72.6% - 92% for sample sizes ranging from 1% to 100% of the training set.

---

[1] I2B2: Informatics for Integrating Biology and the Bedside. http://www.i2b2.org/

J.L. Oliveira et al. (Eds.): ISBMDA 2005, LNBI 3745, pp. 193–201, 2005.
© Springer-Verlag Berlin Heidelberg 2005

Our work focused on comparing the performance of three learning algorithms (Naïve Bayes, Support Vector Machines and Decision Trees) in terms of classification accuracy in a medical domain. The total number of available cases is ~8500. For the purpose of evaluating the classification performance of these learning algorithms, we defined our training sets with sizes ranging from ~150 to ~8500 cases (2% -100% of the total number of cases in the dataset). Each training set containing excerpts referring to smoking status of inpatients extracted from discharge reports. It is worth mentioning that our smallest training set –with 150 cases- is similar in size to the total number of cases reported by Wilcox [13] and Chapman and Haug[14] with 200 and 150 chest x-ray reports respectively.

## 2  Background

The current project is part of Core 2 of I2B2 (Informatics for Integrating Biology and the Bedside), an on-going research to develop a scalable informatics framework that will link clinical research data and the vast data banks arising from basic science research in order to better understand the genetic bases of complex diseases. This knowledge will facilitate the prediction of clinical outcomes for individual patients with common diseases such as Asthma, Diabetes Mellitus and Hypertension. The I2B2 Center is funded as a Cooperative agreement with the National Institutes of Health.

The main goal of I2B2 Core 2 is to develop a robust architecture for the "clinical research chart" that is a necessary foundation for utilization of clinical data. The chart will be based on a proposed architecture called the I2B2Hive that uses modular cells of functionality to flexibly and consistently organize diverse research data that are base sequences coming from high-throughput machines. As a result, Core 2 will provide clinical investigators with the biocomputational tools necessary to collect and manage project-related clinical research data as a cohesive entity – to construct the modern clinical research chart.

## 3  Materials and Methods

In this section we describe the three classification techniques presented in this document, as well as the data source and acquisition and evaluation methods.

### 3.1  Naïve Bayes

Naïve Bayes classifiers involve the formal combination of a priori estimates, and the posterior probability for each category given the feature values of a particular instance or example. This classifier assumes that all features are independent of each other given the context of a class. This assumption, although false in most real-world tasks, does not have a great impact in the performance of Naïve Bayes classifiers. It actually simplifies the learning process because each attribute can be learned separately, especially when the number of attributes is large – as is the case in document classification [9]. In Bayes Theorem, the probability of each category given a document is defined by:

$$P(d_i \mid \vec{x}) = \frac{P(\vec{x} \mid d_i)P(d_i)}{P(\vec{x})} \tag{1}$$

Where $P(d_i \mid \vec{x})$ denotes the conditional probability of a document $d_i$ given a feature vector $\vec{x}$.

### 3.2 Support Vector Machines

Proposed by Vapnik [10], support vector machines are linear classifiers that seek to maximize the distance (margin) between a hyperplane $u$ and the nearest training vectors based on the equation:

$$u = \vec{w} \cdot \vec{x} - b \tag{2}$$

Where $\vec{w}$ is the normal vector to the separating hyperplane, and $\vec{x}$ is the input vector. In the simplest –linear- case, the margin is defined as the distance between the positive and negative examples. Figure 1 shows, the normal vector $\vec{w}$ to the hyperplane separating the two classes.

### 3.3 Decision Tree

Decision Tree algorithms are based on a "divide-and-conquer" strategy. Starting from the root (top node in the tree) the algorithm works recursively seeking at each stage an attribute $x$ that best separates the classes into more specific subsets with similar attributes striving to maximize the information gain that –ideally- will distinguish them from the rest of the classes (Figure 2).

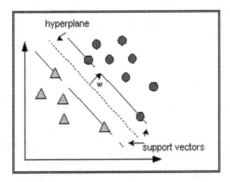

**Fig. 1.** Schematic view of a linear support vector machine separating two classes

### 3.4 Data

Data for these experiments were extracted automatically from narrative discharge reports from the Brigham & Women's Hospital, Boston, USA. Each excerpt is a free-text reference to the smoking status of a patient as one of four possible options: current smoker, past smoker, never a smoker or, denies smoking. The dataset consists of

~8500 excerpts (see Table 1). Each excerpt was tokenized and represented as a vector of the frequency of words:

$$\vec{x} = \{0,0,1,2,0,3...\} \tag{3}$$

Where $\vec{x}$ is the feature vector and $x_i$ is the frequency of the feature (word) $i$ in the excerpt. The number of features in the feature space was determined by their relevance and fit into categories. Given the fact that the task at hand was to classify excerpts of about 50 words or less, as opposed to full documents with hundreds of words, we considered the top 200 one-gram features for our first sample set, and the top 500 bi-grams for the second. Bi-grams were extracted using a 2-word, window size=3 setting.

We did not explore rigorously the optimal number of features for these two experiments. However, these numbers provided acceptable results as we present below.

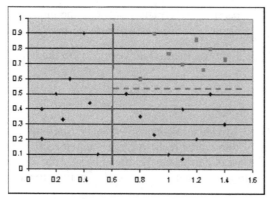 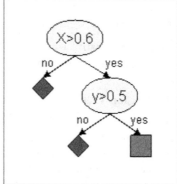

**Fig. 2.** On the left, sample set divided into subgroups by the decision tree on the right. Top node X>0.6 divides the set into two subsets (red solid line), the node below, y>0.5 divides the subset from the right of the chart into two subsets divided by the blue dotted line

### 3.5 Evaluation

We evaluated the classification accuracy of three classification algorithms against two datasets described in Table 1 below:

**Table 1.** Datasets used to evaluate the classification accuracy of the three learning algorithms

|  | Dataset 1 | Dataset 2 |
|---|---|---|
| No. cases from free-text discharge reports | 8467 | 8296 |
| No. features sorted in descending order | top 194 | top 417 |
| Token size | onegram | bigram |
| No. classes | Four:<br>- S: currently smoking;<br>- P: past smoker;<br>- N: never a smoker;<br>- D: denies any smoking | Four:<br>- S: currently smoking;<br>- P: past smoker;<br>- N: never a smoker;<br>- D: denies any smoking |

Our approach consisted of evaluating each learning algorithm independently to assess its classification accuracy when trained with datasets of varying sizes. We conducted our experiments using Weka, a public domain set of data mining tools written in Java. Evaluation was carried out as follows:

```
For each classifier
    Until all cases in dataset are used
        Define training set as percentage of the total
            number of cases in the dataset (range 2-100%)
        Train classifier with current set using four-fold
            cross-validation
        Evaluate the performance of the classifier as the
            average correct classification rate
    % end Until
% end For
```

As shown in Tables 2 and 3, the number of cases used at each stage was selected as a percentage of the total number of cases – ranging from 2 to 100%. We performed 4-fold cross validation instead of the more commonly used 10-fold, since, by definition, every example gets to be in the test set exactly once and in the training set $k-1$ – in this case 3- times regardless of the number of folds. Further, given the large number of cases, both the training and testing sets contained enough examples.

For each classifier, 22 datasets were defined independently, and used for training and testing. Each dataset was divided into 4 subsets. Each time one of the 4 subsets is used for testing and the remaining 3 are used for training, until all four subsets are used for testing. Once this process is over, the average error across all four runs is computed.

## 4  Results

Table 2 summarizes experimental results for the three classification algorithms evaluated using the onegram data set. The curves in Figure 3 show the classification accuracy as a function of the number of cases in the onegram sample set.

Table 3 summarizes experimental results for the three classification algorithms evaluated using the bigram data set. The curves in Figure 4 show the classification accuracy as a function of the number of cases in the bigram sample set.

## 5  Discussion

We explored the effect of the size dataset on the classification accuracy of three learning algorithms. Our results suggest that data set size and classification rate are indeed correlated. These empirical results confirm previous evaluations [1][8][11][12]. A practical implication, specifically in a clinical domain, is the reliability and robustness of the algorithms trained with small vs. large data sets. In both Figures 3 and 4 we can observe a marked fluctuation in the classification accuracy within the first thousand samples. However, as we increase the sample size, performance overcomes local minima –and in some cases, local maxima- and stabilizes, becoming more reliable and accurate. Although classification rate as a measure of the performance of a classifier is influenced by the inherent difficulty of the classification problem at hand, there are however situations when the performance of a classifier is degraded as a result of small number of training cases. Specifically in the clinical domain, [13][14] reported promising results, though highlighting the desirability of larger sample sets to confirm their results. Even though it is impossible to estimate the size of the dataset prior to training a learning algorithm, one can identify a point where performance stabilizes, reducing the possibility of suboptimal classification performance.

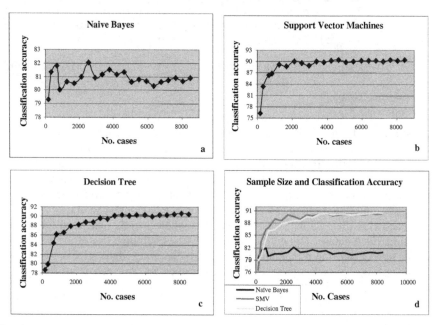

**Fig. 3.** The effect of the size of the training set on the classification accuracy for a) Naïve Bayes; b) SVM; c) Decision Tree; and d) comparison of the three learning algorithms. One-gram dataset

Our findings can be summarized as follows: First, experiments show that the three classifiers perform reasonably well with a small number of cases, although Naïve Bayes tend to perform better than SMV and Decision Trees with small datasets given their tolerance against noisy data (Figures 3a-3c and 4a-4c).

Second, as the number of samples increases, both SMV and Decision Trees show a substantial improvement in their classification accuracy, outperforming Naïve Bayes. This suggests a more consistent learning process where both SMV and Decision Trees are capable of distinguishing relevant from spurious features (Figures 3d and 4d).

Differences in performance can be attributed to the fact that Naïve Bayes is a simple generative classification algorithm that tends to fit the distribution of the data [15][16]. So, as the number of cases, and in this case, variability increase, the accuracy of the classification rate diminishes. On the other hand, SVM and Decision Trees are discriminative algorithms. These algorithms attempt to maximize the classification rate of labeled and unlabeled data, while classifying labeled data as correctly as possible [17].

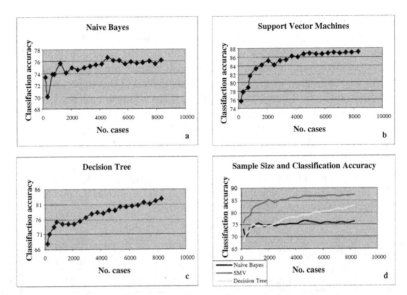

**Fig. 4.** The effect of the size of the training set on the classification accuracy for a) Naïve Bayes; b) SVM; c) Decision Tree; and d) comparison of the three learning algorithms. Bigram dataset

**Table 2.** Classification accuracy for each classification algorithm as a function of the sample size for the "onegram" dataset. Column marked as "% of total" indicates the number of cases in the sample size as a percentage of the total number of cases

| % of total | Classification Accuracy | | | % of total | Classification Accuracy | | |
|---|---|---|---|---|---|---|---|
| | Naïve Bayes | SMV | Decision Tree | | Naïve Bayes | SMV | Decision Tree |
| 2 | 79.2899 | 76.3314 | 78.6982 | 50 | 81.1717 | 90.1488 | 90.1252 |
| 4 | 81.3609 | 83.432 | 79.8817 | 55 | 81.3359 | 90.378 | 90.378 |
| 8 | 81.8316 | 86.4106 | 84.3427 | 60 | 80.6102 | 89.7835 | 90.1181 |
| 10 | 80.0236 | 86.8794 | 86.1702 | 65 | 80.7741 | 90.0781 | 90.2417 |
| 15 | 80.6299 | 89.0551 | 86.5354 | 70 | 80.6615 | 90.297 | 90.297 |
| 20 | 80.508 | 88.7183 | 87.9504 | 75 | 80.3465 | 90.1816 | 89.9528 |
| 25 | 80.9546 | 90.0284 | 88.1853 | 80 | 80.629 | 90.1816 | 90.2259 |
| 30 | 82.0472 | 89.6063 | 88.7008 | 85 | 80.7254 | 90.0639 | 90.2168 |
| 35 | 80.9315 | 89.0314 | 88.6939 | 90 | 80.9055 | 90.4331 | 90.4856 |
| 40 | 81.1577 | 89.9291 | 89.5452 | 95 | 80.654 | 90.2524 | 90.6503 |
| 45 | 81.5223 | 89.7375 | 89.4751 | 100 | 80.9141 | 90.3508 | 90.4925 |

Third, the effect of replacing single tokens (onegram dataset) for more complex bigram features can be seen in Figures 3 and 4. Intuitively, one can argue that bigrams contain more context information than onegrams. However, bigrams increase the complexity of the feature set impacting the performance of the classifiers –as observed in Figure 4.

## 6  Future Work

As part of the I2B2 project, our ultimate goal is to provide clinical investigators with the biocomputational tools that will allow them to link clinical research data and genetic and genomic data from basic science research to better understand and predict the genetic bases of complex diseases and their clinical course. In future work we will evaluate these three algorithms for text classification using other benchmark datasets, e.g. Reuters, to determine whether our empirical results are generalizable in other domains. As a continuation of the experiments described in this paper, we will further investigate the correlation between variability of distribution and classification rate, since, as observed, is an interesting phenomenon. Also, we will evaluate the performance of other classification algorithms that could be integrated into the set of biocomputational tools currently under development.

**Table 3.** Classification accuracy for each classification algorithm as a function of the sample size for the "bigram" dataset. Column marked as "% of total" indicates the number of cases in the sample size as a percentage of the total number of cases

| % of total | Naïve Bayes | SVM | Decision Tree | % of total | Naïve Bayes | SVM | Decision Tree |
|---|---|---|---|---|---|---|---|
| 2 | 73.3333 | 75.7576 | 67.8788 | 50 | 75.5304 | 85.8968 | 78.1581 |
| 4 | 70.0906 | 77.9456 | 70.997 | 55 | 76.6331 | 86.6287 | 78.9566 |
| 8 | 73.9065 | 78.8839 | 73.454 | 60 | 76.2307 | 86.8596 | 79.0838 |
| 10 | 73.8239 | 81.544 | 75.1508 | 65 | 76.1313 | 86.5912 | 80.1001 |
| 15 | 75.7235 | 83.2797 | 74.4373 | 70 | 75.5295 | 86.6885 | 80.1274 |
| 20 | 74.141 | 84.1471 | 74.3822 | 75 | 75.9081 | 86.7406 | 80.5529 |
| 25 | 74.9277 | 85.1013 | 74.5419 | 80 | 75.7233 | 86.9801 | 80.8017 |
| 30 | 74.5579 | 84.1238 | 75.4823 | 85 | 75.8616 | 86.7962 | 81.648 |
| 35 | 74.9914 | 85.1188 | 76.6104 | 90 | 76.0916 | 87.0212 | 81.2349 |
| 40 | 75.1658 | 85.2321 | 77.7878 | 95 | 75.5995 | 87.0448 | 82.2104 |
| 45 | 75.4085 | 86.1505 | 78.2213 | 100 | 76.1933 | 87.2348 | 82.7628 |

# Acknowledgements

Support for this project has been provided by NIH-funded National Center for Biomedical Computing based at Partners HealthCare System.

# References

1. Manning CD, Schutze H. Foundations of Statistical Natural Language Processing. MIT Press, Cambridge Massachusetts, 1999.
2. McKay M, Fitzgerald MA, Beckman, RJ. Sample Size Effects When Using R2 to Measure Model Input Importance. Technical Report LA-UR-99-1357. Los Alamos National Laboratory, Los Alamos, NM, USA.
3. Yang Y. Expert network: Effective and efficient learning from human decisions in text categorization and retrieval. *Proceedings of the 17th Annual International ACM SIGIR Conference on Research and Development in Information Retrieval*, 13-22, 1994.
4. Lewis DD, Ringuette M. A comparison of two learning algorithms for text categorization. *Third Annual Symposium on Document Analysis and Information Retrieval*, 81-93, 1994.
5. Wiener E, Pedersen JO, Weigend AS. A neural network approach to topic spotting. *Proceedings of the 4th Annual Symposium of Document Analysis and Information Retrieval*, 1995.
6. Cohen WW, Singer Y. Context-sensitive learning methods for text categorization. *Proceedings of the 19th Annual International ACM SIGIR Conference on Research and Development in Information Retrieval*, 307-315, 1996.

7. Joachims T. Text categorization with support vector machines: Learning with many relevant features. *Proceedings of the 10^{th} European Conference on Machine Learning.* Springer Verlag 1998.
8. Dumais S, Platt J, Heckerman D, Inductive Learning Algorithms and Representations for Text Categorization. *In Proceedings of the 7th International. Conference on Information and Knowledge Management,* 1998.
9. McCallum A, Nigam K. A Comparison of Event Models for Naïve Bayes Text Classification. *In AAAI-98 Workshop on Learning for Text Categorization,* 1998. 1286.
10. Vapnik VN. *The Nature of Statistical Learning Theory.* Springer, 1995.
11. Ghani R. Using Error-Correcting Codes for Text Classification. Workshop on Text Mining at the First IEEE Conference on Data Mining (2001).
12. Raudys SJ, Jain AK. Small Sample Size Effects in Statistical Pattern Recognition: Recommendations for Practitioners. IEEE Transactions on Pattern Analysis and Machine Intelligence, Vol. 13, No. 3, March 1991.
13. Wilcox A, Hripcsak G. Classification algorithms applied to narrative reports. *Proceedings of the American Medical Informatics Association (AMIA) Symposium.* 1999;:455-9.
14. Webber Chapman W, Haug PJ. Comparing Expert Systems for Identifying Chest X-ray Reports that Support Pneumonia. *Proceedings of the American Medical Informatics Association (AMIA) Symposium* 1999;:216-20.
15. McCallum A, Nigam K. A comparison of event models for Naive Bayes text classification. AAAI-98 Workshop on Learning for Text Categorization, 1998.
16. Ng A, Jordan M. On discriminative vs. generative classifiers: a comparison of logistic regression and naive Bayes. NIPS 14, 2002.
17. Gilles G. Semi-supervised learning. Tech. Rep. Dept. Informatique et Recherche Operationnelle, Universite de Montreal, Montreal, QC, Canada H3C 3J7.

# Influenza Forecast: Comparison
# of Case-Based Reasoning and Statistical Methods

Tina Waligora and Rainer Schmidt

Universität Rostock, Institut für Medizinische Informatik und Biometrie,
Rembrandtstr. 16 / 17, D-18055 Rostock, Germany
{tina.waligora,rainer.schmidt}@medizin.uni-rostock.de

**Abstract.** Influenza is the last of the classic plagues of the past, which still has to be brought under control. It causes a lot of costs: prolonged stays in hospitals and especially many days of unfitness for work. Therefore many of the most developed countries have started to create influenza surveillance systems. Mostly statistical methods are applied to predict influenza epidemics. However, the results are rather moderate, because influenza waves occur in irregular cycles. We have developed a method that combines Case-Based Reasoning with temporal abstraction. Here we compare experimental results of our method and statistical methods.

## 1 Introduction

Since influenza results in many costs, e.g. for delayed stays in hospital and especially for an increased number of unfitness for work, many of the most developed countries have started to generate influenza surveillance systems (e.g. US: www.flustar.com, France [1], and Japan [2]). The idea is to predict influenza waves or even epidemics as early as possible and to indicate appropriate actions like starting vaccination campaigns or advising high-risk groups to stay at home.

Mostly statistical methods are applied to predict influenza epidemics. However, the results are rather moderate, because influenza waves occur in irregular cycles and Farrington pointed out that statistical methods have difficulties to cope with infectious diseases characterised by irregular cyclic behaviour [3].

So, we have developed a method that combines Case-Based Reasoning with temporal abstraction. Before we explain our method and subsequently present comparative results of our method and of statistical methods, we discuss the question which data should be used.

### 1.1 Data

It is well known that a couple of factors can be responsible for influenza outbreaks. One of them is the weather. It is often assumed that a strong winter increases the spread of influenza. However, studies could reveal only an extremely small relation between temperatures and spread of influenza [4]. Dowell even suggests that the occurrence of influenza in winter is not related to the temperature but to the annual light/dark pattern [5]. Other influences are the mutations of the virus and influenza outbreaks in foreign countries, even as far as Hongkong. Unfortunately, no exact knowledge about these influences is available. So far, only seasonal behaviour of influenza is well known from observations [3, 5]. Therefore all surveillance systems focus on observed numbers of infected people, especially on their increase.

J.L. Oliveira et al. (Eds.): ISBMDA 2005, LNBI 3745, pp. 202–210, 2005.

Considerations about how to forecast influenza begin with questions about which data should be used and which data are available. The answer varies with the country and its health organisation.

In some countries with rather private health systems like Germany, research groups interested in predicting influenza have started to develop surveillance nets based on voluntary participation of general practitioners. They felt that the data collected and provided by official health centres were insufficient, because they are usually available with a delay of two or even three weeks. These surveillance nets are based on general practitioners who give once a week some sort of standardised reports. It needs a huge effort to initiate and organise such nets and rural areas are very often not adequately represented, because it is difficult to find doctors willing to participate. Since the reports are always subjective, misjudgements and data interpretation errors may occur, which may lead to false assessments – especially in areas with low density of participating doctors.

The alternative means to use official data from health centres. In Germany, these data are more objective, because they contain reports and laboratory results of all occurrences of notifiable diseases. Unfortunately, because of the hierarchical and bureaucratic organisation of the health centres, the availability of these data is delayed for at least two weeks. In countries with more public health systems, sometimes the situation seems to be much better, e.g. in Japan [2].

However, we have chosen another alternative. Since 1997 we receive data for our federal state Mecklenburg-Western Pomerania from the main health insurance scheme. These data are sick certificates of employees and of people who receive unemployment benefit. Fortunately we get the data daily. Of course there is a short delay between doctors writing the certificates and the insurance scheme receiving them by mail from their policyholders. We do not recur on the days when the certificates have been issued by doctors, but on the daily data sets received by the insurance scheme. Since there are some daily fluctuations by chance, influenza surveillance systems usually use weekly aggregated data.

The disadvantage of using insurance data is their superficiality, because the certificates usually contain just the first diagnoses, which might be refined or changed later on. However, for influenza this is only a minor problem, because the symptoms of influenza, acute bronchitis, etc. are so similar that most surveillance groups use a superficial category anyway, namely all acute respiratory diseases to infer influenza.

## 2  Prognostic Methods

All influenza surveillance systems make use of developments in the past. Most of them have tried statistical methods. The usual idea is to compute mean values and standard courses based on weekly incidences of former influenza seasons (from October till March) and to analyse deviations from a statistic normal situation.

Influenza waves usually occur only once a season, but they start at different time points and have extremely different intensities. Since Farrington pointed out that statistical methods are inappropriate for diseases like influenza that are characterised by irregular cyclic temporal spreads [3], we have developed a method that uses former influenza seasons more explicitly. We apply the Case-Based Reasoning idea: that means to determine the most similar former courses of weekly incidences and to use them to decide whether a warning is appropriate. Viboud [6] from the group that is responsible for the French Surveillance net has developed a method that is very similar to our one. However, both methods differ in their intentions. Viboud attempts to predict incidences few weeks in advance, while we are interested in more practical results, namely in the computation of appropriate warnings.

## 2.1 Case-Base Reasoning to Forecast Influenza

Inspired by our former program for the prognosis of kidney function courses [7], we have developed a method to decide about the appropriateness of warnings against approaching influenza waves (figure 1).

Every influenza season consists of 26 weeks (from October till March). Since we consider weekly incidences, seasons are represented as sequences of 26 numeric values. Each week it has to be decided anew whether a warning is appropriate or not. For this decision, just the recent development is important. So, we consider only a sequence of the four most recent weeks. When an influenza season is finished, it is separated into 23 four-week courses; all of them are stored as cases in the case base.

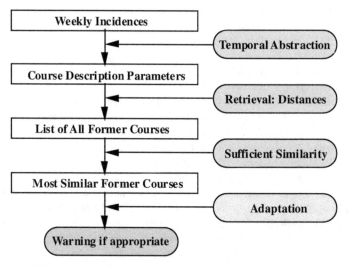

**Fig. 1.** Case-Based Reasoning method to forecast influenza

The first step of our method is a temporal abstraction of a sequence of four weekly incidences into three trend descriptions that assess the changes from last week to this week, from last but one week to this week and so forth. Secondly, these three assessments and the four weekly incidences are used to determine similarities between a current query course and all four-weeks courses stored in the case base. Our intention for using these two sorts of parameters is to ensure that a query course and an appropriate similar course are on the same level (similar weekly incidences) and that they have similar changes on time (similar assessments). More details about these first two steps of our method can be found in [8].

The result of computing distances is a very long list of all former four-week courses sorted according to their distances in respect to the query course. For the decision whether a warning is appropriate, this list is not really helpful, because most of the former courses are rather dissimilar to the query course. So, the next step means to find the most similar ones. We decided to filter the most similar cases by applying two explicit similarity conditions. First, the difference concerning the sum of the three trend assessments between a query course and a similar course has to be below a threshold X. This condition guarantees similar changes on time. And secondly, the difference concerning the incidences of the current weeks must be below a threshold Y. This second condition guarantees an equal level of the current week of a similar case and the current week of the query course. We have learned good settings for the threshold parameters X and Y by taking in turn one season out of the case base and comparing the results when varying the settings.

The result of this third step usually is a very small list containing only the most similar former courses. As in compositional adaptation [9] we take the solutions of a couple of similar cases into account, namely of all courses in this small list.

In retrospect, we have marked those time points of the former influenza seasons where we believed a warning would have been appropriate; e.g. in the 4th week of 2001, which is the 17th week of the 2000/2001 season (marked as square in fig.2).

For the decision to warn, we split the list of the most similar courses in two lists. One list contains those courses where a warning was appropriate; the second list gets the other ones. For both of these new lists we compute their sums of the reciprocal distances of their courses to get sums of similarities. Subsequently, the decision about the appropriateness of a warning depends on the question which of these two sums is bigger.

## 2.2  Statistical Methods to Forecast Influenza

A couple of statistical tests are available. Under the assumption of binomial distribution, we have tested whether the observed weekly count of infected people is significant. With the following formula the probability of exactly k insured people being infected:

$$P(X=k)=\binom{n}{k}p^k(1-p)^{n-k} \text{ for } k=0,...,n$$

where

k = number of observed insured people being infected
n = number of insured people

and

$$p = \frac{Number\ of\ infected\ people\ on\ average}{Number\ of\ insured\ people}$$

Here, "on average" means the average number of infected people per week concerning the specific months. Concerning the whole time period of three years or just the influenza seasons would be too vague, because influenza mainly occurs in some months, and concerning specific calendar week might be randomly.

If the sum of the probabilities of k people being infected, k+1 people being infected etc. is 5% or more, an influenza wave can be assumed:

$$P(X\geq i)=\sum_{k=1}^{n} P(X=k) \text{ for } i=0,...,n$$

## 3    Experimental Results

We have performed some experiments to compare both methods. However, we used different data. For doctors it is extremely difficult to distinguish between a real influenza infection and other acute respiratory infections, especially at the first diagnosis. So, in most influenza surveillance systems influenza is usually inferred from the counts of all acute respiratory infections.

For the Case-Based Reasoning method we used acute respiratory infections (ICD9: 460 to 487 and ICD10: J00 to J99, except very few chronic diseases) for the influenza seasons from 1997 to 2002.

For the statistical tests we just used data of real influenza infections (ICD10: J10 and J11) for three years, namely for 2000, 2001, and 2002, because the ICD10 code was introduced in 2000 and for 1997-1999 (ICD9: 487) the counts did not fit together the later ICD10 counts.

For both experiments, we have not used more up-to-date data, because in Mecklenburg-Western Pomerania the last influenza wave occurred in early 2001 and the course of the 2001/2002 season is typical for the following ones.

### 3.1    Case-Based Reasoning Method

First, we have marked those time points where we, in retrospect, believed a warning would have been appropriate (the three squares in figure 2). Later on we assumed that these warnings might be a bit late. So, we have additionally attempted earlier desired warnings (the three circles in figure 2).

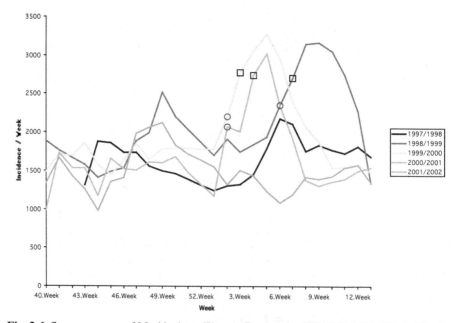

**Fig. 2.** Influenza seasons of Mecklenburg-Western Pomerania, ICD10: J00-J99, ICD9: 460-487

### 3.1.1  First Experiment

For our first tests, we used the five seasons shown in figure 2 with the desired warnings depicted as squares. In turn we used one season as query course. Furthermore, we wanted to discover how much the results are improved by the number of seasons stored in the case base. So, for every query season we varied the case base, and we did not only put the seasons in chronological order into the case base, but attempted every combination. That means, for each query season we made four attempts with one season in the case base, six attempts with two seasons etc.

The results are shown in table 1. Sensitivity means proportion of computed warnings to desired warnings; specificity means proportion of computed "non-warnings" to desired "non-warnings".

**Table 1.** Sensitivity and specificity of our first experiment

|  | Sensitivity | Specificity |
|---|---|---|
| 1 season in case base | 50 % | 100 % |
| 2 seasons in case base | 83 % | 100 % |
| 3 seasons in case base | 100 % | 100 % |
| 4 seasons in case base | 100 % | 100 % |

At first glance the results seem to be very good: there are no false warnings and to exactly compute the desired warnings, for every season it is sufficient to use just three of the four remaining seasons as case base. However, since for every query season 23 decisions have to be made, most of them are obvious "non-warnings", a few are follow-up warnings (determined by a simple heuristic when the week before a warning or a follow-up warning was computed), and only few decisions are really crucial.

### 3.1.2  Earlier Warnings

Since we imagined that the desired warnings of our first experiment might be a bit late, we tried earlier ones in a second experiment, in figure 2 depicted as circles. We made the same experiment again and the results are shown in table 2.

**Table 2.** Sensitivity and specificity of our second experiment: with earlier warnings

|  | Sensitivity | Specificity |
|---|---|---|
| 1 season in case base | 45 % | 95 % |
| 2 seasons in case base | 69 % | 96,7 % |
| 3 seasons in case base | 80 % | 97,2 % |
| 4 seasons in case base | 80 % | 96,1 % |

Now it is more difficult to compute the new desired warnings. However, the problems are mainly caused by the peak in the 49th week of 1998, which is the 10th week of the 1998/1999 season. Since the incidences of this peak are higher than the incidences of the desired warnings and the developments are similar too, consequently a warning is computed. Only in retrospect it becomes clear that this was not the beginning of an influenza wave. And since this peak is marked as not worth for a warning, it prevents our program from computing desired warnings for other seasons.

However, this is not so much a problem of the method, but rather a question of the availability of appropriate data. Since we use health insurance data, we do not have access to laboratory results, which often indicate causes. In fact, concerning the analysis of such data by the Robert-Koch Institute [11], the peak in the 49<sup>th</sup> weak of 1998 was probably (but this was never definitely proved) the result of a pathogen (respiratory syntactical virus) that causes similar symptoms as influenza. Unfortunately, such data from health centres even the Robert-Koch Institute gets only delayed (about two weeks).

## 3.2  Statistical Tests

For the statistical tests we did not consider all acute respiratory infected people but only those with real influenza infections (ICD10: J10 and J11). Since the counts of ICD9 and ICD10 did not fit together, we used only data for 2000, 2001, and 2002 (figure 3).

Though the considered infections differ, in both figures (see figure 2 and figure 3) the increase of seasons 2000 and 2001 occurs obviously at the same moment, namely in the second calendar week. The fact that influenza waves start in two following seasons in the same week is poor chance. Sometimes influenza waves start much later, see e.g. the 1998/1999 season (figure 2). We performed tests under the assumption of binomial distribution (see section 2.2). Parts of results, namely for the first four calendar weeks, are shown in table 3. A significant count of infected people is indicated, when the value is below 5%.

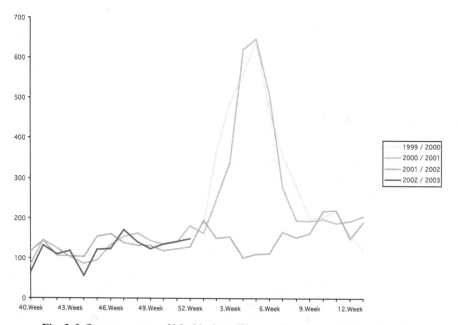

**Fig. 3.** Influenza seasons of Mecklenburg-Western Pomerania, ICD10: J10+J11

For 2000 table 3 shows the desired result, namely the beginning of an influenza wave in the second calendar week. For 2001 the obviously desired result is a start of an influenza wave in the second week too, but the statistical test discovers the start not until the fourth week. Of course, one problem is that we considered just three seasons. However, the results illuminate a problem of statistical methods applied to diseases with cyclical behaviour. In two of the three considered seasons influenza waves occurred which increase the average value very much. So, instead of using a general average value only those weeks without influenza waves should be considered as "normal" and should be used for computing average values. However, sometimes it is difficult to decide whether a week is "normal" and for specific calendar weeks there may be just very few "normal" ones.

**Table 3.** Results of tests under assumption of binomial distribution. NIL means that no computation was necessary, because the observed count is even below the average value

|         | Counts 2000 | % | Counts 2001 | % | Counts 2002 | % |
|---------|-------------|---|-------------|---|-------------|---|
| 1.Week  | 184 | 41 | 163 | Nil | 195 | |
| 2.Week  | 366 | 0,1 | 246 | Nil (32) | 151 | Nil |
| 3.Week  | 489 | 0,1 | 340 | 23 | 154 | Nil |
| 4.Week  | 559 | 0,1 | 620 | 0,1 | 102 | Nil |

## 4 Conclusion

In contrast to most medical diagnostic problems, we cannot ask experts about the correctness of the computed warnings. Instead, nobody knows in which week a first warning should be computed. However, the simultaneous increase in both sorts of data indicates that for the Case-Based Reasoning method the earlier warnings of the second experiment are probably ideal moments for first warnings against approaching influenza waves.

So far, it is difficult to assess the quality of our Case-Based Reasoning method. However, the results are at least as good as with statistical tests that assume binomial distribution. Unfortunately for our research, there has not occurred an influenza wave since 2002 in Mecklenburg-Western Pomerania.

Furthermore, we believe that the results of influenza surveillance depend more on the data than on the method. This does not only mean the a priori quality of the data and the speed of their availability, but additionally the quality for discriminating risky situations. The a priori quality of our health insurance data is rather poor, especially the diagnoses are often superficial, but there is only a very short delay concerning their availability. Official data from German health centres are more profound, but for bureaucratic reasons there availability is delayed for too long.

## References

1. Prou, M., Long, A., Wilson, M., Jacquez, G., Wackernagel, H., Carrat, F.: Exploratory Temporal-Spatial Analysis of Influenza Epidemics in France. In: Flahault, A., Viboud, C., Toubiana, L., Valleron, A.-J. (eds.): Abstracts of the 3rd International Workshop on Geography and Medicine, Paris, October 17-19 (2001) 17
2. Shindo, N. et al.: Distribution of the Influenza Warning Map by Internet. In: Flahault, A., Viboud, C., Toubiana, L., Valleron, A.-J. (eds.): Abstracts of the 3rd International Workshop on Geography and Medicine, Paris, October 17-19 (2001) 16

3. Farrington, C.P., Beale, A.D.:,The Detection of Outbreaks of Infectious Diseases. In: Gierl L. et al. (eds.): International Workshop on Geomedical Systems, Teubner, Stuttgart (1997) 97-117

4. Rusticucci, M., Bettolli, M.L., De Los Angeles, Harris, M., Martinez, L., Podesta, O.: Influenza outbreaks and weather conditions in Argentina. In: Toubiana, L., Viboud, C., Flahault, A., Valleron, A.-J. (eds.): Geography and Health, Inserm, Paris (2003) 125-140.

5. Dowell, S.F.: Seasonal Variation in host susceptibility and cycles of certain infectious diseases. Emerg Inf Dis 7 (3) (2001): 369-374

6. Viboud, C. et al.: Forecasting the spatio-temporal spread of influenza epidemics by the method of analogues. In: Abstracts of 22nd Annual Conference of the International Society of Clinical Biostatistics, Stockholm, August 20-24 (2001) 71

7. Schmidt, R., Gierl, L.: Prognoses for Multiparametric Time Courses. In: Brause, R.W., Hanisch, E. (eds.): Medical Data Analysis. Proceedings of ISMDA 2000, Springer-Verlag, Berlin (2000) 23-33

8. Schmidt, R., Gierl, L.: A prognostic model for temporal courses that combines temporal abstraction and case-based reasoning. Int J Medical Informatics 74 (2005) 307-315

9. Wilke, W., Smyth, B., Cunningham, P.: Using Configuration Techniques for Adaptation, In: Lenz, M., et al. (eds.): Case-Based Reasoning Technology, From Foundations to Applications. Springer-Verlag, Berlin (1998) 139-168

# Tumor Classification
# from Gene Expression Data:
# A Coding-Based Multiclass Learning Approach

Alexander Hüntemann[1,*], José C. González[2], and Elizabeth Tapia[3]

[1] Katholieke Universiteit Leuven, Celestijnenlaan 300B,
3000 Leuven, Belgium
Alexander.Huntemann@mech.kuleuven.be
[2] E.T.S.I. Telecomunicación, Universidad Politécnica de Madrid,
Ciudad Universitaria s/n, 28040 Madrid, Spain
jgonzalez@gsi.dit.upm.es
[3] Facultad de Ciencias Exactas, Ingeniería Agromesura, Escuela de Ingeniería
Electrónica, Riobamba 245 bis, 2000 Rosario, Argentina
etapia@eie.fceia.unr.edu.ar

**Abstract.** The effectiveness of cancer treatment depends strongly on an accurate diagnosis. In this paper we propose a system for automatic and precise diagnosis of a tumor's origin based on genetic data. This system is based on a combination of coding theory techniques and machine learning algorithms. In particular, tumor classification is described as a multiclass learning setup, where gene expression values serve the system to distinguish between types of tumors. Since multiclass learning is intrinsically complex, the data is divided into several biclass problems whose results are combined with an error correcting linear block code. The robustness of the prediction is increased as errors of the base binary classifiers are corrected by the linear code. Promising results have been achieved with a best case precision of 72% when the system was tested on real data from cancer patients.

## 1   Introduction

Effective cancer treatment depends strongly on an accurate diagnosis of the type of tumor. Nowadays, the diagnosis of such malignancy relies strongly on histopathological and clinical data. Molecular tests have not yet been widely exploited to predict the type of cancer since molecular markers have not been identified for all possible tumors.

A promising technology has been introduced in molecular biology called microarrays, where thousands of genes can be analyzed simultaneously. On the surface of these devices, fragments of DNA or RNA are deposited. Then, a dyed sample of tissue is applied to the microarray for analysis. Hybridization occurs

---

* This work has been done while the author was at the Department of Telematic Systems Engineering, Universidad Politécnica de Madrid, Spain

J.L. Oliveira et al. (Eds.): ISBMDA 2005, LNBI 3745, pp. 211–222, 2005.

at the spots where the sample genetic material matches the DNA/RNA on the surface of the microarray. The output of such an experiment is a colormap where the color intensity is related to the degree of expression of the genes under consideration. Since microarrays are built in a controlled fashion, it is possible to identify gene expression patterns from the obtained colormaps. This technique opens new ways for researchers to further investigate the existing relationships between gene expression patterns and cancer.

Following current trends in bioinformatics, the present article studies the classification of cancer tissues based on genetic information provided by a microarray. The analyzed data was obtained from the Whitehead Institute of Cancer Research in the USA [3]. It consists of two independent databases, i.e. one training set of 144 classified instances and one test set of 54 classified instances. Each instance is composed of 16063 gene expression values. The database is small in size since it is an extremely costly process to obtain sample tissues valid for later analysis. The employed samples correspond to primary biopsy tissues enriched by 50% in malignant cells in order to make the analysis of the data easier. They represent the 14 most common classes of cancer in human beings, i.e. breast, prostate, colon, lung, uterus, renal, ovary, bladder, pancreas, central nervous system cancer, leukaemia, lymphoma and mesothelioma. In every case independent medical experts of different cancer research centers in the USA verified the initial diagnosis twice. Once the samples were classified, a high throughput technique was used to profile the data genetically: Affymetrix's GeneChips.

The problem in the presented research is to train classifiers on a multiclass dataset of 144 instances and validate the results on the 54-instance test set. This problem's complexity is due to two reasons: first, there exist much more attributes than classified instances and second, multiclass machine learning is intrinsically more complex than biclass learning because most of the available algorithms are designed for the biclass case.

## 1.1   Related Work

Previous work has been published on the creation of a diagnosis system for cancer based on gene expression data and machine learning algorithms. We can compare for instance the results of Yeang et al. [15], Ramaswamy [9] and Golub [4]. Yeang and Ramaswamy have developed a system for multiclass diagnosis and evaluated it on the same dataset. They are considered, therefore, an important reference for the current research and their results establish a goal to achieve. Golub introduces in his article the statistical Signal to Noise Ratio (SNR) that we will also use for feature selection and correlation measurements. The approach followed by Yeang and Ramaswamy is similar to our system since they also decompose the multiclass learning problem into $N$ binary and simpler problems. Their approach differs from ours on how to combine the individual predictions of the binary learners. They use a *One-Versus-All* code where every classifier is trained to distinguish each class from the rest. A disadvantage of their system is that contradictions may exist when several classifiers have positive output.

Instead, in our approach, linear codes are employed to combine the output of the individual binary learners to create a better prediction. Not every linear block code is suited for machine learning. We rely on the good properties of low density parity check codes (LDPC codes), as the Gallager code [6], [7]. An important reference for our work is Dietterich et al. [1]. Their publication explains how error correcting output codes can be used in multiclass learning as opposed to standard multiclass learning algorithms like, for instance, ID3. The foundations of how to apply LDPC codes in machine learning can be found in the Ph.D. dissertation of Tapia [12].

An important choice in multiclass learning based on linear coding theory is the underlying binary learning algorithm. As it is normally done in bioinformatics when dealing with gene expression datasets, an algorithm is selected capable of handling a number of attributes that is higher than the number of samples. This algorithm is called *Support Vector Machines* (SVM). In addition to the powerful SVM learners, the performance of the binary classifiers is improved by *boosting* them. A very good introduction to *boosting* and the employed *boosting* algorithm, AdaBoost.M1, can be found in the publication by Freund et al. [2].

The present article is structured in seven sections. After the brief introduction, the feature selection process dealing with the dimensionality problem of the data is explained. Then, an overview over all preprocessing steps is given. Later, the selected machine learning approach is detailed. In the fifth section, the results of the classification on a real dataset are presented. A conclusion and an acknowledgement section finish the article.

## 2    Feature Selection

The complexity of the problem is addressed partially by a feature selection process. The objective is to find out which genes are most correlated with the class distinction and use only these genes in the classification. Attribute selection is performed according to a parameter called Signal to Noise Ratio (SNR) [4] for its similarity to the SNR parameter used in communication theory.

$$SNR\left(\overrightarrow{gen}, \overrightarrow{class}\right) = \frac{\mu_{class+}\left(\overrightarrow{gen}\right) - \mu_{class-}\left(\overrightarrow{gen}\right)}{\sigma_{class+}\left(\overrightarrow{gen}\right) + \sigma_{class-}\left(\overrightarrow{gen}\right)} . \tag{1}$$

The main advantages of the SNR statistic are that it respects the correlation structure of the data and it does not assume any hypothesis about the statistical distribution of the samples, which would have to be verified [9].

Another added value is that it can be computed empirically if the selected attributes are meaningful in a statistical sense by a hypothesis contrast. The goal of the hypothesis contrast is to state, with a given significance level, if a hypothesis is valid or not. In our case the null hypothesis is that the SNR ratio does not select the most correlated genes with the class distinction. This hypothesis must be rejected for the machine learning process to remain valid.

A hypothesis contrast can have two possible outcomes: the null hypothesis can be either true or false. The significance level $\alpha$ is the probability to reject a correct null hypothesis $H_0$. We call the error of rejecting a correct null hypothesis Type I error. Another possible scenario is to accept a false $H_0$. This type of error is called Type II and its probability is given by the power of the contrast, $\beta$. If we reduce the significance level $\alpha$ of a hypothesis contrast, we reduce the probability to reject a valid $H_0$ hypothesis. A *p-value* is the lowest significance level at which it is possible to reject the null hypothesis. The *p-value* also gives the lowest $\alpha$ for which the observed statistic is meaningful. It is possible to use the *p-value* in a decision rule for accepting or rejecting the null hypothesis.

$$p - value \leq \alpha \iff \text{reject } H_0; \ p - value > \alpha \iff \text{accept } H_0 . \qquad (2)$$

In our case it is possible to calculate *p-values* experimentally[1] and compare them with the significance level of the null hypothesis in order to contrast its validity. The *p-value* of a gene is the probability that the SNR hypothesis contrast of a random permutation of the class labels is greater than or equal to the SNR observed. We calculate in some experiments the number of genes that exceed the real SNR ratios, when the class labels are permutated. Formally:

$$p - value \left( \overrightarrow{gen_i} \right) = \sum_{B=1}^{N_{PERM}} \frac{\# \{ j \in \{1, 2, \ldots, N_{genes}\} : |C_j| \geq |A_i| \}}{N_{genes} \cdot N_{perm}} \qquad (3)$$

$$A_i = SNR_{\text{real}} \left( \overrightarrow{gen_i}, \overrightarrow{class} \right) ; \quad C_j = SNR_{\text{rand}}^{B} \left( \overrightarrow{gen_j}, \overrightarrow{class^*} \right) .$$

where $C_j$ is the $SNR$ statistic in random permutation $B$ of the class labels, $\overrightarrow{class^*}$. If the *p-value* calculated in the permutation test is lower than the significance level $\alpha$, the null hypothesis is rejected.

In figure 1 we can see that, even in the worst case, the null hypothesis can be rejected since there are enough genes with a small *p-value*. We can conclude that the SNR ratio is a good correlation measure with the class distinction.

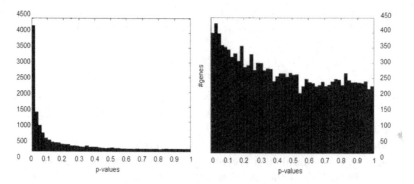

**Fig. 1.** Histograms representing the number of genes as function of the *p-values*. The figure shows on the left the best case (bladder cancer) and on the right the worst case (ovary cancer) scenario for the hypothesis contrast. In both cases the null hypothesis can be rejected since there are enough genes with small *p-values*

[1] For further explanation please refer to [11]

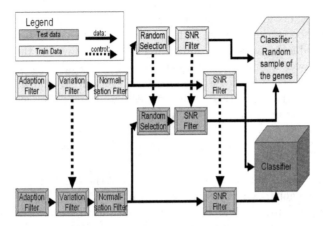

**Fig. 2.** Preprocessing and filtering stages. Two paths of data processing are shown: one affecting the training data and the other referring to the test set. Both databases must be compatible after modifying the attribute set in order to achieve meaningful results

## 3   Preprocessing

In figure 2 all stages of preprocessing and filtering are summarized. On top, the flow for the training set is represented and below the steps relative to the test set are shown. The process starts with an adaptation filter that transforms the data from its original format into an understandable format for the employed machine learning library, WEKA [14]. The adaptation of the data is followed by a variation filter that eliminates those genes with not enough dynamic variation among samples, i.e. the genes without marked difference in expression across different classes. Genes selected in the training dataset must also be selected in the test data by the variation filter for both datasets to be compatible. After the variation filter, the data is normalized to mean zero and standard deviation one. The data is normalized in order to avoid that genes with higher absolute expression values mask other genes with smaller values. At this point in the process, the tracks for both the test and training dataset split. With a random selection filter, attributes are chosen from the total set of attributes to verify if the attributes of the feature selection stage are truly marker genes. Both paths of training and test set contain as last stage of preprocessing a SNR filter that chooses the most correlated sets of attributes out of the total set. Note that again full compatibility between training and test set is maintained.

## 4   Machine Learning Approach

In order to deal with multiclass classification using biclass base algorithms, a novel approach is used related to coding theory. Binary classifiers are trained transforming the original fourteen class-learning task in $N$ binary learning problems. The number of learners is higher than the amount of bits needed to code fourteen classes, i.e. four bits. This way, if one of the classifiers fails, the original class may still be retrieved if the codeword is decoded correctly using the properties of linear block codes. The underlying model employed is the transmission of codewords over a binary memoryless channel with additive white gaussian noise. The received codeword is decoded bearing in mind that it might have a small amount of errors. In this setting the errors are modeled according to the training error performance of the binary classifiers.

$$\overrightarrow{r} = \left(\overrightarrow{t} + \overrightarrow{e}\right) \bmod 2 = \left(G_{nxk}^T \cdot \overrightarrow{s_k} + \overrightarrow{e}\right) \bmod 2 . \tag{4}$$

In the above equation $\overrightarrow{s_k}$ represents a vector of the $k$ possible source symbols and $G_{nxk}^T$ is the generator matrix of the linear block code $C(n, k)$. The received codeword is the sum modulo two of the transmitted codeword and an error vector. At the moment of reception, both the transmitted and the error vectors are unknown. One of them has to be guessed based on the available data, i.e. the received vector and the properties of the linear block code. An optimum decoder estimates the transmitted vector with the maximum a posteriori probability of having transmitted this vector given the received vector and $G_{nxk}^T$. Formally:

$$\widehat{t} = \arg \max_t p\left(\overrightarrow{t} \mid \overrightarrow{r}, G_{nxk}^T\right) . \tag{5}$$

The main disadvantage of optimum decoding is its computational complexity that makes it impractical. Normally, the problem of optimum decoding is NP-complete [7].

Closely related to the generator matrix $G_{nxk}^T$ is the parity check matrix $H_{(n-k)xn}$ that has the property of being orthogonal to the generator matrix. If this property is applied to the equation of reception in a memoryless channel, the syndrome relation is obtained:

$$syndrome : z = H_{(n-k)xn} \cdot \overrightarrow{e} \bmod 2 . \tag{6}$$

The syndrome of a vector can be used in the decoding process as it is done in the sum-prod algorithm that is used in the presented research.

## 4.1   Gallager Codes

A special kind of linear code that can be decoded iteratively is used as composing scheme. This code was developed by MacKay and Neal [7], [6]. Gallager codes have a parity check matrix with a very low density of '1's. Let's denote by $m = n - k$ the number of parity bits (rows of $H_{(n-k)xn}$) and by $t$ the number of '1's of a column. The parity check matrix of the code is constructed by choosing randomly its bits. The number of '1's per column is constrained to $t$ and the number of '1's per row is as uniform as possible.

Based on the syndrome relation, the following sets are defined:

$$L(m) \equiv \{l : H_{ml} = 1\}; \; M(l) \equiv \{m : H_{ml} = 1\} . \tag{7}$$

$L(i)$ represents the set of bits that participate in the syndrome equation $z_i$ . $M(l)$ is the set of indexes of the parity check equations, which involve the bit $l$ of the vector whose syndrome is calculated.

The sum-prod algorithm is a two-step process consisting of a horizontal and a vertical step. During these stages two parameters related to $H_{(n-k)xn}$, $q_{ml}^x$ and $r_{ml}^x$, are updated iteratively until the syndrome condition (equation 6) is satisfied. The value $q_{ml}^x$, with $x = 1$ or $x = 0$, represents the probability that bit $l$ of the vector whose syndrome is calculated is equal to $x$, given all the information obtained from all parity check equations except equation $m$. The value $r_{ml}^x$, with $x = 1$ or $x = 0$, gives the probability that syndrome equation $m$ is satisfied if the bit $l$ has the value $x$ and the rest of the bits have a separable probability distribution given by:

$$\{q_{ml'} : l' \in L(m) \setminus l\} \ . \tag{8}$$

During the initialization stage, initial values are assigned to the variables $q_{ml}^x$ with $H_{mxn} = 1$. These initial values are calculated as the a priori probabilities of the error vector. The horizontal step is devoted to calculate the $r_{ml}^x$ variables for all bits of $L(m)$ iterating through the rows of the parity check matrix. During the vertical step the $q_{ml}^x$ probabilities are updated using the $r_{ml}^x$ values obtained in the horizontal step. For the details of how these values are calculated, we refer to the publications of MacKay and Neal [6], [7]. With $q_{ml}^x$ it is possible to calculate the values of the a posteriori probabilities that are used to estimate the bits of the error vector. If the estimated error vector satisfies the syndrome condition (equation 6), the process is stopped. Otherwise, the horizontal and vertical steps are repeated, updating variables with the values of the previous iteration. The decoding is accomplished by setting a one on position $l$ of the error vector if the a posteriori probability exceeds 0.5. As part of the decoding, the syndrome condition is always verified to see if all the constrains are met. An error occurs if a maximum number of iterations is exceeded. Nevertheless, the final values of an erroneous decoding process can serve as initial values for the next run of the sum-prod algorithm. Undetected errors can appear if the estimated error vector satisfies the syndrome equation (equation 6) but does not correspond to the actual transmission error pattern.

## 4.2  Relation Between Gallager Codes and Multiclass Learning

The learning system consists of the elements depicted in figure 3: channel coder, binary symmetric memoryless transmission channel and channel decoder. A binary source that produces binary symbols represents the training set. The supervisor is a novel element in the transmission model. This supervisor determines if the output of the system is correct. In the case errors are present, the system is updated accordingly. The supervisor calculates error probabilities of the binary classifiers, which later determine the behavior of the discrete memoryless channel of the model. The supervisor does not change the channel coder. Once obtained, the linear code, parity check and generator matrices remain fixed.

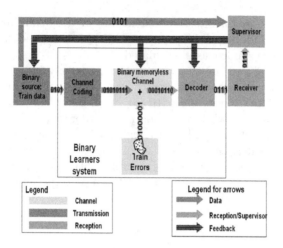

**Fig. 3.** Model for the multiclass learning problem. The original multiclass setting is transformed into $N$ binary problems. The output of the $N$ binary classifiers is later combined with a linear block code

The class of a new sample is obtained evaluating the output of all binary base classifiers. After all base learners have provided their output, a decoding process is started where the error probabilities are estimated according to the training errors of the binary classifiers. It is important to note that the system takes the performance of all base classifiers into account when doing a prediction of a new sample's class. This is achieved because the individual training error probabilities are used during the decoding process. Globally, the system adapts to the learning task by training binary learners and applying later a decoding approach.

Low Density Parity Check codes (LDPC codes), as for example the Gallager Codes, have an interesting property that makes them attractive for machine learning problems. There is a threshold for the crossover probability, $p_0^*$, that defines the maximum value at which the number of erroneous messages tends to zero as the number of iterations tends to infinity. If the number of '1's per column of $H_{(n-k) x n}$ is greater than three, the components of the error probability vector decrease exponentially. From this it is possible to infer that LDPC codes behave better if they are very long. The above condition can be satisfied more easily as $p_0^*$ decreases with the channel rate. Therefore, if the performance of a code is not satisfactory, it is only necessary to reduce the channel rate to improve the results [12].

### 4.3   Support Vector Machines

So as to keep the training error low, strong binary classifiers are used: *boosted Support Vector Machines* (SVM). SVM is a novel machine learning algorithm widely employed for genomic data, where there are much more attributes than instances. The main advantage of SVM is that they perform a classification in high dimensionality feature spaces and do the computations in the original feature space by kernel functions. The main goal of the algorithm is to find the maximum margin hyperplane that separates the data, i.e. the best conceivable separation of samples so as to obtain the smallest possible error [8]. In the presented research, the SVM implementation of the WEKA machine learning library was used with polynomial kernel functions.

### 4.4   Boosting: *AdaBoost*

SVM is a powerful classification scheme but another algorithm called *boosting* is added to further improve the results. *Boosting* is a technique that minimizes the training error by performing several iterations on the training data. Normally, it is used in combination with weak learning algorithms in order to improve their performance. A *boosted* classifier is trained on different distributions of the initial training set. Every sample is assigned a probability related to the error probability when classifying it. On every iteration, a stronger effort is made on wrongly classified samples because the probability distribution of the samples is modified according to classification results. At the end all hypothesis are combined having a higher weight those hypothesis with smaller error. It is important to notice, however, that a small training error does not necessary imply small test errors. This is only true if the training and test databases have similar valued attributes for the same classes [2].

The *boosting* implementation employed in the presented research is called *AdaBoost.M1* [2] and forms part of the WEKA machine learning libraries [14]. In *AdaBoost.M1* a base classifier is trained for a fixed number of iterations on the training set. This base classifier returns in the $i$-th iteration a hypothesis that classifies the data minimizing the training error. The training error is calculated according to the probability distribution $D_i$ that describes the difficulty to classify each sample. $D_i$ is updated after every iteration proportionally to the training error. At first, $D_0$ is uniform for all the samples of the training set. $D_{i+1}$ is calculated from $D_i$ and the weak hypothesis $h_i$ multiplying the weight of the sample by a number related to the training error. If a sample was classified correctly in the previous iteration, its weight is left unchanged for the next step. The dependency of the number that multiplies the weight distribution is such that erroneously classified samples get a higher probability in the next iteration. At the end of the process, the final hypothesis is obtained as a weighted sum of the weak hypothesis, $h_i$, being the weight related to the training error.

## 5   Results

In this section the results of the machine learning approach are presented. On two graphs the precision, i.e. the relation among correctly classified positives and the number of instances classified as positives, is displayed. Figure 4(A,B) clearly shows that with increasing decoding iterations of the LDPC recursive Gallager code, the error of the classifier is lower in average. The minimum error obtained

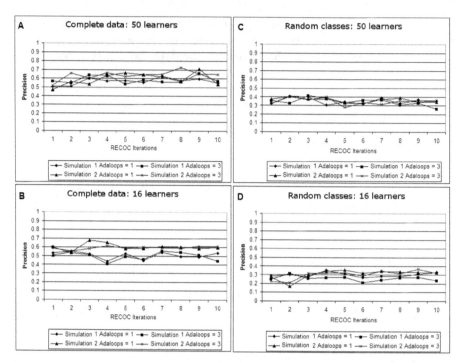

**Fig. 4.** Results of the classification process. The precision is plotted in function of the number of decoding and *boosting* iterations. Higher number of learners improve the performance of the classifiers (A,C). A best case precision of 72% is achieved (A). The results for random permutation data (C,D) are significantly worse than for the complete dataset (B,D)

in 10 simulations is of 40%, which is a little bit higher than the error achieved by Ramaswany et al. on the same database. It is important to mention however, that reducing the channel rate from 0.25 to 0.08, i.e. increasing the number of learners from 16 to 50, significantly improves the quality of the results. With a higher number of learners, the error correcting capability of a linear code increases and thus the overall error is reduced. Computational constraints only permitted to use at maximum 50 learners. Exceeding this number of classifiers significantly increased the computation time of the simulations to a non-tolerable limit given the available computers resources. With more powerful processors or even clusters we expect to further reduce test errors. Nevertheless, the results are comparable to other publications as for example [9] and [15] that worked on the same dataset and achieved a best-case precision of 78%.

In order to verify that the performance of the system is satisfactory, the obtained results are compared with the classification of randomly selected attribute sets. This way it can be proved that the genes selected by high SNR ratio are truly marker genes for the analyzed pathologies. From figure 4(C,D) it can be concluded that the classification provided by the most correlated genes in the sense of the SNR ratio is meaningful. In the case of the randomly selected genes, the best-case precision does not exceed 40%.

## 6    Conclusions

The addressed problem bears a high complexity due to the following reasons: first, the relation between number of attributes and instances is very low (curse of dimensionality) and second, multiclass machine learning is intrinsically very complicated[2]. The small size of the available sample databases increases the difficulty of training proper classifiers. In particular, only eight instances per category are available. Nevertheless, our results are comparable to the results published by other research groups [9], [15]. This is achieved by transforming the multiclass learning problem via a coding approach into several simpler biclass learning settings. The results of the binary classifiers are combined to a joint prediction with a linear block code that allows a small number of classification errors to be present. The learning problem is thus similar to a transmission through a binary symmetric memoryless channel. The output of each base binary classifier represents a bit of the received codeword and the training errors of the classifiers can be assimilated to the channel's error probability. Decoding is done with a recursive Gallager code with excellent performance. Due to the iterative decoding, error rates can be limited with sufficiently high number of learners.

It is also necessary to face the problem of dimensionality in the present research, which means that the number of attributes exceeds largely the number of available classified samples. This problem is common to gene expression datasets since modern high throughput techniques allow analyzing thousands of genes simultaneously. Still, it is a costly process to obtain many classified samples and, therefore, the size of the databases never exceeds a few hundred instances. This problem has been solved in our investigation using statistical feature selection algorithms based on the SNR ratio. It is verified statistically by a hypothesis contrast that the SNR ratio correctly measures the correlation to a class distinction before training the classifiers with the filtered datasets. The importance of the performed feature selection is not only related to machine learning requirements, it also offers insight into biological processes by identifying possible marker genes for a particular kind of tumor. As the number of attributes is still too high even after SNR filtering, a base learning algorithms is chosen able to deal with a huge number of features: *Support Vector Machines.*

Presently it is being investigated how to improve the computational efficiency in order to reduce the error rate of the process even more. The results obtained up to now encourage to continue researching on molecular diagnosis systems that may improve further the treatment of patients. The combined use of techniques from the areas of information and coding theory along with machine learning algorithms represent a new and encouraging approach to the use of gene expression data for medical diagnosis.

---

[2] In multiclass learning the random guessing probability is $\frac{1}{k}$ for $k$ classes in comparison to the much higher random guessing probability for biclass learning of $\frac{1}{2}$

## Acknowledgments

The authors wish to thank the Spanish Ministry of Culture and Education for funding this research that was carried out at the Universidad Politécnica de Madrid under a Research Collaboration Grant.

## References

1. T. Dieterich and G. Bakiri: *Error-correcting output codes: A general method for improving multiclass inductive learning programs.* Proceedings of the $9^{th}$ National Conference on Artificial Intelligence (AAAI-91), AAAI Press, pp. 572-577, 1991.
2. Y. Freund and R. R. Schapire. *Experiments with a new boosting algorithm.* In Machine Learning: Proceedings of the Thirteenth International Conference on Machine Learning. Morgan Kaufmann, 1996.
3. www-genome.wi.mit.edu/MPR/GCM.html .
4. T.R. Golub et al., *Molecular Classification of Cancer: Class Discovery and Class Prediction by Gene Expression.* Science 1999 286: pp. 531-537, 1999.
5. S. Lin and D. J. Costello, Jr., *Error Control Coding: Fundamentals and Applications.* Englewood Cliffs, NJ: Prentice-Hall, 1983.
6. D. J. C. MacKay, R. M. Neal, *Good Codes based on Very Sparse Matrices*; Cryptography and Coding the IMA Conference; 1995.
7. D. J. C. MacKay, R. M. Neal, *Good Error-Correcting Codes based on Very Sparse Matrices*, IEEE transactions on Information Theory, 1999.
8. S. Mukherjee, *Classifying Microarray Data Using Support Vector Machines.* In: A Practical Approach to Microarray Data Analysis, D. P. Berrar, W. Dubitzky and M. Granzow (Eds.), Kluwer Academic Publishers, pp. 166-185, 2003.
9. S. Ramaswamy et al., *Multi-Class Cancer Diagnosis Using Tumor Gene Expression Signatures*, PNAS 98: pp. 15149-15154, 2001.
10. B. Schölkpof, A. Smola, *Learning with Kernels Support Vector Machines, Regularization, Optimization and Beyond*, MIT Press, 2001.
11. J. Storey and R. Tibshirani, *Statistical Significance for Genome-Wide Experiments* http://www-stat.stanford.edu/~tibs/ftp/fdringenomics.pdf, 2003.
12. E. Tapia, *New learning models based on recursive error correcting codes*, Doctoral Thesis, ETSI de Telecomunicación Universidad Politécnica de Madrid, Spain, 2001.
13. E. Tapia, J. C. González, A. Hüntemann, J. García-Villalba *Beyond Boosting: Recursive ECOC Learning Machines* In Multiple Classifier Systems, MCS 2004, Lecture Notes in Computer Science Vol. 3077, pp. 62-71. Springer 2004.
14. I. H. Witten, E. Frank, *Data Mining: Practical Machine Learning Tools and Techniques with Java Implementations*, Morgan Kaufmann, 1999.
15. C. H. Yeang et al. *Molecular classification of multiple tumor types*; Bioinformatics 17 (Suppl. 1): pp. 316-322, 2001.

# Boosted Decision Trees
# for Diagnosis Type of Hypertension

Michal Wozniak

Chair of Systems and Computer Networks, Wroclaw University of Technology
Wybrzeze Wyspianskiego 27, 50-370 Wroclaw, Poland
michal.wozniak@pwr.wroc.pl

**Abstract.** The inductive learning algorithms are the very attractive methods generating hierarchical classifiers. They generate hypothesis of the target concept on the base on the set of labeled examples. This paper presents some of the decision tree induction methods, boosting concept and their usefulness for diagnosis of the type of hypertension (essential hypertension and five type of secondary one: fibroplastic renal artery stenosis, atheromatous renal artery stenosis, Conn's syndrome, renal cystic disease and pheochromocystoma). The decision on the type of hypertension is made only on base on blood pressure, general information and basis biochemical data.

## 1 Introduction

Machine learning [1] is the attractive approach for building decision support systems. For this type of software, the key-role plays the quality of the knowledge base. In many cases we can find following problem:

- the experts can not formulate the rules for decision problem, because they might not have the knowledge needed to develop effective algorithms (e.g. human face recognition from images),
- we want to discover the rules in the large databases (data mining) e.g. to analyze outcomes of medical treatments from patient databases; this situation is typical for designing telemedical decision support system, which knowledge base is generated on the base on the large number of hospital databases,
- program has to dynamically adapt to changing conditions.

Those situations are typical for the medical knowledge acquisition also. For many cases the physician can not formulate the rules, which are used to make decision or set of rules given by expert is incomplete.

In the paper we present two type of decision tree induction algorithm and we discus if boosting methods can improve the quality of decision tree for the real medical problem.

The content of the work is as follows. Section 2 introduces idea of the inductive decision tree algorithms. In Section 3 we describe mathematical model of the hypertension's type. Next section presents results of the experimental investigations of the algorithms. Section 5 concludes the paper.

J.L. Oliveira et al. (Eds.): ISBMDA 2005, LNBI 3745, pp. 223–230, 2005.

## 2  Algorithms

### 2.1  Decision Tree Induction

The most of algorithm as C4.5 given by R. J. Quinlan [3] or ADTree (Alternative Decision Tree)[13] are based on the idea of "Top Down Induction of Decision Tree". Therefore let us present the main idea of it

```
Create a Root node for tree
IF all examples are positive
      THEN return the single node tree Root with label
yes and return.
IF all examples are negative
      THEN return the single node tree Root with label no
and return.
IF set of attributes is empty
      THEN return the single node tree Root with label =
      most common value of label in the set of examples
      and return
Choose "the best" attribute A from the set of attributes.
   FOR EACH possible value vi of attribute
      1. Add new tree branch bellow Root, corresponding
         to the test A=vi.
      2. Let Evi be the subset of set of examples that
         has value vi for A.
      3. IF Evi is empty
            THEN bellow this new branches add a leaf node
            with label = most common value of label in
            the set of examples
            ELSE below this new branch add new subtree
            and do this function recursive.
   END
   RETURN Root
```

The central choice in the TDIDT algorithm is selecting "the best" attribute (which attribute to test at each node in the tree). Family of algorithm based on ID3 method [4] (e.g. C4.5) uses the information gain (or its' modification gain ratio) that measures how well the given attribute separates the training examples according to the target classification. This measure based on the Shanon's entropy of learning set $S$:

$$Entropy(S) = \sum_{i=1}^{M} - p_i \log_2 p_i \qquad (1)$$

where $p_i$ is the proportin of $S$ belonging to klas $i$ $(i \in M, M = \{1, 2, ..., M\})$.

The information gain of an attribute $A$ relative to the collection of examples $S$, is defined as

$$Gain(S, A) = Entropy(S) - \sum_{c \in values(A)} \frac{|S_v|}{|S|} Entropy(S_v), \qquad (2)$$

where $values(A)$ is the set of all possible values for attribute A and $S_v$ is the subset of S for which $A = v$. The future implementations of decision tree induction algorithm use measure based on defined in (2) information gain (e.g. information ratio [6]).

## 2.2  Boosting

Boosting is general method of producing an accurate classifier on base of weak and unstable one[9-10]. The boosting often does not suffer from overfitting. AdaBoost is the most popular algorithm introduced in 1995 by Freund and Shapire [1]. Pseudocode of AdaBoost.M1 (one of the version of AdaBoost algorithm) is presented below[2]:

```
Input:
1. sequence of m examples {(x₁, y₁), (x₂, y₂), ..., (xₘ, yₘ)} with la-
bels yᵢ ∈ Y = {1, ..., k}
2. weak learning algorithm WeakLearn
3. integer T specifying number of iterations
Initialize D₁(i) = 1/m for all I
do for t = 1, 2, ..., T :
1. Call WeakLearn, providing it with the distribution Dₜ
2. Get back a hypothesis hₜ : X → Y .
3. Calculate the error of hₜ :
```

$$\varepsilon_t = \sum_{i: h_t(x_i) \neq y_i} D_t(i).$$

```
4. If εₜ > 1/2, then set T = t-1 and abort loop.
5. Set βₜ = εₜ/(1-εₜ).
6. Update distribution Dₜ :
```

$$D_{t+1} = \frac{D_t}{Z_t} \times \begin{cases} \beta_t & \text{if } h_t(x_i) = y_i \\ 1 & \text{otherwise} \end{cases}$$

where $Z_t$ is a normalization constant (chosen so that $D_{t+1}$ will be a distribution).

Output the final hypothesis: $h_{fin}(x) = \arg\max_{y \in Y} \sum_{t: h_t(x) = y} \log \frac{1}{\beta_t}$ .

# 3  Model of Type of Hypertension (HT) Diagnosis

During the hypertension's therapy is very important to recognize state of patient and the correct treatment. The physician is responsible for deciding if the hypertension is of an essential or a secondary type (so called the first level diagnosis). The senior physicians from the Broussais Hospital of Hypertension Clinic and Wroclaw Medical Academy suggest 30% as an acceptable error rate for the first level diagnosis.

The presented project was developed together with Service d'Informatique Médicale from the University Paris VI. All data was getting from the medical database ARTEMIS, which contains the data of the patients with hypertension, whose have been treated in Hôpital Broussais in Paris.

The mathematical model was simplified. Hover the experts from the Broussais Hôspital, Wroclaw Medical Academy, regarded that stated problem of diagnosis as very useful.

It leads to the following classification of type of hypertension:

1. essential hypertension (abbreviation: essential),
2. fibroplastic renal artery stenosis (abbreviation: fibro),
3. atheromatous renal artery stenosis (abbreviation: athero),
4. Conn's syndrome (abbreviation: conn),
5. renal cystic disease (abbreviation: poly),
6. pheochromocystoma (abbreviation: pheo).

Although the set of symptoms necessary to correctly assess the existing HT is pretty wide, in practice for the diagnosis, results of 18 examinations (which came from general information about patient, blood pressure measurements and basis biochemical data) are used, whose are presented in table 1.

**Table 1.** Clinical features considered

| No | Feature |
|---|---|
| 1 | Sex |
| 2 | body weight |
| 3 | High |
| 4 | Cigarette smoker |
| 5 | limb ache |
| 6 | Alcohol |
| 7 | Systolic blood pressure |
| 8 | Diastolic blood pressure |
| 9 | Maximal systolic blood pressure |
| 10 | Effusion |
| 11 | Artery stenosis |
| 12 | Heart failure |
| 13 | Palpitation |
| 14 | carotid or lumbar murmur |
| 15 | Serum creatinine |
| 16 | Serum potassium |
| 17 | Serum sodium |
| 18 | Uric acid |

## 4  Experimental Investigation

All learning examples were getting from medical database *ARTEMIS*, which contains the data of 1425 patients with hypertension (912 with essential hypertension and the rest of them with secondary ones), whose have been treated in Hôpital Broussais.

We used  WEKA systems [11] and our own software for experiments e.g. [15]. Quality of correct classification was estimated using 10 folds cross-validation tests.

### 4.1  Experiment A

The main goal of experiment was to find quality of recognition the C4.5 algorithm and its' boosted form. The obtained decision tree is shown in Fig.1.

The frequency of correct classification of this tree is 67,79% and the confusion matrix looks as follow

**Table 2.** Confusion matrix for decision tree

| Real diagnosis | | | | | | Recognized class |
|---|---|---|---|---|---|---|
| Athero | conn | essent | fibro | Pheo | Poly | |
| 4 | 3 | 54 | 16 | 0 | 0 | athero |
| 0 | 44 | 92 | 11 | 0 | 0 | Conn |
| 2 | 25 | 878 | 7 | 0 | 0 | Essent |
| 4 | 5 | 64 | 40 | 0 | 0 | Fibro |
| 2 | 3 | 80 | 2 | 0 | 0 | Pheo |
| 0 | 2 | 81 | 6 | 0 | 0 | Poly |

We rejected the classifier because his quality did not satisfy expert. But we have to note that advantage of this tree is that the essential hypertension was recognized pretty good (96,26%).

We tried to improve quality of obtained classifier using boosting concept. Unfortunately new classifier had worse quality than original one (59,30%). The confusion matrix of the boosted C4.5 is presented in Tab.2

**Table 3.** Confusion matrix for boosted decision tree

| Real diagnosis | | | | | | Recognized class |
|---|---|---|---|---|---|---|
| Athero | conn | essent | fibro | pheo | Poly | |
| 8 | 4 | 52 | 7 | 2 | 4 | athero |
| 2 | 22 | 106 | 10 | 4 | 3 | Conn |
| 19 | 47 | 790 | 34 | 12 | 10 | Essent |
| 3 | 6 | 86 | 9 | 8 | 1 | Fibro |
| 1 | 3 | 73 | 5 | 4 | 1 | Pheo |
| 1 | 4 | 70 | 0 | 2 | 12 | Poly |

## 4.2 Experiment B

Physician-experts did not accept classifiers obtained in Experiment A. After the discussion we simplified the problem once more. We were trying to construct classifiers which would point at essential type of hypertension or secondary one. We used two methods to obtain the classifiers:

1. Alternative Decision tree (ADTree),
2. C4.5 algorithm.

For each of classifier we check its' boosted form also. The results of tests is shown in Fig.2.

As we see the frequency of correct classification of ADTree algorithm is 79,16%. Unfortunately the quality of recognition the secondary hypertension is only 58,48%. We tried improve the quality of classifier by AdaBoost.M1 procedure and we obtained new classifier based on ADTree concept (we use 10 iterations), which frequency of correct classification grew to 83,30% and 72,90% of correct classified secondary type of hypertension. This results satisfied our experts.

Additionally we check the quality of C4.5 for the same dichotomy problem. We obtained the decision tree similar to tree in Fig.1 which quality is 74,74% and frequency of correct recognized secondary type of hypertension is 51,85%. The boosting procedure did not improve the average quality of C4.5 (72,42%) but strongly improved the recognition secondary type of hypertension (60,81%).

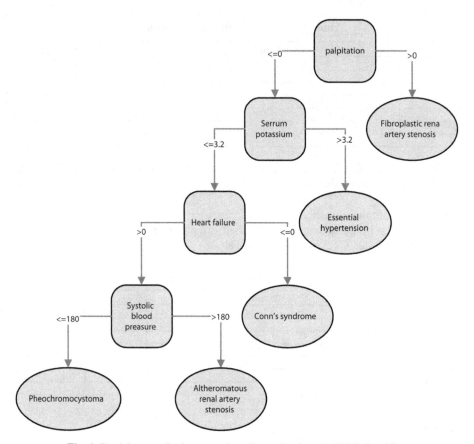

**Fig. 1.** Decision tree for hypertension diagnosis given by C4.5 algorithm

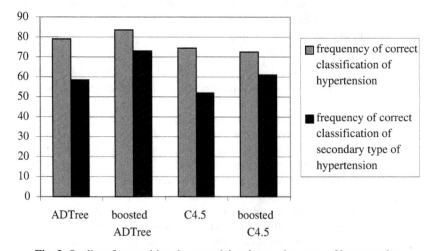

**Fig. 2.** Quality of recognition the essential and secondary type of hypertension

## 5   Discussion and Conclusion

The methods of inductive learning were presented. The classifiers generated by those algorithms were applied to the medical decision problem (recognition of the type of hypertension). The general conclusion is that *boosting* does not improve each classifier for each decision task. For the real decision problem we have to compare many classifiers. The similar observations were described by Quinlan in [7] where he did not observe quality improvements of boosted C4.5 for some of databases.

Most of obtained classifier (especially based on C4.5 method) did not satisfy experts. The best classifier (obtained for simplified decision problem) satisfied our expert. Now we want to construct classifier ensemble on the based on stacked classifier concept [5, 14] which idea is depicted in Fig.3. Obtained boosted ADTree classifier can be use for the first stage of recognition. Now we are working on the classifier of HT for patient with diagnosed secondary type of hypertension.

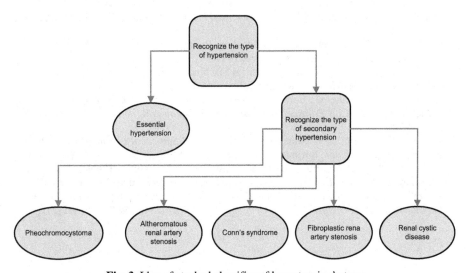

**Fig. 3.** Idea of stacked classifier of hypertension's type

The similar problem of computer-aided diagnosis of hypertension's type was described in [12] but authors used another mathematical model and implement Bayes decision rule. They obtained slightly better classifier than our, its' frequency of correct classification of secondary type of hypertension is about 85% (our 83,30%). Advantage of our proposition is simplified and cheaper model than presented in [12] (we use 18 features, authors of [12] 28 ones).

Advantages of the proposed methods make it attractive for a wide range of applications in medicine, which might significantly improve the quality of the care that the clinician can give to his patient.

This work is supported be The Polish State Committee for Scientific Research under the grant which is realizing in years 2005-2007.

# References

1. Freund Y., Schapire R.E., A decision-theoretic generalization of on-line learning and application to boosting, *Journal of Computer and System Science,* 55(1),1997, pp. 119-139.
2. Freund Y., Schapire R.E., Experiments with a New Boosting Algorithm, *Proceedings of the International Conference on Machine Learning*, 1996, pp. 148-156.
3. Jain A.K., Duin P.W., Mao J., Statistical Pattern Recognition: A Review, *IEEE Transaction on Pattern Analysis and Machine Intelligence*, vol 22., No. 1, January 2000, pp. 4-37.
4. Mitchell T., *Machine Learning*, McGraw Hill, 1997.
5. Opitz D., Maclin R., Popular Ensemble Methods: An Empirical Study, *Journal of Artificial Intelligence Research*, 11 (1999) 169-198
6. Quinlan J.R., *C4.5: Programs for Machine Learning*, Morgan Kaufmann, 1993.
7. Quinlan J.R., Bagging, Boosting, and C4.5, *Proceedings of the 13th National Conference on Artificial Intelligence and Eighth Innovative Applications of Artificial Intelligence Conference, AAAI 96, IAAI 96*, Volume 1, Portland, Oregon, August 4-8, 1996, pp. 725-230.
8. Schapire R. E., The boosting approach to machine learning: An overview. *Proc. Of MSRI Workshop on Nonlinear Estimation and Classification*, Berkeley, CA, 2001.
9. Shapire R.E., The Strength of Weak Learnability, *Machine Learning*, No. 5, 1990, pp. 197-227.
10. Schapire R.E., A Brief Introduction to Boosting, *Proceedings of the Sixteenth International Joint Conference on Artificial Intelligence*, 1999.
11. Witten I.H., Frank E., Data Mining: Practical Machine Learning Tools and Techniques with Java Implementations, Morgan Kaufmann Pub., 2000.
12. Blinowska A., Chatellier G., Bernier J., Lavril M., Bayesian Statistics as Applied to Hypertension Diagnosis, *IEEE Transaction on Biomedical Engineering*, vol. 38, n0. 7, July 1991, pp. 699-706.
13. Hastie T., Tibshirani R., Friedman J., *The Elements of Statistical Learning. Data Mining, Inference, and Prediction*, Springer Series in Statistics, Springer Verlag, New York 2001.
14. Puchala E., A Bayes Algorithm for the Multitask Pattern Recognition Problem – direct and decomposed approach, *Lecture Notes in Computer Science*, vol. 3046, 2004, pp. 39-45.
15. Koszalka L., Skworcow P., Experimentation system for efficient job performing in veterinary medicine area, *Lecture Notes in Computer Science*, vol. 3483, 2005, pp. 692-701.

# Markov Chains Pattern Recognition Approach Applied to the Medical Diagnosis Tasks

Michal Wozniak

Chair of Systems and Computer Networks, Wroclaw University of Technology,
Wybrzeze Wyspianskiego 27, 50-370 Wroclaw, Poland
michal.wozniak@pwr.wroc.pl

**Abstract.** In many medical decision problems there exist dependencies between subsequent diagnosis of the same patient. Among the different concepts and methods of using "contextual" information in pattern recognition, the approach through Bayes compound decision theory is both attractive and efficient from the theoretical and practical point of view. Paper presents the probabilistic approach (based on expert rules and learning set) to the problem of recognition of state of acid-base balance and to the problem of computer-aided anti-hypertension drug therapy. The quality of obtained classifier are compared to the frquencies of correct classification of three neural nets.

## 1 Introduction

Nowadays, the pattern recognition algorithms are widely used in supporting the medical diagnosis. In most of such applications we have to make decision about patient's state only once, but there are problems in which such decisions should be made in a sequential way [2]. For example in the recognition of the equilibration of acid-base state we recognize the sequence of patient's state using the same features vector. In this case we should take into account that after observation and decision about patient's state, the corresponding treatment is applied. Another example, which leads us to the problem of sequential diagnosis, is concerned with the computer-aided anti-hypertension drug therapy. In this situation, after observing the patient's medical data we should decide about the diagnosis and next to decide what kind of anti-hypertension drug therapy should be applied. As usual, in both mentioned above examples the information about patient's state is represented by feature vector and the patient's state is a class, which have to be recognized. Additionally, we should consider the set of possible treatment. In such medical cases, our mathematical task consists on recognizing the sequences of patterns, one by one. Of course, in such cases of sequential medical diagnosis there are exists dependences among the patterns to be recognized. Moreover, it is reasonable to assume, that the current patient's state depends on the previous state and applied treatment.

Among the different concepts and methods using "contextual information" in pattern recognition, an attractive from theoretical point of view and efficient approach is Bayes' compound decision theory [1] in which a classifying decision is made on one pattern at a time, using additional information from the entire past. Furthermore, the Markov dependence among the patterns to be recognized is assumed. There is a great deal of available papers dealing with the recognition problems under assumption of first-order or higher-order Markov dependence, but we should take into account that between the observations the patient is treated. It means that, the recognized process

J.L. Oliveira et al. (Eds.): ISBMDA 2005, LNBI 3745, pp. 231–241, 2005.
© Springer-Verlag Berlin Heidelberg 2005

is controlled as well. Consequently, it is assumed the mathematical model of decision process that the sequence of states of recognized process forms first-order controlled Markov chain [1]. Based on this model and using Bayes' approach in the case of complete probabilistic information the pattern recognition algorithms for controlled Markov chains are derived. Next, the algorithms with learning are proposed in the case when there is a lack of full probabilistic information. In this paper also rule-based algorithm [4, 13] and its medical application is presented. Another possibility, which also was taken into account, uses neural networks [8] approach. In the end, some results of medical application concerning with two above mentioned medical problems are presented. For presented medical examples frequencies of correct classifications of different approaches are compared.

## 2  Problem Statement

Let us consider the classical problem of pattern recognition that is concerned with the assignment of a given pattern to one of $m$ possible classes. Let $x_n$, which takes values in $r$-dimensional Euclidean space, denote the vector of measured features of the $n$-th pattern to be recognized and let $j_n$ denote the label of the class to which the pattern in question belongs. In medical application it means that the same medical tests are repeated several times and each time the patient's state should be recognized as one of possible states from the set $M = \{1, ..., m\}$. In such situation, in which the medical investigations are repeated, it is necessary to take into account that between each observation the patient can be treated. The idea of this process is presented in Fig.1.

Let us denote the set of all possible treatments by $K = \{1, ..., k\}$. Thus

$$\bar{x}_n = \{x_1, x_2, ..., x_n\}, \ \bar{j}_n = \{j_1, j_2, ..., j_n\}, \ \bar{u}_n = \{u_1, u_2, ..., u_n\}, \tag{1}$$

state the sequence of feature vectors, true identities and used treatments, respectively. In this paper it is assumed that $x_n, j_n, u_n$ are observed values of random variables $X_n, J_n, U_n$ for $n = 1, 2, ...$ . It is also assumed that the sequence of classifications forms a controlled Markov chain:

$$P\big(J_n = j_n \big| J_{n-1} = j_{n-1}, ..., J_1 = j_1, U_{n-1} = u_{n-1}, ..., U_1 = u_1\big) =$$
$$= P\big(J_n = j_n \big| J_{n-1} = j_{n-1}, U_{n-1} = u_{n-1}\big) = p_{j_n j_{n-1}}(u_{n-1}) \tag{2}$$

This assumption means, that the current patient's state depends on his previous state and applied medical treatment. Note that the controlled Markov chain is described by the set of initial probabilities:

$$P\big(J_1 = j_1\big) = p_{j_1}, \qquad j_1 \in M, \tag{3}$$

and the set of transition probabilities:

$$p_{j_n j_{n-1}}(u_{n-1}), \qquad j_n, j_{n-1} \in M, \ u_{n-1} \in K, \ n = 2, 3, ... \tag{4}$$

Let $f(x_n | j)$ be the conditional density function of random variable $X_n$ given that $J_n = j, j \in M$, identically for all natural $n$ and independently of used controls. For simplicity assume the conditional independence among variables $X_n$, $n = 1, 2, ...$ , which implies that: ·

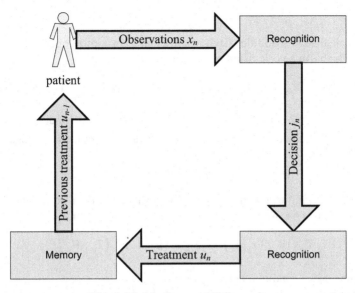

**Fig. 1.** Model of sequential diagnosis

$$\bar{f}_n\left(\bar{x}_n\mid\bar{j}_n\right)=\prod_{\alpha=1}^{n} f\left(x_\alpha\mid j_\alpha\right),\quad n=1,2,\ldots\;. \tag{5}$$

In this work, two cases of control algorithms are considered. First, simple probabilistic control algorithm is considered i.e. the treatments are chosen according to the stationary probabilities of control

$$p_c\left(u_n\mid j_n\right)=P\left(U_n=u_n\mid J_n=j_n\right) \tag{6}$$

where for $n=1,2$, $u_n\in K$ and $j_n\in M$ are observed values of discrete random variables $U_n$ and $J_n$ respectively. Next, deterministic algorithm of control is assumed. It means, that after decision about patient's state we make corresponding, deterministic decision about treatment, which should be applied. In this case the set of classes $M$ is the same as the set of controls $K$.

## 2.1 Case-Based Reasoning

The pattern recognition algorithm for first-order controlled Markov chains in the case of complete statistical information and for the algorithm of control (6)

$$i_n=\Psi_n^*\left(\bar{x}_n,\bar{u}_{n-1}\right),\quad n=1,2,\ldots, \tag{7}$$

is presented in the paper. For the special case of a 0-1 loss function, i.e.

$$L\left(i_n,j_n\right)=0\;\;if\;\;i_n=j_n\;\;or\;\;1\;\;otherwise\;\;for\;n=1,2,\ldots, \tag{8}$$

the rule (7) assigns the $n$-th pattern to the class $i_n$ with the highest posterior probability after observing $\bar{x}_n$ and $\bar{u}_{n-1}$ for all natural n:

if for $s\in M$, $s\neq i_n$,

$$p_{1,n}\left(i_n \mid \bar{x}_n, \bar{u}_{n-1}\right) > p_{1,n}\left(s \mid \bar{x}_n, \bar{u}_{n-1}\right). \tag{9}$$

Instead of calculate probabilities (9) it is sufficient to maximize only the discriminate functions, which can be calculated recursively as follows [2,5]:

$$d_{j_n}\left(\bar{x}_n, \bar{u}_{n-1}\right) = f\left(x_n \mid j_n\right) \cdot \sum_{j_{n-1}=1}^{m} p_{j_n j_{n-1}}\left(u_{n-1}\right) \cdot p_c\left(u_{n-1} \mid j_{n-1}\right) \cdot d_{j_{n-1}}\left(\bar{x}_{n-1}, \bar{u}_{n-2}\right) \tag{10}$$

for all natural $n \geq 2$ and for every $j_n \in M$ with the initial condition obtained by very well known method of minimization of statistical risk at the first stage of classification

$$d_{j_1}\left(x_1\right) = p_{j_1} \cdot f\left(x_1 \mid j_1\right), \quad j_1 \in M. \tag{11}$$

In the case of deterministic control, it is easy to show that the discriminate functions of algorithm (9) can be simpler calculated according to the formulae:

$$d_{j_n}\left(\bar{x}_n, \bar{u}_{n-1}\right) = f\left(x_n \mid j_n\right) \cdot \sum_{j_{n-1}=1}^{m} p_{j_n j_{n-1}}\left(u_{n-1}\right) \cdot d_{j_{n-1}}\left(\bar{x}_{n-1}, \bar{u}_{n-2}\right) \tag{12}$$

with the same initial condition (11).

These algorithms require complete statistical information about recognized process. It means, that all conditional density functions $f\left(x_n \mid j\right), j \in M$, and all probabilities (3), (4), describing the first-order controlled Markov chain must be known. In medical applications, there is a lack of the exact knowledge of probabilities (3), (4) and conditional density functions in all classes, whereas only partial information is available. In considered case, we have the set of N learning sequences i.e. the set of correctly classified sequences of samples:

$$\bar{S}_N = \left\{S_1, S_2, ..., S_N\right\} \tag{13}$$

where the sequence $S_i$ refers to one object (e.g. in medical decision problems it refers to the observation of the one patient). It contains a sequence of states that occurred at the successive moments $\left(j_{i1}, j_{i2}, ..., j_{in_i}\right)$, a corresponding sequences of the feature vector values $\left(x_{i1}, x_{i2}, ..., x_{in_i}\right)$ and of the applied control $\left(u_{i1}, u_{i2}, ..., u_{in_i}\right)$.

In this case one obvious and conceptually simple method is to estimate probabilities (3), (4) and densities $f\left(x_n \mid j\right), j \in M$, from the set of learning sequences (13) and then to use these estimators to calculate discriminate functions according to (10), (11) as though they were correct [2, 3]. Then the pattern recognition algorithm for first-order controlled Markov chain with learning:

$$i_n = \Psi\left(\bar{S}_N, \bar{x}_n, \bar{u}_{n-1}\right), \quad n = 1, 2, ..., \quad i_n \in M \tag{14}$$

classifies the n-th recognized pattern $x_n$ to the class $i_n$, for which the discriminate function (10), (11) is maximal. These functions can be calculated from the same recursive formulas, in which the true probabilities (3), (4), (6) are replaced by their frequency estimators obtained from (13), and the values of density functions in classes in every needed points $x_n$ are evaluated using nonparametric estimators like Parzen estimator for example [1].

## 2.2  Expert Rules

We have two kinds of rules: initial and transition ones:

$$R = R_{in} \cup R_{tr}. \tag{15}$$

Where each set consists of the subset which point in the same state index and each of them consists of the single rules:

$$R_i = \left\{ r_i^{(1)}, r_i^{(2)}, \ldots, r_i^{(N_i)} \right\} \qquad R_{ij} = \left\{ r_{ji}^{(1)}, r_{ij}^{(2)}, \ldots, r_{ij}^{(N_{ij})} \right\}. \tag{16}$$

The analysis of different practical examples leads to the following forms of rules:

### Initial Rule $r_i^{(k)}$

**if** at the first moment, $x_1 \in D_i^{(k)}$ **then** at this moment the state of object is $J_1 = i$ **with** *posterior* probability $\beta_i^{(k)}$ greater than $\underline{\beta}_i^{(k)}$ and less than $\overline{\beta}_i^{(k)}$

where

$$\beta_j^{(k)} = \int_{D_i^{(k)}} p(i|x)dx. \tag{17}$$

### Transition Rule $r_{tri}^{(k)}$

**if** at the given moment $n$, $x_n \in D_{ij}^{(k)}$ **and** the previous state $J_{n-1} = j$ **and** the last applied therapy was $u \in U$ **then** at this moment the state of object is $J_n = i$ with *posterior* probability $\beta_{ij}^{(k)}$ greater than $\underline{\beta}_{ij}^{(k)}$ and less than $\overline{\beta}_{ij}^{(k)}$

where

$$\beta_{ij}^{(k)} = P(J_n = i | J_{n-1} = j, x_n \in D_{ij}^{(k)}, u_{n-1} = u) = p_n(i/j, D_{ij}^{(k)}, u). \tag{18}$$

We can thus get rule which inform us about *posterior* probability for the given decision area. Our objective is to construct the rule-based recognition algorithm with learning $\psi_n^{(SR)}(\overline{x}_n, \overline{u}_{n-1})$, which uses the sets $S$ and $R$ to recognize the pattern at the $n$-step of classification on the basis of all available information.

For that form of knowledge we can formulate the decision algorithm $\Psi_n^{(R)}(x)$ which pointed at the class $i$ if the *posterior* probability estimator obtained from the rule set has the biggest value. The knowledge about probabilities given by expert estimates the average *posterior* probability for the whole decision area. Then for decision making it is very important to calculate the exact value of the *posterior* probability for given observation. For the logical knowledge representation the rule with the small decision area can be over fitting the training data [7] (especially if the training set is small). For our proposition we respect this danger for the rule set obtained from learning data. For the estimation of the *posterior* probability from rule we assume the constant value of for the rule decision area. Therefore lets propose the relation "more specific" between the probabilistic rules pointed at the same class.

**Definition.** Rule $r_i^{(k)}$ is "more specific" than rule $r_i^{(l)}$ if

$$\left(\overline{\beta}_i^{(k)} - \underline{\beta}_i^{(k)}\right)\left(\int_{D_i^{(k)}} dx \Big/ \int_X dx\right) < \left(\overline{\beta}_i^{(l)} - \underline{\beta}_i^{(l)}\right)\left(\int_{D_i^{(l)}} dx \Big/ \int_X dx\right) \tag{19}$$

Hence the proposition of the *posterior* probability estimator $p^{(R)}\!\left(i_1|x_1\right)$ is as follow. From subset of initial rules $R_i(x) = \left\{r_i^{(k)} : x \in D_i^{(k)}\right\}$ choose the "most specific" rule $r_i^{(m)}$:

$$p^{(R)}\!\left(i_1|x_1\right) = \left(\overline{\beta}_i^{(m)} - \underline{\beta}_i^{(m)}\right) \Big/ \int_{D_i^{(m)}} dx \tag{20}$$

We can use similar procedure to obtain *posterior* probabilities estimators from the transition rules. We can get the estimator of *posterior* probability obtained from rule set using following formulae

$$p^{(R)}(j_n / \overline{x}_n, \overline{u}_{n-1}) = \sum_{j_{n-1}=1}^{M} \sum_{j_{n-2}=1}^{M} \cdots \sum_{j_1=1}^{M} p_n^{(R)}(j_n / j_{n-1}, x_n, u_{n-1})$$
$$\times p_{n-1}^{(R)}(j_{n-1} / j_{n-2}, x_{n-1}, u_{n-2}) \times \cdots \times p_1^{(R)}(j_1 / x_1). \tag{21}$$

We thus obtain the rule-based algorithm

$$\psi_n^{(R)}(\overline{x}_n, \overline{u}_{n-1}) = i_n \text{ if } p^{(R)}(i_n|\overline{x}_n, \overline{u}_{n-1}) = \max_{k \in M} p^{(R)}(k|\overline{x}_n, \overline{u}_{n-1}). \tag{22}$$

### 2.3  The Neural Network Approach

Two types of neural networks were applied to the considered decision problems: back propagation and counter propagation one [8]. The first one consisted of two trained layers. The neurons were described by sigmoid transition function and were trained according to back propagation method. Each layer consisted of 5 neurons. Number of inputs to the first layer varied and was dependent on the length of input vector. The second network consisted of three layers. First one was a normalization layer, and scaled the data so they received a value from the 0 to 1 range. This layer was not trained. Second one was trained according to Kohonen's rule (without teacher). The output layer was trained with Grossberg Outstar rule. Number of neurons in Kohonen's layers was empirically estimated and varied from 50 to 70 depended on input vector length and in the output layers was equal to 5.

Two different methods of the data presentation were applied. For the first one (method A) there were only the vectors of features and the therapy, used as the information needed for the recognition which leads to the following object form in learning sets:

$$X_k = (x_{k1}, u_{k1}, x_{k2}, u_{k2}, \ldots, x_{k_nk}, u_{k_nk}) \ \ Y_k = j_{k_nk}, \tag{23}$$

where

$X_k$ input of the net,

$Y_k$ classification (desired output of net).

For the second method (B) there were also medical diagnoses from previous states given as the information. Hence:

$$X_k = (x_{k1}, j_{k1}, u_{k1}, x_{k2}, j_{k2}, u_{k2}, ..., x_{k_nk}, u_{k_nk})   Y_k = j_{k_nk} \qquad (24)$$

Each of 5 neurons in the output layer were attributed to the one of class number states and during training was set to 1 if it represented diagnosed lines or to 0 in the other cases for both networks. During recalling neuron, which received the biggest stimulation pointed the recognition result. The result was considered correct if it was in agreement with the medical diagnosis.

# 3 Mathematical Models of Decision Problems

## 3.1 Acid-Base Balance

In the course of many diseases there occur disorders in the human acid-base balance state (ABB). It is effect of anomalies in production and elimination of hydrogen ions and carbon dioxide in the patient's organism. These anomalies can cause acidosis or alkaloses (each of them can be of metabolic or respiratory origin and additional the ABB state can be acute or chronic). For simplicity the set of classes $M$ consists of five classes: {1-respiratory acidosis, 2-respiratory alkalosis, 3-metabolic acidosis, 4-metabolic alkalosis, 5-normal state}. In ABB state recognition the most important are blood gas results, although the set of symptoms is pretty wide. Then the features vector $x = [x^{(1)}, x^{(2)}, x^{(3)}]^T$ consists of: $x^{(1)}$ - denotes pH of blood, $x^{(2)}$ - denotes carbon dioxide pressure, $x^{(3)}$ - denotes actual bicarbonate concentration .

Acid-base balance states quickly change in time, depending on the previous state and the applied treatment, and require frequent repetition of examinations. Between the observations the patient usually is treated. In this case the therapies can be comprised in three categories, so the set of treatments $K$ = {1- respiratory treatment, 2-pharmacological treatment, 3- no therapy}. In the Neurosurgery Clinic of Wrocław Medical Academy the set of data has been gathered, which contains $N$ = 78 learning sequences (13) of different length from 10 to 25. The medical tests were taken twice a day, mornings and evenings. In order to apply the pattern recognition algorithm for controlled Markov chain with learning (15) first we should estimate the probabilities describing the Markov chain (3), (4) and the probabilities of control (6) using frequency estimators and the learning set of medical data. Next, assuming the independence between the features we must calculate the values of density functions $f(x_n | j)$ in every class and in every tested point. In this paper estimator k-NN is used. More details can be found in [2, 14].

During the experiment, from the same set of learning sequences (13) the set of 55 of probabilistic rules (16), (17) has been gathered. The exemplary rules are presented bellow:

```
IF 5.5≤pH≤6.0 AND 32≤pCO₂
AND respiratory alkalosis was previously diagnosed
AND pharmacological treatment was applied
```

```
THEN at this moment metabolic alkalosis
WITH posterior probability 0.6


IF 4.5≤pH AND concentration of HC₂≤30
AND respiratory alkalosis was previously diagnosed
AND pharmacological treatment was applied
THEN at this moment metabolic alkalosis
WITH posterior probability 0.7
```

### 3.2  Hypertension Drug Therapy

During the hypertension treatment there is very important to recognize state of patient, the progress or regression of disease. On the base on this state (it has not the typical medical interpretation sometimes) and previous treatment the physician chooses the therapy. On the base on its results the physician can decide if therapy is correct or not, if it is necessary to change the dose of the drug or to change, to add or to remove the drug. We simplified the problem. Our computer system point on the one of six group of drug which has to be applied. In this case we say nothing about changing, removing drug or manipulation of doses. Our experts confirmed that that formulated problem is still interesting.

The presented project is developed together with Service d'Informatique Médicale from the University Paris VI. All data are getting from the medical database *ARTEMIS*, which contains the data of the patients with hypertension, whose have been treated in Hôpital Broussais in Paris.

The mathematical model of problem is as follow:

**Features**

| | | | |
|---|---|---|---|
| 1 | alcohol | 16 | sex |
| 2 | pain of legs | 17 | tail |
| 3 | cerebral haemorrhage | 18 | weight |
| 4 | pregnancy | 19 | systolic blood-pressure |
| 5 | coronary heart disease | 20 | diastolic blood-pressure |
| 6 | cardio vascular disease | 21 | maximal systolic blood-pressure |
| 7 | palpitation | 22 | calcium |
| 8 | aorta's souffle | 23 | potassium |
| 9 | lumbar's souffle | 24 | sodium |
| 10 | stenoses | 25 | creatinine |
| 11 | perspiration | 26 | HDL cholesterol |
| 12 | headache | 27 | cholesterol total |
| 13 | smoking | 28 | uricemia |
| 14 | tumour | 29 | sokolow |
| 15 | age | | |

**Classes M (Controls K)**

1-diuretics, 2- $\beta$-blockers, 3 - $\alpha$-$\beta$-blockers, 4 - ACE inhibitors, 5 - calcium antagonists, 6 – other.

The learning set was selected from Artemis database from the histories of patient with primary hypertension whose were treated in Hôpital Broussais in Paris since 1992. Those records were verified by our experts. This set consists of 168 histories of patient, where each of them was observed five times at least. The set of rules was generated by experts from Service d'Informatique Médicale (department of the University Paris VI) and Clinic of Pediatric Nephrology (Medical Academy of Wroclaw). It consists of 15 initial rules and 21 transition rules.

The examples of rules are presented bellow:

```
IF pain of legs = yes AND HDL cholesterol = high
THEN ACE inhibitors WITH average posterior probability
greater than 0,70 and less than  0,80
```

```
IF previous ACE inhibitor was taken
AND age > 40 AND age < 50
AND coronary heart disease = yes
THEN other WITH average posterior probability greater
than 0,75 and less than  0,85
```

## 4  Experimental Investigations

In order to study the performance of the proposed recognition concept and evaluate its usefulness to the computer-aided diagnosis of human acid-base balance states and to the problem of anti-hypertension drug therapy some computer experiments were made. The tested methods were as follows:

1) K-NN - k(N)-Nearest Neighbors algorithm,
2) RB – probabilistic rule-based algorithm,
3) CP A - Counter Propagation network with method A of data presentation,
4) CP B - Counter Propagation network with method B of data presentation,
5) BP A&B - Back Propagation network with methods A and B of data presentation.

The results of experimental investigations are presented below:

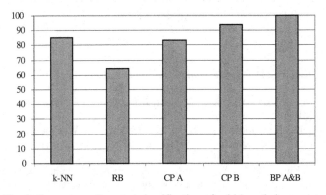

**Fig. 2.** Percentage of correct classification of acid-base balance state

The results of experimental investigations for the presented medical diagnosis task are similar. The neural nets always gave the better results than classifier based on Bayes formulae. Note that nets taught by data which respects context gave the best results. The quality of classifier based on the k-NN estimator of density function has similar quality as neural net which did not respect context of information (CP A).

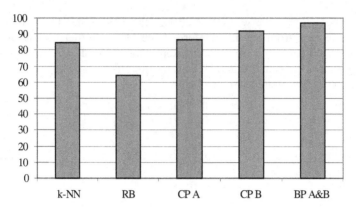

**Fig. 3.** Frequency of correct classification for anti-hypertension drug therapy

# 5  Conclusion

The paper proposed the algorithms whose can be used to help the clinician to make his own diagnosis. Of course he or she can use own expertise and algorithms' outputs to make the final decision. This observation induce to test in the future concept of classifiers ensembles where the final diagnosis is made on the base on the group of recognizers' decisions[9-12].

The superiority of the presented empirical results for the case-based classifiers over heuristic one demonstrates the effectiveness of the proposed concepts in such computer-aided medical diagnosis problems. Advantages of the proposed methods make it attractive for a wide range of applications in medicine, which might significantly improve the quality of the care that the clinician can give to his patient.

This work is supported be The Polish State Committee for Scientific Research under the grant which is realizing in years 2005-2007.

# References

1. Devijver P., Kittler J., *Pattern Recognition- A Statistical Approach*, Prentice Hall, London (1982)
2. Haralick R. M., Decision Making in Context, *IEEE Trans. on Pattern Anal. Machine Intell.*, vol. PAMI-5, (1983)
3. Duda R.O., Hart P.E., Stork D.G., Pattern Classification, John Wiley and Sons, 2001
4. Giakoumakis E., Papakonstantiou G., Skordalakis E., Rule-based systems and pattern recognition, Pattern Recognition Letters, No 5, 1987.
5. Jackowski K., Kurzynski M., Wozniak M., Zolnierek A., Different approaches to the sequential diagnosis problem: a comparative study, Computers in Medicine, 1997, vol.1, pp. 220-225.
6. Mitchell T., *Machine Learning*, McGraw Hill, 1997.
7. Lin T.Y., Wildberger A. (red.), Soft Computing: Rough Sets, Fuzzy Logic, Neural Networks, Uncertainty Management, Knowledge Discovery, San Diego, Simulation Councils Inc., 1995.
8. Xu L., Krzyżak A., Suen Ch.Y. (1992), Methods of Combining Multiple Classifiers and Their Applications to Handwritting Recognition, *IEEE Transactions on Systems, Man, and Cybernetics,* vol.22, no. 3, pp.418-435.

9. Ji Ch., Ma S. (1997), Combination of Weak Classifiers, *IEEE Transaction on Neural Networks*, vol 8, no.1, pp.32-42.
10. Kittler J., Alkoot F.M. (2003), Sum versus Vote Fusion in Multiple Clasifier Systems, *IEEE Transaction on Pattern Analysis and Machine Intelligence*, vol. 25, no. 1, pp. 110-115.
11. Lam L., Suen Ch.Y. (1997), Application of Majority Voting to Pattern Recognition: An Analysis of Its Bechavior and Performance, *IEEE Transaction on Systems, Man, and Cybernetics-Part A: Systems and Humans*, vol. 27, no. 5, pp. 553-567.
12. Puchala E., A Bayes Algorithm for the Multitask Pattern Recognition Problem – direct approach, *Lecture Notes in Computer Science*, vol. 2659, 2003, pp. 3-10.
13. Burduk R., Decision Rules for Bayesian Hierarchical Classifier with Fuzzy Factor. Soft Methodology and Random Information Systems, *Advances in Soft Computing*, Springer, Berlin 2004, pp. 519-526.

# Computer-Aided Sequential Diagnosis Using Fuzzy Relations –
# Comparative Analysis of Methods

Marek Kurzynski[1,2] and Andrzej Zolnierek[1,2]

[1] Wroclaw University of Technology, Faculty of Electronics, Chair of Systems and
Computer Networks, Wyb. Wyspianskiego 27, 50-370 Wroclaw, Poland
{marek.kurzynski,andrzej.zolnierek}@pwr.wroc.pl
[2] The Witelon University of Applied Sciences,
Ul. Sejmowa 5A, 59-220 Legnica, Poland

**Abstract.** A specific feature of the explored diagnosis task is the dependence between patient's states at particular instants, which should be taken into account in sequential diagnosis algorithms. In this paper methods for performing sequential diagnosis using fuzzy relation in product of diagnoses set and fuzzified feature space are developed and evaluated. In the proposed method first on the base of learning set fuzzy relation is determined as a solution of appropriate optimization problem and next this relation in the form of matrix of membership grade values is used at successive instants of sequential diagnosis process. Different algorithms of sequential diagnosis which differ with as well the sets of input data as procedure are described. Proposed algorithms were practically applied to the computer-aided recognition of patient's acid-base equilibrium states where as an optimization procedure genetic algorithm was used. Results of comparative experimental analysis of investigated algorithms in respect of classification accuracy are also presented and discussed.

## 1 Introduction

Multiple diagnoses of the patient's state based on results of successive examinations is one of the most frequent and typical medical diagnosing tasks. Such a task, henceforth called the sequential diagnosis, involves dealing with a complex decision problem. This is caused by the dependence of the patient's state at a given time on the preceding states and on the already applied treatment. Although there remains no doubt as to the very existence of this dependence, it may be of a diversified nature and range; its simplest instance can be a one-instant-backwards dependence to so complex arrangements as those in which the current state depends on the whole former course of the disease.

We have to take into account these sequential diagnosing dependencies when we intend to support diagnosing tasks using a computer. In other words, when constructing an appropriate decision algorithm we must not limit our approach

J.L. Oliveira et al. (Eds.): ISBMDA 2005, LNBI 3745, pp. 242–251, 2005.

to only the narrow information channel that concerns just the current symptoms but we have to consider all the available measurement data instead, as they may contain important information about the patient's state at a given instant. The measurement data comprise all the examination results obtained so far, the applied treatment procedures, as well as the diagnoses formulated at previous diagnosis instants. Thus the bulk of data is very rich and grows over time from one instant to another; this can be viewed at as both an advantage and disadvantage, depending on the viewpoint. Every medical practitioner will recognize the great usefulness of such data. Performing the sequential diagnosis he/she will inevitably ask the patient not only about the former symptoms and applied treatment but about the previous diagnostic statements. On the other hand, however, the lavishness and incremental nature of the available data make it impossible to comprehend them completely, as a result of which fact various simplifications and compromises must be made.

A specific feature of the explored diagnosis task is the dependence between patient's states at particular instants, which dependence is, specifically again, taken into account in sequential diagnosis algorithms. The dependence can be included at an as early stage as that of formulating a mathematical model for the diagnosis task, or as late as at the stage of selecting the appropriate input data set in the decision algorithm which otherwise does not differ from the classical diagnosis task. An example for the former case can be the probabilistic approach which offers the effective, as it has turned out, description of the dependences and actual treatment, in the form of a controlled Markov chain ([4], [5]). We call it the effective description because it leads to a constructive algorithm. However, its form is quite complex and depends on the context range that is taken into account in the actual model.

The other case occurred when the approximated inference engine was applied based on a fuzzy rule system, and when artificial neural networks were applied ([1]). Both methods deal with a well-know procedure used either for fuzzy rule construction based on empirical data, or for neural network training. The specificity of the investigated diagnostic task reveals itself here exclusively in the form of input data which are not associated only with the direct symptoms that manifest the current state, but comprise up to an extent the "historic" information that regards the preceding course of the disease. For this case we do not know how far backwards the examined input data should spread into the past; the "the more the better" rule need not necessarily be true here. As far now, there are no analytical evidence to be used in this issue, whilst any attempts to answer the question are under way of experimental research.

In the paper we propose a novel approach to the problem of algorithmization of sequential diagnosis task based on a concept of fuzzy relation given by matrix of membership grade values. The paper is a sequel to the author's earlier publications ([1], [2], [3], [4]) and it yields new results dealing with the application of fuzzy inference systems to the decision making at the successive instants of sequential diagnosis procedure. The contents of the work are as follows. In section 2 necessary background is introduced and the sequential diagnosis method is described and formalized. In section 3 algorithms for sequential diagnosis are discussed. In the proposed approach fuzzy relation in the product of set of decisions and feature space is determined as a solution of appropriate optimization problem. In the presented example the genetic algorithm was applied to find optimal solution. In section 4 we discuss the results of application of proposed fuzzy decision systems to the computer-aided recognition of acid-base equilibrium states.

## 2    Preliminaries and the Problem Statement

We will treat the sequential diagnosis task as a discrete dynamical process. The patient (object) is at the $n$-th instant in the state $j_n \in \mathcal{M}$, where $\mathcal{M}$ is an $M$-element set of possible states numbered with the successive natural numbers. Thus

$$j_n \in \mathcal{M} = \{1, 2, ..., M\}. \tag{1}$$

Obviously, the notion of instant has no specific temporal meaning here, as its interpretation depends on the character of the case under consideration. The actual measure used may be minutes, hours, days, or even weeks.

The state $j_n$ is unknown and does not undergo our direct observation. What we can only observe is the indirect symptoms (also called features or tokens) by which a state manifests itself. We will denote a $d$-dimensional symptom value vector by $x_n \in \mathcal{X}$, for symptoms measured at the $n$-th instant (thus $\mathcal{X}$ is the symptom space); let us also denote by $u_n$ the therapy chosen from the therapy set $\mathcal{U}$ to be applied at the $n$-th instant.

As already mentioned, the patient's current state depends on the history and thus in the general case the decision algorithm must take into account the whole sequence of the preceding symptom values, $\bar{x}_n = \{x_1, x_2, \ldots, x_n\}$, and the sequence of applied therapies, $\bar{u}_{n-1} = \{u_1, u_2, \ldots, u_{n-1}\}$. It must be underlined here that sometimes it may be difficult to include all the available data, especially for bigger $n$. In such cases we have to allow various simplifications (e.g. make allowance for only several recent values in the $\bar{x}_n$ and $\bar{u}_{n-1}$ vectors), or compromises (e.g. substituting the whole disease history segment that spreads as far back as the $k$-th instant, i.e. the $\bar{x}_k$ and $\bar{u}_{k-1}$ values, with data processed in the form of a diagnosis established at that instant, say $i_k$).

Apart from the data measured for a specific diagnosed patient we need some more general information to take a valid diagnostic decision, namely the a priori knowledge concerning the general associations that hold between diagnoses on the one hand, and symptoms and the applied treatment schemata, on the other. This knowledge may have multifarious forms and various origins. From now on we assume that it has the form of a so called training set, which in the investigated decision task consists of $m$ training sequences:

$$\mathcal{S} = \{S_1, S_2, ..., S_m\}. \tag{2}$$

A single sequence:

$$S_k = ((x_{1,k}, u_{1,k}, j_{1,k}), (x_{2,k}, u_{2,k}, j_{2,k}), ..., (x_{N,k}, u_{N,k}, j_{N,k})) \tag{3}$$

denotes a single-patient disease course that comprises $N$ symptom observation instants, the applied treatment, and the patient's state.

Analysis of the sequential diagnosis task implies that, when considered in its most general form, the explored decision (diagnostic) algorithm can in the $n$-th step make use of the whole available measurement data (perhaps partly substituted with former diagnoses), as well as the knowledge included in the training set. In consequence, the algorithm is of the following form:

$$\Psi_n(\bar{x}_n, \bar{u}_{n-1}, \mathcal{S}) = i_n. \tag{4}$$

The next chapter describes in greater detail the construction of the diagnostic algorithm (4) using concept of fuzzy relation in product of feature space for various input data and decision set.

## 3   Algorithms of Sequential Diagnosis

In the presented method of sequential diagnosis, first on the base of learning sequences (2) we find fuzzy relation between fuzzified feature space and class number set as a solution of appropriate optimization problem. Next this relation expressed as matrix which elements represent intensity of fuzzy symptoms for each diagnosis (patient's state) can be used to make decision in particular steps of the whole sequential diagnosis process.

Two procedures has been proposed which differ exclusively with the relevant selection of input data. The first algorithm includes $k$-instant-backwards-dependence ($k < N$) with full measurement data. It means, that decision at the $n$th instant is made on the base of vector of features

$$\bar{x}_n^{(k)} = (x_{n-k}, x_{n-k+1}, ..., x_{n-1}, x_n). \tag{5}$$

In the second approach however, we also include $k$-instant- backward- dependence, but using the previous diagnoses in lieu of the previous symptom values. In the both approaches information about the sequence of applied therapies

$$\bar{u}_n^{(k)} = (u_{n-k}, u_{n-k+1}, ..., u_{n-1}). \tag{6}$$

plays the role of peculiar "switch" which allows to select appropriate fuzzy relation (matrix).

Before we will describe both algorithms let first introduce sets $\mathcal{S}^{\bar{u}^{(k)}}$ and $\mathcal{S}_{\bar{j}^{(k)}}^{\bar{u}^{(k)}}$ denoting sequences of $(k+1)$ learning patterns from $\mathcal{S}$ in which at the first $k$ position the sequence of therapies $\bar{u}^{(k)} \in \mathcal{U}^k$ (Cartesian product of $\mathcal{U}$) appears and additionally sequence of classes $\bar{j}^{(k)} \in \mathcal{M}^k$ appears, respectively.

### 3.1   Algorithm with $k$th Order Depencence (AkD)

The algorithm with the full measurement features can be expressed in the following points:

1. Cover the space of features $\mathcal{X}$ with fuzzy regions. In the further example we use triangular fuzzy numbers with 3 regular partitions [2]. Obtained fuzzy sets correspond to the linguistic values of features which state fuzzified feature space $\mathcal{X}_F$.

2. For each sequence of therapies $\bar{u}^{(k)} \in \mathcal{U}^k$:

   - Determine observation matrix $O^{\bar{u}^{(k)}}$, i.e. fuzzy relation in the product of Cartesian product of fuzzified feature space $\mathcal{X}_F^k$ and learning subset $\mathcal{S}^{\bar{u}^{(k)}}$. The $i$th row of observation matrix contains grades of memberships of features $\bar{x}^{(k)}$ of $i$th learning sequence from $\mathcal{S}^{\bar{u}^{(k)}}$ to the fuzzy sets of space $\mathcal{X}_F^k$.

   - Determine decision matrix $D^{\bar{u}^{(k)}}$, i.e. relation in product of learning sequences $\mathcal{S}^{\bar{u}^{(k)}}$ and the set of decisions (diagnoses) $\mathcal{M}$. Elements of $i$th row of decision matrix are equal to zero with except the position corresponding to the last class number of $i$th sequence in the set $\mathcal{S}^{\bar{u}^{(k)}}$ which is equal to one.

   - Find matrix $E^{\bar{u}^{(k)}}$, so as to minimize criterion

$$\rho(O^{\bar{u}^{(k)}} \circ E^{\bar{u}^{(k)}}, D^{\bar{u}^{(k)}}), \tag{7}$$

where operator $\circ$ denotes max-min-norm composition of relations, i.e. multiplication of matrices O and E with $\times$ and $+$ operators replaced by min and max operators (more general by $t$-norm and $s$-norm operators) ([7]). Criterion (7) evaluates difference between matrices $A$ and $B$, i.e. $\rho(A,B) \geq 0$ and $\rho(A,B) = 0$ iff $A = B$. In the further experimental investigations we adopt

$$\rho(A,B) = \sum_{i,j}(a_{ij} - b_{ij})^2 \tag{8}$$

and as method of minimization (7) the genetic algorithm will be applied.

Matrix $E^{\bar{u}^{(k)}}$ is a fuzzy relation in product of decision set $\mathcal{M}$ and feature space $\mathcal{X}_F^k$ for objects (patients) subject to therapies $\bar{u}^{(k)}$ in which reflects knowledge contained in the learning sequences.

The manner of utilize the matrices $E$ for decision making is obvious. At the $n$th step of sequential diagnostic process, first for known sequence (6) we select matrix $E^{\bar{u}_n^{(k)}}$ and for sequence of feature observations (5) the row-matrix of fuzzy observation $O(\bar{x}_n^{(k)})$ is determined. Next we calculate the row-matrix of soft decisions:

$$O(\bar{x}_n^{(k)}) \circ E^{\bar{u}_n^{(k)}} = D(\bar{x}_n^{(k)}, \bar{u}_n^{(k)}) \tag{9}$$

and final diagnosis is made according to the maximum rule.

It must be emphasized that proposed procedure leads to the very flexible sequential recognition algorithm due to optional value of $k$. In particular the value of $k$ need not be constant but it may dynamically change from step to step. It means next, that choice $k = n - 1$ for $n$th instant of sequential diagnosis denotes the utilization of the whole available information according to the general form of diagnostic rule (4). On the other side however, such concept - especially for bigger $n$ - is rather difficult for practical realization.

## 3.2   Reduced Algorithm with $k$th Order Depencence and Crisp History (RkDC)

In this approach for diagnosis at the $n$th instant, we substitute the whole disease history segment which - as previously - covers the $k$ last instances, i.e. $(x_{n-k}, x_{n-k+1}, ..., x_{n-1})$ values with data processed in the form of diagnoses established at these instances, say

$$\bar{i}_n^{(k)} = (i_{n-k}, i_{n-k+1}, ..., i_{n-1}). \tag{10}$$

Such a concept significantly simplifies the computational procedure since - the sequence of previous diagnoses can play exactly the same role as a sequence of previous therapies in the algorithm AkD. Thus, we get identical procedure for determining matrices $E$ in which set $\mathcal{S}^{\bar{u}^{(k)}}$ is replaced with set $\mathcal{S}_{\bar{j}^{(k)}}^{\bar{u}^{(k)}}$ and fuzzified product feature space $\mathcal{X}_F^k$ with simply $\mathcal{X}_F$.

As a consequence for each sequence of possible therapies (6) and diagnoses $\bar{j}^{(k)}$, on the base of learning sequences (2) via optimization procedure (7) the matrix $E_{\bar{j}^{(k)}}^{\bar{u}^{(k)}}$ is determined, which applied in the formula

$$O(x_n) \circ E_{\bar{i}_n^{(k)}}^{\bar{u}_n^{(k)}} = D(x_n, \bar{u}_n^{(k)}, \bar{i}_n^{(k)}). \tag{11}$$

leads to the vector of soft decision at the $n$th instant and next after maximum defuzzification procedure, to the crisp diagnosis.

## 3.3   Reduced Algorithm with $k$th Order Depencence and Soft History (RkDS)

In the RkDC algorithm with crisp history, matrix $E$ for given sequence of therapies $\bar{u}_n^{(k)}$ was univocally determined by observed sequence of previous diagnoses (10). In the concept of algorithm with soft history however, we take into account soft decisions at previous instances, i.e. sequence of diagnoses for previous instances before defuzzification procedure (row-matrices $D$ containing grades of membership for particular diagnoses) instead of sequence (10) of crisp decisions.
Let

$$D_{n-i} = (d_{n-i}^{(1)}, d_{n-i}^{(2)}, ..., d_{n-i}^{(M)}), \tag{12}$$

be the vector of membership grades for all classes produced by diagnostic algorithm at the $(n-i)$th instant ($i = 1, 2, ..., k$).

In the RkDS algorithm at the $n$th instant we replace in (11) matrix $E_{\bar{i}_n^{(k)}}^{\bar{u}_n^{(k)}}$ for observed sequence of previous diagnoses (10) with the weighted sum of matrices for all possible sequences $\bar{i}_n^{(k)} \in \mathcal{M}^k$, viz.

$$O(x_n) \circ \sum_{\bar{i}_n^{(k)} \in \mathcal{M}^k} w_{\bar{i}_n^{(k)}} \times E_{\bar{i}_n^{(k)}}^{\bar{u}_n^{(k)}} = D(x_n, \bar{u}_n^{(k)}, ), \tag{13}$$

where weight coefficients are equal to product of elements of vectors (12) corresponding to the elements of vector (10), namely

$$w_{\bar{i}_n^{(k)}} = d_{n-k}^{(i_{n-k})} \cdot d_{n-k+1}^{(i_{n-k+1})} \cdot \ldots \cdot d_{n-1}^{(i_{n-1})}. \tag{14}$$

All the decision algorithms that are depicted in this chapter have been experimentally tested as far as the decision quality is concerned. Measure for the decision quality is the frequency of correct diagnoses for real data that are concerned with recognition of human acid-base equilibrium states. The purpose of our research and associated tests was not only the comparative analysis of the presented algorithms but also answering the question whether including the inter-state dependence (whatever its form would be) would yield a better decision quality as compared to algorithms that did not take into account such a dependence. The next chapter describes the performed tests and their outcome.

## 4  Practical Example: Sequential Diagnosis of Acid-Base Equilibrium States

In the course of many pathological states, there occur anomalies in patient organism as far as both hydrogen ion and carbon dioxide production and elimination are concerned, which leads to disorders in the acid-base equilibrium (ABE). Thus we can distinguish acidosis and alkalosis disorders here. Either of them can be of metabolic or respiratory origin, which leads to the following ABE classification: metabolic acidosis, respiratory acidosis, metabolic alkalosis, respiratory alkalosis, correct state.

In the process of treatment, correct recognition of these anomalies is indispensable, because the maintenance of the acid-base equilibrium, e.g. the pH stability of the fluids is the essential condition for correct organism functioning. Moreover, the correction of acid-base anomalies is indispensable for obtaining the desired treatment effects.

In medical practice, only the gasometric examination results are made use of to establish fast diagnosis, although the symptom set needed for correct ABE estimation is quite large. The utilized results are: the pH of blood, the pressure of carbon dioxide, the current dioxide concentration.

An anomalous acid-base equilibrium has a dynamic character and its changes depend on the previous state and the therapy applied, and in consequence they require frequent examinations in order to estimate the current ABE state. It is clear now that the sequential decision methodology presented above well suits the needs of computer aided ABE diagnosing.

The current formalization of the medical problem leads to the task of the ABE series recognition, in which the classification basis in the $n$-th moment constitutes the quality feature consisting of three gasometric examinations and the set of diagnostic results $\mathcal{M}$ is represented by 5 mentioned acid-base equilibrium states. This model can be completed also with therapeutic possibilities (controlling) which patient might undergo. Assuming the certain simplification, these therapies could be divided into three following categories: respiratory treatment, pharmaceutical treatment, no treatment.

In order to study the performance of the proposed concept of sequential diagnosis, the algorithms presented in previous section were applied for the ABE state sequential recognition. Experiments have been worked out on the basis of evidence material that was collected in Neurosurgery Clinic of Medical Academy of Wroclaw and constitutes the set of training sequences (2). The material comprises 78 patients (78 sequences) with ABE disorders caused by intracranial pathological states for whom the following data were regularly put down on the 12-hour basis: 1. gasometric examination results, 2. the correct ABE state diacrisis, and 3. the decision concerning the therapy to be applied. There were around 20 examination cycles for each patient, yielding the total of 1486 single examination instances.

In order to find matrices $E$ for different therapies the genetic algorithm was applied, which is a popular method in optimization and can improve the search procedure ([8]). The genetic algorithm proceeded as follows.

- *Coding method* – the values of elements of matrix $E$ were directly coded to the chromosome.
- *The fitness function* – was defined as follows:

$$Fit = Q - \rho(A, B),\tag{15}$$

where $\rho$ is as in (8) and $Q$ is suitably selected positive constant.
- *Initialization* – the initial population of chromosomes with which the search begins was generated randomly. The size of population – after trials – was set to 40.
- *Reproduction* – roulette wheel with elitism.
- *Crossover and mutation* – a two-point crossover was used and probability of mutation was 0.05.
- *Stop procedure* – evolution process was terminated after 1000 generations.

The outcome is shown in Table 1. It includes the frequency of correct diagnoses for the investigated algorithms (their names being explained in previous section). Additionally, the algorithm A0D includes neither inter-state-dependences nor the influence the applied therapy has exerted on a patient's

**Table 1.** Frequency of correct diagnosis for various diagnostic algorithms

| Algorithm | A0D | A1D | A2D | R1DC | R2DC | R1DS | R2DS |
|---|---|---|---|---|---|---|---|
| Result | 80.3% | 88.4% | 91.3% | 85.9% | 87.7% | 85.6% | 88.2% |

state but utilises only the current symptom values instead. Thus it will be obtained by putting $k = 0$ in the algorithm AkD.

These results imply the following conclusions:

1. Algorithm A0D that does not include the inter-state dependences and treats the sequence of states as independent objects is worse than those that were purposefully designed for the sequential medical diagnosis task, even for the least effective selection of input data. This confirms the effectiveness and usefulness of the conceptions and algorithm construction principles presented above for the needs of sequential diagnosis.
2. There occurs a common effect within each algorithm group: the model of the second order dependency (A2D, R2DC, R2DS) turns out to be more effective than the first order dependence approach (A1D, R1DC, R1DS).
3. Algorithms A1D and A2D that utilize the original data (i.e. symptoms along with therapy) yield always better results than those which substitute the data with diagnoses.
4. There is no essential difference between the algorithms with crisp and soft history.

## 5   Conclusions

This work presents fuzzy approach to the algorithmization of the sequential diagnosis task and the implied decision algorithms which take into account a specific feature of the sequential diagnosis, i.e. dependence between patient's states at particular instants. The fuzzy methods which have been already applied to the computer-aided sequential diagnosis, use fuzzy rules generated from the learning set and Mamdani inference engine ([1]). Recognition method presented in this work however, deals with fuzzy relation concept developed to the sequential diagnosis process.

The comparative analysis presented above for the sequential diagnosis algorithms is of the experimental nature. We have carried out a series of experiments on the basis of a specific exemplar that concerns acid-base unequilibrium state diagnosing using rich enough set of real-life data. The objective of our experiments was to measure quality of the tested algorithms that was defined by the frequency of correct decisions. The algorithm-ranking outcome cannot be treated as one having the ultimate character as a that of a law in force, but it has been achieved for specific data within a specific diagnostic task. However, although the outcome may be different for other tasks, the presented research may nevertheless suggest some perspectives for practical applications. All the experiments show that algorithms which are appropriate for sequential diagnosis, i.e. ones that include - in whatever form - the inter-state dependences are much more effective as far as the correct decision frequency is concerned than algorithms which do not include the actual associations. This testifies that the proposed conceptions are correct, and the constructed algorithms effective, for computer aided sequential medical diagnosis.

# References

1. Kurzynski, M.W.: Benchmark of Approaches to Sequential Diagnosis. In: Lisboa, P., Ifeachor, J., Szczepaniak P.S. (eds.): Perspectives in Neural Computing. Springer-Verlag, Berlin Heidelberg New York (1998) 129-140
2. Kurzynski, M.W.: Multistage Diagnosis of Myocardial Infraction Using a Fuzzy Relation. In: Rutkowski, L., Tadeusiewicz, R. (eds): Artificial Intelligence in Soft Computing, Lecture Notes in Artificial Intelligence, Vol. 3070, Springer-Verlag, Berlin Heidelberg New York (2004) 1014-1019
3. Kurzynski, M.W., Zolnierek, A.: A Recursive Classifying Decision Rule for Second-Order Markov Chains, Control and Cybernetics **18** (1990) 141-147
4. Zolnierek, A.: The Empirical Study of the Naive Bayes Classifier in the Case of Markov Chain Recognition Task. In: Kurzynski, M.W., Wozniak M. (eds.): Computer Recognition Systems CORES 05, Springer-Verlag, Berlin Heidelberg New York (2005) 329-336
5. Devroye, L., Gyorfi, P., Lugossi, G.: A Probabilistic Theory of Pattern Recognition. Springer-Verlag, Berlin Heidelberg New York (1996)
6. Duda, R., Hart, P., Stork, D.: Pattern Classification. John Wiley and Sons, New York (2001)
7. Czogala, E., Leski, J.: Fuzzy and Neuro-Fuzzy Intelligent Systems, Springer-Verlag, Berlin Heidelberg New York (2000)
8. Goldberg, D.: Genetic Algorithms in Search, Optimization and Machine Learning. Adison-Wesley, New York (1989)

# Service Oriented Architecture
# for Biomedical Collaborative Research

José Antonio Heredia[1], Antonio Estruch[1], Oscar Coltell[2], David Pérez del Rey[3], Guillermo de la Calle[3], Juan Pedro Sánchez[4], and Ferran Sanz[5]

[1] Departamento de Tecnología, Universitat Jaume I, Castellón
[2] Systems Integration and Re-Engineering Research Group, Universitat Jaume I
[3] Biomedical Informatics Group, Department of Artificial Intelligence,
Universidad Politécnica de Madrid
[4] Área de Bioinformática y Salud Pública, Instituto de Salud Carlos III,
Ministerio de Sanidad y Consumo, Madrid
[5] GRIB, IMIM/UPF Barcelona, Spain

**Abstract.** Following a systems engineering approach we have identified the information system requirements for biomedical collaborative research. We have designed a Service Oriented Architecture following a dynamic and adaptable to change approach, using technology and specifications that are being developed in an open way, utilizing industry partnerships and broad consortia such as W3C and the Organization for the Advancement of Structured Information Standards (OASIS), and based on standards and technology that are the foundation of the Internet. The design has been translated in a pilot implementation infrastructure (INBIOMED) that is now been populated with web services for data and images analysis and collaborative management.

## 1 Towards the Collaborative Management
## of Biomedical Research

Publishing and sharing the research results is something inherent to the scientific method and has been performed since the dawn of science by means of written publications. Thanks to the possibilities offered by electronic formats it has become possible to reduce the time that elapses before they see the light, and use of the network makes them far more easily accessible. At the same time, in many disciplines the academic journals require researchers to make the data they used to reach their conclusions publicly available and some communities have even set up large pools of experimental data. These data, as a manifestation of the communities that generated them, are unconnected and the fact that they were produced by different technologies makes it difficult to integrate them. This basic level collaboration can be understood as a **first level of collaboration** in which the model of information exchange is limited to making both the data and the findings publicly available (table 1).

**Table 1.** Levels of collaboration and models of information exchange

| Level of collaboration | Model of information exchange |
|---|---|
| Work in isolation | Publication of data and findings |
| Coordinated projects | Sharing applications |
| Collaboration processes | Sharing information system |

J.L. Oliveira et al. (Eds.): ISBMDA 2005, LNBI 3745, pp. 252–261, 2005.

Yet, it is becoming increasingly clearer that isolated biomedical research does not lead to any significant solutions in clinical practice. Thus, the current trend is for public institutions to back research projects involving increasingly large numbers of research groups in order to be able to make more efficient use of the scarce resources available. At the same time, these research groups participate in networks where they try to coordinate with other groups by sharing knowledge as well as tangible resources. This approach to research is very recent and we are at present in a phase in which cultural changes are taking place, and structures and mindsets are therefore still being adjusted to the new environment. This means that, although in general terms research groups appear to be assuming this new situation and are willing to adapt to it, for the time being many shortcomings can be observed in the incipient processes of collaboration. Indeed, collaboration is often essentially limited to sharing funding and findings. In this phase, which is the one that currently predominates, we would be in a **second level of collaboration** where the models of information are characterised by sharing applications, as well as data and findings.

The challenges biomedical research faces in the medium and long term require large-scale coordinated projects involving collaborative research processes that really break down the walls between different departments. These projects will be fostered and will require the use of suitable infrastructures for computing and communication over the Internet – *cyberinfrastructures* [1] – that will allow both hardware and software resources to be shared in a network.

The fact is that reaching an understanding of the molecular mechanisms underlying diseases, such as the different types of cancer, is now seen as an increasingly more distant objective. Just as different research groups deal with their work in depth, the gap between what they currently know and what they actually need to know appears to be getting bigger. We think that required experimental technology appears to be already available, although it will be continually improved and new techniques will be proposed. However, the real challenge is to manage the complexity and scale of the problem being tackled: the organisational complexity that entails cultural changes related to managing research, and the huge amount of data generated by technologies used in molecular scale experiments. Handling such information requires sophisticated database management systems, in addition to complex data mining algorithms for its analysis. For certain applications, it is believed that it will be necessary to use Grid-based computing services [2].

The processes involved in disease research include research projects that are interrelated on different scales and which focus on a variety of aspects (molecular, cellular, organs, individuals, family, community, population). These kinds of research are usually carried out by researchers from different disciplines, with different languages and cultures, and this will also make it more difficult to integrate the research processes.

Hence, as happens in other scientific and technological disciplines, it appears that we are entering a new age in which it will be necessary to work in a large-scale collaborative manner with many other groups. We will therefore become part of a management process with common final goals and which shares large amounts of resources that enable extensive, costly and well designed experiments to be conducted in order to reach conclusive outcomes. This is what many call *Big Science*. At first, perhaps large emblematic projects will be implemented (by means of platforms), but alongside such work individual research will undoubtedly continue to be carried out. These studies will also have to benefit from collaborative work with other groups that are dealing with the same issues.

A paradigmatic example of the collaborative management of a large project was the Cooperative Human Linkage Center (CHLC) consortium set up as part of the genome project with the intention of creating gene maps [3]. The CHLC acted as a geographically distributed virtual centre that connected small highly specialised laboratories by means of an Internet-based computer system. Each group worked within the context it was familiar with and the different parts were put together to make maps using predefined workflows. To do so, when necessary, they also used distributed processing over a network of computers. The data, intermediate analyses and maps were distributed over the Internet by web servers.

This example is a proof-of-concept that the key objectives of collaborative management can be reached even with previous-generation computer technology [1]. The design and management of the research processes are even more important than specific computer solutions. The computer systems, although insufficient, like any technological tool, are necessary.

Collaboration during the research process requires the sharing of information systems, that is to say, working on the same information system or on systems that are truly interoperable so that the information flows from one group to another in the most effective and efficient manner.

## 2  The Service Oriented Architecture

The information technology industry is basing on web services technology the construction of the latest generation of distributed computer systems. This technology allows the internal resources to be encapsulated within the service, thus providing a logical application layer between those resources and their consumers. The owners of the services can modify the resources as time goes by without the need to carry out changes in the messages used to communicate with other consumer services, which means that they can continue to interoperate as they did before the modifications were introduced. This independence among the parts that go to make up a system makes it possible to build very robust systems that are open and in continual development, without the need for anyone to be in control of the system.

The specific way that certain information architects design the basic system or infrastructure that is going to be built with web services is called, as is only natural, Service Oriented Architecture (SOA). The services in this architecture are loosely coupled [4], since one application does not necessarily have to know the technical details of another – in fact, it does not even need to know which platform or language it has been programmed in. In contrast, however, if they are to be able to communicate with one another they must have well and clearly defined communication interfaces. Although Web services technology is not the only approach to realizing an SOA, it is one that the IT industry as a whole has enthusiastically embraced. With Web services, the industry is addressing yet again the fundamental challenge that distributed computing has provided for some considerable time: to provide a uniform way of describing components or services within a network, locating them, and accessing them. The difference between the Web services approach and traditional approaches (for example, distributed object technologies such as the Object Management Group – Common Object Request Broker Architecture (OMG CORBA), or Microsoft Distributed Component Object Model (DCOM) ) [5] lies in the loose coupling aspects of the architecture. Instead of building applications that result in tightly integrated collections of objects or components, which are well known and understood at development time, the whole approach is much more dynamic and adaptable to change. Another key difference is that through Web services, the IT industry is tackling the problems using technology and specifications that are being developed in an open way, utilizing industry partnerships and broad consortia such as W3C and the Organization for the Advancement of Structured Information Standards (OASIS), and based on standards and technology that are the foundation of the Internet.

Many developers, architects, managers and academics still see web services as the next episode in the continued saga of distributed object technologies such as CORBA, DCOM and RMI. Web services and distributed objects systems are both *distributed systems* technologies, but that is where the common ground ends. They have no real relation to each other, except maybe for the fact that web services are now sometimes deployed in areas where in the past the application of distributed objects has failed. If we look for relationships within the distributed technology world it is probably more appropriate to associate web services with messaging technologies, as they share a common architectural view, but address different types of applications.

Web services are based on XML documents and document exchange, and as such one could call the technological underpinning of web services *document-oriented computing*. Exchanging documents is a very different concept from requesting the instantiation of an object, requesting the invocation of a method on the specific object instance, receiving the result of that invocation back in a response, and after a number of these exchanges, releasing the object instance [6].

## 2.1 Service Oriented Architecture Technologies: XML and WEB Services

The foremost technology for the description and interaction of services is based on XML and HTTP, although extended with a set of specifications. From which the most widely used are SOAP and WSDL. SOAP (Simple Object Access Protocol) is a "binding protocol" utilised to coordinate the query/response interactions among Web services. WSDL (Web Services Description Language) specifies how to describe a service using a document with a special XML format. The big advantage of using XML standards for Web services is that all the computer platforms (such as Microsoft, IBM, Oracle, BEA, Apache) support them.

The World Wide Web Consortium (W3C), which has managed the evolution of the SOAP and WSDL specifications, defines Web services as follows: "*A software system designed to support interoperable machine-to- machine interaction over a network. It has an interface described in a machine-processable format (specifically WSDL). Other systems interact with the Web service in a manner prescribed by its description using SOAP messages, typically conveyed using HTTP with XML serialization in conjunction with other Web-related standards*"[7]

Along with the standards for describing and the transfer of Web services, it is also necessary to use methods that enable communication to be performed with high levels of security by encrypting the messages. The standard that is currently becoming the most popular is the VPN (Virtual Private Network). The VPN (Virtual Private Network) concept was to produce the virtual "dedicated circuit", pump it over the internet, and use cryptography to make it secure providing protections against eavesdropping and active attacks. Different alternatives are available to achieve secure networking in VPNs. IPSec was the first major effort to develop a standard. Traditional IPSec implementations required a great deal of kernel code, complicating cross-platform porting efforts. By contrast, other standards like SSL (Secured Sockets Layer) matured quickly, due to heavy usage on the web. SSL runs in user space, simplifying implementation and administration.

## 2.2  Layers of the Service Oriented Architecture

The basic functionalities of a SOA on a conceptual level are usually classified in layers. We propose the following three layers.

On the lowest level there are the functions and components that allow the infrastructure to connect with other services and sources of data, which is why this is usually called the *connectivity layer*. On top of the connectivity layer, the *orchestration layer* allows us to organise the different services and data sources that are accessed previously to enable the composition of the workflows. These workflows are the ones that allow us to reach the intended goal by implementing the information system. These may include the different workflows needed for collaborative management, knowledge management, project management, querying distributed databases, grid computing [8], or transactional processes.

The services made available through the infrastructure constitute a catalogue that must be administered so that service suppliers and customers can locate and communicate with each other. In addition to the services, the other resources that are managed by the infrastructure, such as data sources and users, also need to be administered. The architecture must therefore include capabilities that allow users to subscribe and unsubscribe, as well as the management of roles and the access privileges to the different resources. This is the basic function of the *administration layer*.

The distinct layers can be made functional by means of specialised services and/or through a low level language of the infrastructure itself that glues the different layers together. This SQL-type language makes the system very strong and flexible.

The infrastructure can be considered to be middleware, since it can be seen how it is situated between the interfaces with the people and the services and sources of data that supply the logic and the data (Figure 1).

The middleware infrastructure provides a common set of tools and services that enable both people and other applications to behave as if the computation, the data pools and other distributed resources were all part of one large virtual system. This common service infrastructure simplifies the development of applications, enhances robustness and interoperability, and increases efficiency in the development and maintenance of information systems.

# 3  The INBIOMED Infrastructure

Based on this Service Oriented Architecture approach, we have developed an infrastructure suited to biomedical research that is currently being enhanced in a pilot

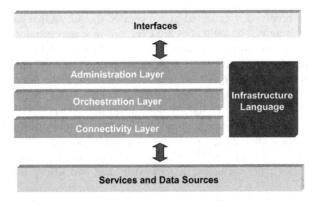

**Fig. 1.** Diagram of the Service Oriented Architecture

experience by the incorporation of data and imaging analysis services, collaborative management services [9], distributed databases (CaCore [10], SRS[11]), interfaces, administration applications, ontology managers, workflow tools, and so forth.

The software development is guided by the objectives and requirements identified by the INBIOMED research network [12] and the Web Services Architecture Working Group [13]. The INBIOMED research network is made up of several bioinformatics research groups that, in turn, collaborate with prestigious biomedical research groups both in Spain and around the world, from whom requirements have been gathered through personal interviews, dedicated workshops and surveys.

The system has as nucleus a Site Web Service that serves as a point of entry to the client applications. It receives the query and processing requests in a language oriented to the batch execution called IQL (Inbiomed Query Language). This language is the key to communicate with the platform. A more detailed description of how the infrastructure was implemented and the language definition can be found in [14].

## 3.1 The INBIOMED Query Language

Every request to the Site Web Service is made by calling a unique method (see figure 2), where must be specified, at least, the sentences that have to be executed, in a string format. In that way, it is consolidated a basic communication interface that will not be modified by the addition of new functionality to the IQL language.

```
String Login(String a_sLoginName, String a_sPassword)
   Starts a new session for the specified Login.
String ExecuteQuery(String a_sGuidSession, String a_sQuery,
String[] a_dsArguments, Boolean a_bWait)
   Executes an IQL script with the specified arguments,
   instructing the Executing Engine to wait the end of the
   script execution to return his result, or return the Guid of
   the associated thread to query his result later.
String Logout(String a_sGuidSession)
   Close the specified session from the client.
```

**Fig. 2.** Site Web Service calling method description

Every request received is executed in a separated thread, so that can be controlled its execution separately. The sentences are executed in a sequential way, allowing in one request ask for data from several data sources and call a data processing service that processes the data obtained, delegating all the processing load in the server. Input data included as part of the request are sent to the server with the query sentences that reference these input data as parameters.

All the data manipulation in IQL is done using an ADO.NET data type called DataSet [15]. This data type can contain traditional relational information including tables and relations in memory, or scalar information implemented like one table with one column with one row of data. A lot of operators had been implemented in IQL to work with this data type, including queries with joins and filters, mathematical and string operators.

To make easy the data manipulation, during the execution of the IQL sentences, variables can be defined to store temporally partial results and be reused during the execution of a sequence of instructions. Once the sentences execution is finished, the variables are removed. In case that you want to store the results to use them later, there is a Persistent Data Repository, which has an analogue structure to a conventional file system with folders and subfolders and with archives that store the result of the execution of IQL sentences. This sentences can be saved and load whenever, whether the user has the proper rights.

As the execution of a sequence of IQL sentences can take a long time, the system can leave running the query in the server and return an ID to the client, that can be used by the client to interrogate to the system about the state of the query, and when the query is finished to recover the result. When the client waits until the end of the execution the result is returned to the client that made the request.

For complex tasks, IQL Procedures can be declared containing other IQL sentences to access to Data Sources located in several nodes, do basic data manipulation, process data calling Web Services located in other nodes, do flow control if necessary and return only the final result. These Procedures are stored and can be executed lately when is needed or be called from other stored Procedures.

The language IQL contains basic sentences to query data from local data sources with ADO.NET or from remote data sources managed by other Site Web Services. ADO.NET grants the accessibility to data through .NET managed providers, OLE DB Providers or widely used ODBC drivers. To call Data Processing Web Services, SOAP is used to invoke the standard method passing all data in strings with XML serialized data to ensure maximum interoperability from different development platforms.

The client applications only have to implement the User Interface with their input forms and the presentation results, delegating in the infrastructure the responsibility to query, store and process the information as much as possible.

The communication between the client applications and the Site Web Service must be made using the SOAP protocol, so the applications have to be implemented using a programming language that is able to consume Web Services. The common Data Model will establish the data structures used in the data flows between the several layers.

As part of the platform development, among others, an administrative tool has been developed named INBIOMED Manager (figure 4), that facilitates the connection with the platform, throwing the execution of IQL sentences and has the necessary options to manage all the elements in the platform such as users, roles, data source connections, or the Data Process Services Catalogue, among others.

```
SET @var = expr;
IF (boolexpr)
{
truesentences;
}
ELSE
{
falsesentences;
}
WHILE boolexpr DO
{
loopsentences;
BREAK;
CONTINUE;
}
CREATE PROCEDURE myproc (@arg1, …)
{
bodysentences;
RETURN expr;
}
SET @var = CALL myproc (expr, …)
```

**Fig. 3.** Complex Procedure structure example with IQL

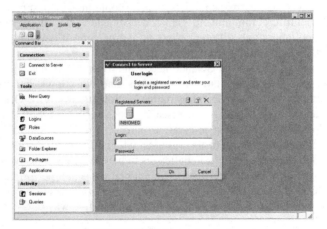

**Fig. 4.** INBIOMED Manager typical screen

## 4   Conclusions: Challenges for Collaborative Research in Biomedicine

After developing the computer infrastructure and beginning to use it in the first pilot trials, one of the fundamental conclusions we reached was that adopting collaboration processes in biomedical research has to overcome two basic obstacles, one of a technological nature and the other involving organisational aspects.

Infrastructures developed following a SOA facilitate interoperability among computer systems but do not completely resolve the problem. If a service is to operate as part of an infrastructure it will have to comply with the standards languages that have been specified for that particular system. Therefore the bioinformatics community will have to define its own standards for collaborative management processes. Such standards will have to go beyond one-off developments such as MAGE-ML [16] (for representing microarray data). These specifications will represent yet another step towards the development of software that is designed to be interoperable from its earliest stages, but until such applications become available the services and databases that are developed for other infrastructures will have to be adapted manually.

Moreover, research process models must also be elaborated to guide the development of specifications. Unlike the field of business management, where work has been carried out for decades on the development of a process perspective (which was initially internal and more recently has become focused on inter firm collaboration), very little has been done to define the processes involved in biomedical research. The term *processes* is understood to mean all the activities that are needed to accomplish a particular objective, their interrelations and the resources that are used at each stage [17]. In order to manage processes we need an information system and an organisational structure. To date, the need to carry out research into suitable models of organisational structures and research processes and their representation in formalised languages has not been covered. The progressive definition of biomedical research process models will provide us with a clearer idea of what scientific workflows [18] are more relevant in biomedicine, as well as the priorities in the development of specifications. Moreover, the reference models, which should include best practices, can be used by the scientific community to gain a deeper and easier understanding of the advantages of collaboration, different ways of approaching it and the organisational structures that are best suited to each project.

The collaborative research in Science needs interoperable information systems to share data, applications, computational resources, and workflows, but technology is not enough. We have develop an infrastructure following a Service Oriented Architecture that can be used to integrated research processes as a proof of concept that technology does not represent a integration barrier. The Web Services technology facilitates interoperability and foster systems integration with flexibility and efficiency. Although there are technical challenges to be solved with this new technology (mainly interoperability issues between different platforms and performance), what is most needed is to develop research process models and organizational approaches to guide the cultural change, from an isolated to a collaborative way of doing science.

# References

1. Kenneth H. Buetow. Cyberinfrastructure: Empowering a "third way" in Biomedical Research. Science. Vol 308. pag. 821-824. 6 May 2005.
2. Oliveira IC, Oliveira JL, Sanchez JP, Lopez-Alonso V, Martin-Sanchez F, Maojo V, Sousa Pereira A. Grid requirements for the integration of biomedical information resources for health applications. Methods Inf Med. 2005;44(2):161-7
3. J.C. Murray et al. A comprehensive human linkage map with centimorgan density. Science 265, 2049-2054. 1994.4. D. Booth et al., Web Services Architecture (WC, Working draft, 2003; /www.w3.org/TR/ws-arch/)

4. Sanjiva Weerawarana, Francisco Curbera, Frank Leymann, Tony Storey, Donald Ferguson. Web Services Platform Architecture: SOAP, WSDL, WS-Policy, WS-Addressing, WS-BPEL, WS-Reliable Messaging, and More. Prentice Hall PTR.2005
5. Werner Vogels. Web services are not distributed objects. IEEE Internet Computing. 59-66.Nov-Dec 2003.
6. Hugo Haas, W3C Web services Working Group. Web services glossary. http://www.w3.org/TR/2004/NOTE-ws-gloss-20040211/
7. I. Foster. The grid: computing without bounds. Sci Am. 2003 Apr; 288(4):78-85
8. Manuel Pastor, Paolo Benedetti, Angelo Carotti, Antonio Carrieri, Carlos Diaz, Cristina Herraiz, Hans-Dieter Höltje, M. Isabel Loza, Tudor Oprea, Fernando Padin, Francesc Pubill, Ferran Sanz, Friederike Stoll & the LINK3D Consortium. Distant collaboration in drug discovery: The LINK3D project. Journal of Computer-Aided Molecular Design, 16: 809–818, 2002.
9. Cacore. http://ncicb.nci.nih.gov/NCICB/core
10. SRS. www.lionbioscience.com
11. Jose A. Heredia. Plataforma Inbiomed: Objetivos y Arquitectura. 21-31. Informática Biomédica. Madrid. 2004. Editorial: Inbiomed.
12. Daniel Austin, Abbie Barbir, Christopher Ferris, Sharad Garg, Web Services Architecture Requeriments. http://www.w3.org/TR/wsa-reqs
13. Estruch, A., Heredia J.A., "Technological platform to aid the exchange of information and applications using web services" Lecture Notes in Computer Science, 3337, 458-468, Nov. 2004.
14. NET Framework Class Library. DataSet Class. http://msdn.microsoft.com/library/default.asp?url=/library/en-us/cpref/html/frlrfsystemdata datasetclasstopic.asp
15. Mage-ML. www.mged.org
16. José A. Heredia. Modelo de información de los procesos de análisis en la investigación de patologías.223-247. Informática Biomédica. Madrid. 2004. Editorial: Inbiomed.
17. William W. Stead, Randolph A. Miller, Mark A. Musen, and William R. Hersh Integration and Beyond: Linking Information from Disparate Sources and into Workflow. J Am Med Inform Assoc. 2000 Mar–Apr; 7(2): 135–145.

# Simultaneous Scheduling of Replication and Computation for Bioinformatic Applications on the Grid*

Frédéric Desprez[1], Antoine Vernois[1], and Christophe Blanchet[2]

[1] LIP, UMR CNRS-INRIA-UCBL 5668
ENS Lyon, France
{Frederic.Desprez,Antoine.Vernois}@ens-lyon.fr
[2] LBRS Laboratory / IBCP Institute
UMR 5086 CNRS, Univ. Claude Bernard Lyon 1
7 passage du Vercors, 69007 Lyon, France
Christophe.Blanchet@ibcp.fr

**Abstract.** One of the first motivations of using grids comes from applications managing large data sets like for example in High Energy Physic or Life Sciences. To improve the global throughput of software environments, replicas are usually put at wisely selected sites. Moreover, computation requests have to be scheduled among the available resources. To get the best performance, scheduling and data replication have to be tightly coupled which is not always the case in existing approaches.

This paper presents an algorithm that combines data management and scheduling at the same time using a steady-state approach. Our theoretical results are validated using simulation and logs from a large life science application (ACI GRID GriPPS). The PattInProt application searches sites and signatures of proteins into databanks of protein sequences.

## 1 Introduction

One of the first motivations of using grids [11, 21] comes from applications managing large data sets [17, 31] such in Life Science [7, 24, 26] or for example in High Energy Physic [23]. Indeed, life Science is a scientific field that produce continuously lot of data through experiences such as complete genome sequencing projects (1220 projects in november 2004 [12]). These raw bioinformatic datasets come generally from different sources located in different institutes, and need to be analyzed with many different algorithms [29]. Grid is a good mean to solve the equation of analysing such large datasets with a large panel of bioinformatic software. To improve the global throughput of software environments, replicas are usually put at wisely selected sites. Moreover, computation requests have to be scheduled among the available resources. To get the best performance, scheduling and data replication have to be tightly coupled which is not always the case

---

* This work was supported in part by the ACI GRID of the french department of research

in existing approaches. Usually, in existing grid computing environments, data replication and scheduling are two independent tasks. In some cases, replication managers are requested to find best replicas in term of access costs. But the choice of the best replica has to be done at the same time as the schedule of computation requests.

Our motivating example comes from an existing life science application (see Section 2). This kind of application has usually the following characteristics: a large number of independent tasks of small duration (for example pattern scanning, searching for signature or functional site of protein family into a databank), reference databases from some MBs to several GBs which are updated on a daily or weekly basis, several computational servers available on the network, and the size of the overall data set is too important to be replicated on every computational server on the whole. The resolution of such application on the grid leads to solve two problems related to replication: **finding how (and where) to replicate the databases** and **choosing wisely the data to be deleted when new data have to be stored.** On the scheduling side, **computation requests must be scheduled on servers by minimizing some performance metric, taking into account the data location.** This paper presents an algorithm that combines data management and scheduling simultaneously using a steady-state approach. Our theoretical results are validated using simulation and logs from a large life science application.

This paper is organized as follows. In a first section, we present the application that motivated this work. In Section 3, we discuss some previous work around data replication, web cache mapping, data and computation scheduling. In Section 4, we present our model of the problem and the algorithm we designed to solve it. Finally, before some conclusions and our future work, we discuss our experimentation using the OptorSim simulator [9] for replica managers.

## 2   Motivating Example

Our motivation for this work comes from the PattInProt application about the search of sites and signatures of proteins into databanks of protein sequences.

Genomic acquiring programs such as full genomes sequencing projects produce large amounts of data, made available to the community. These raw data have to be understood and annotated in order to be useful for further studies and for cross references to and from other datasets. There is also a large number of bioinformatic tools used to analyze these data, and they come from different fields of Bioinformatic (similiraty and homology, protein function analysis, sequence analysis, etc). But many of them can be modeled as shown in Figure 1. Protein function analysis, such as the PattInProt application studied in a grid context by the GriPPS project, can act as a good model of such a bioinformatic application requiring access to several datasets of various sources and sizes.

Functional sites and signatures of protein are very useful for analyzing these data or for correlating different kinds of existing biological data. Sites and signatures of protein can be expressed using the syntax defined by the PROSITE [14] databank, and written as a regular expression. Then, the search of functional sites or signature into databanks can be very similar to simple pattern matching except that some biological relevant error between search pattern and matching protein can be allowed. These methods can be applied, for example, to identify and find a characterization of the potential functions of new sequenced proteins, or to clusterize the sequences contained into international databanks into families of proteins. Most of the time, this kind of analysis, i.e. searching for a matching protein into a databank, is quite fast and its execution time mainly depends on the size of the databanks, but the number of requests for such analysis can be very high as the number of users increases every day thanks to the Internet.

The difficulties come from the fact that the number of datasets used as reference for this kind of search can be large and of very different sizes. These datasets can be international protein sequence databanks such as Swiss-Prot/TrEMBL [13], PIR [34], etc. But they also can be raised quite directly from genome sequencing projects with the translation of the CDS (CoDing Sequence) extracted from the gene sequence obtained. In this case, the number of datasets are as large as the genome project [12], and as variable as the size of the genomes (e.g. 3 Gpb for the human genome or 120 Mpb for the Saccharomyces cerevisiae genome).

Figure 1 describes the classical architecture of a bioinformatic application. We can notice two kinds of components connected together by the Internet network. On one side, there is a set of clients which submit requests to computational servers. Clients are seen as personal computers that have no knowledge from each others but which are often gathered in some big sites. Usually, these are office computer of researchers in biology or bioinformatics research centers. Computational servers are dedicated to computation. They usually are single processors computers or, sometime clusters of computers. These servers locally store a limited number of reference databanks and algorithms on which they can be applied. Often, they are independent from each other and are located and managed in bioinformatic centers such as EBI [1], NCBI [3], SIB [5], or NPS@ [4]. Clients access computational servers through web portals or directly by asking for an account to servers administrators.

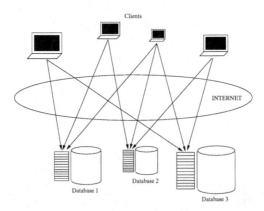

**Fig. 1.** Current view of a Bioinformatic Application

We accessed the logs of such a cluster that provides computational work through a bioinformatic web portal and allows users to apply some well known algorithms to existing databanks. The portal is the "Network Protein Sequence Analysis - NPS@" bioinformatic portal [18], providing around 40 algorithms and 7 databanks to biologists and bioinformaticians for biological queries. NPS@ is up since 1998 and had answered to more than 6 million bioinformatic analyses. It currently computes more than 3 thousands of such analyses per day. The portal and the cluster are located at IBCP [2] in Lyon (France), a research institute on biology and chemistry of proteins, and they are managed by the bioinformatic team of this laboratory.

Input of such requests are user's protein sequences or signatures that usually do not excess a few kilobytes. This is a centralized cluster with limited capacities so only major databanks and algorithms are available. This is currently improved by the GriPPS [22] which aims at distributing its work among a large number of servers made accessible through the grid.

## 3   Related Work

Data replication has attracted much attention over the last decade. Our work is connected to several others: high performance web caches, data replication, and scheduling in grids.

With the rapid growth of the Internet, scalability became in major issue for the design of high performance web services [35]. Several researches have studied how to optimally replace data in distributed web caches [15, 30]. Even if this problem seems to be close to ours, the fundamental difference between the two is that our problem has a non-negligible computation cost that depends upon the speed of the machine hosting a given replica.

In computation grids, some work exist around replication [28] and among them the researches for the Datagrid project from the CERN [6]. OptorSim [9, 19] allows to simulate data replication algorithms over a grid. This tool is more precisely described in Section 5.1. In [8], several strategies are simulated like unconditional replication (oldest file deleted, LRU) and with an economic approach. The target application is the data management of the Datagrid physic application. Simulation shows that the economical model is as fast as classical algorithms. OGSA [27] also proposes a replication service which is currently not connected to request scheduling. In [25], the authors describe Stork, a scheduler for data mapping in grid environments. Data are considered as resources that have to be managed as computation resources. This environment is mainly used to be able to map data close to computation during the scheduling of task graphs in Condor.

The closest researches to the results presented in our paper are the one that aim to schedule computation requests and data mapping on remote sites at the same time. In [32, 33], several strategies are evaluated to manage data and computation scheduling. These strategies are either strongly related to the scheduling of computation or completely disconnected. However, these strategies are highly dynamic and the mapping is not proved close to the optimal. In [16], the authors present an algorithm (Integrated Replication and Scheduling Strategy) in which performance are iteratively improved by working alternatively on the data mapping and the task mapping.

## 4   Joint Data and Computation Scheduling Algorithm

In this section, we present the algorithm we designed that combines data replication and scheduling (Scheduling and Replication Algorithm or SRA).

### 4.1   Model

Our model is based on three kinds of objects: a set of computational servers $P_i$, $i \in [1..m]$, a set of data $d_j$ of size $size_j$, $j \in [1..n]$ and a set of algorithms $a_k$, $k \in [1..p]$ that use one $d_j$ as an input. We call a request, or task, $R_{k,j}$ a couple $(a_k, d_j)$ where $a_k$ is an algorithm and $d_j$ is a data which will be used as an input of the algorithm $a_k$. All algorithms can not be applied on all kind of data, so we define $v_{k,j} = 1$ if $R_{k,j}$ is a request that is possible, otherwise $v_{k,j} = 0$. The complexity of algorithm $a_k$ is linear in time with the size of the data. Thus the amount of computation needed to compute a request $R_{k,j}$ is $\alpha_k \cdot size_j + c_k$, where $\alpha_k$ and $c_k$ are two constants defined for each algorithm $a_k$. For each server, we also introduce $n_i(k, j)$, which is the number of requests $R_{k,j}$ that will be executed on server $P_i$. A server $P_i$ is described by two constants: its computational power $w_i$ and its storage capacity $m_i$. $f_{k,j}$ is the fraction of request of type $R_{k,j}$ in the pool of requests. We suppose that this proportion of request is always the same whatever the interval of time you consider as soon as it is large enough. Our study focus on managing data and their replication taking all these parameters into account to improve the computation time of a set of requests. We also make the assumption that it is possible to store at least one replica of each data.

Our goal is to find a placement of the databanks that maximizes the throughput of the platform. We call $TP$ this throughput. It is the number of requests that can be executed per unit time on the platform. The ratio of each kind of request is defined by $f(j, k)$. So the number of requests of type $R(k, j)$ that is executed is restricted by this ratio in order to avoid to take in account more requests than the number that will be submitted.

The throughput is limited by some constraints due to the specifications of the platform and the requests. First, the space on each server is limited by its storage capacity. So the total size of data stored on server $P_i$ cannot exceed $m_i$ Then number of requests a server $P_i$ can handle is restricted by its computation capacity. Thus the amount of computation that a server will execute cannot exceed $w_i$. To compute a request $R_{k,j}$ on server $P_i$, the data $d_j$ should be stored on this server. If it is not the case, then $n_i(k, j)$ should be equal to 0, otherwise, $n_i(k, j)$ is limited by the maximal number of requests $R_{k,j}$ this server can handle.

Let $\delta_i^j = 1$ if there is a replica of data $d_j$ on server $P_i$, $\delta_i^j = 0$ otherwise. Considering previous constraints, we can define the linear program of Figure 2. The solution of this linear program will give us a placement for the databanks on the servers but also, for each kind of job, on which server they should be

MAXIMISE $TP$,
CONSTRAINT TO

$$\begin{cases}
(1) \; \sum_{j=1}^{n} \delta_i^j \geq 1 \\
\quad 1 \leq i \leq m \\
(2) \; \sum_{j=1}^{n} \delta_i^j . size_j \leq m_i \\
\quad 1 \leq i \leq m \\
(3) \; n_i(k,j) \leq v_{k,j} . \delta_i^j . \frac{w_i}{\alpha_k . size_j + c_k} \\
\quad 1 \leq i \leq m, 1 \leq j \leq n, 1 \leq k \leq p \\
(4) \; \sum_{k=1}^{p} \sum_{j=1}^{n} n_i(k,j)(\alpha_k * size_j + c_k) \leq w_i \\
\quad 1 \leq i \leq m \\
(5) \; \sum_{i=1}^{m} n_i(k,j) = f_{k,j} . TP \\
\quad 1 \leq i \leq m, 1 \leq j \leq n \\
(6) \; \delta_i^j \in \{0,1\} \\
\quad 1 \leq i \leq m, 1 \leq j \leq n
\end{cases}$$

**Fig. 2.** Linear Program Formulation

executed. More precisely, for a kind of request $R_{k,j}$, we know how many job can be executed on the platform and we also know how many requests of this kind should be executed on each server to reach optimal throughput. Thus, with the placement of data, the linear program also gives good information for the scheduling of requests.

## 4.2   A Greedy Solution

Starting with the same platform and algorithm models, we also design a greedy algorithm to solve the mapping problem. The idea behind this algorithm is to try to map data that need the most computational power to the server that has the most computation capacities first.

The algorithm starts by computing the amount of computation needed by each data proportionally to its usage. Then data are sort by decreasing values of this amount. We also sort the list of servers by decreasing computation abilities. We try to map the data that need the most computational power to the server that has the highest computation capacity. If there is not enough space, we try to map the data on the second server and so on and so forth until the data is mapped. Then, we try to place the second data by computation need to the first server. We do this operation for each data. If a data cannot be placed on any server we skip it and try to place the following data item. We restart from the beginning of data list till there is enough space available to place a data on the platform.

## 5   Experiments

To experiment the results of our model, we used OptorSim [9, 19], a simulator of Data Grid environments developed in the Work Package 2 of EU Datagrid project [6]. We have modified OptorSim to exactly match our needs.

## 5.1    Experimental Environment

The target platforms of our studies are distributed and may span multiple administrative domains. Therefore, it may be quite difficult to conduct repeatable experiments for long running applications on such systems. So we choose to make use of simulation to experiments our algorithms. Other advantage to use simulation is the ability to easily test our algorithm with different network configurations which is not possible using experiments on a real platform.

The simulated grids have five major components. Computing Elements (CE) act like gateways, or masters of a batch scheduler system and will distribute jobs that are submitted to them to their Worker Nodes (WN). Worker Nodes execute jobs and are defined by their computation power expressed in Mflops. All Worker Nodes managed by the same CE have the same capacity of computation, but WN from different CE may have different capacities.

The third kind of component is the Storage Element (SE). It is where data are stored and is defined by its storage capacity (in MB). The same file can be stored on different SEs at the same time. To work properly, a CE should have a local SE that is accessible by all of its Worker Nodes. Access time to data located on the local SE by a WN is considered to be null.

The Replica Manager (RM) is in charge of all data movements between sites. And finally, jobs are created and scheduled by the Resource Broker which is able to instantiate communications with CE and RM to get all information needed about SE, network bandwidth, job queues, etc. for scheduling purpose. In our case, a job is defined by an algorithm and a databank on which the algorithm is applied.

For our experiments, we extracted from raw logs all information about data sets and algorithm usage. With external information about data sizes, algorithm computation costs, and a description of the target platform, we generated the concrete instance of the linear program described in Section 4. This linear program is solved using *lp_solve* [10]. The results give us all information about data mapping and job scheduling. These outputs are used, with other configuration files, as inputs for the simulator.

For the experiments, the topology of the simulated platform is inspired from the architecture of the European DataGrid testbed. There are ten clusters of eight nodes with associated SE and seven routers without any storage nor computation abilities.

Requests are submitted to the RB with a frequency around ten per second. This could seems to be a very high rate, but discussions with the biologist and bioinformatic community lead to the conclusion that the more computation power we can give them, the more they will use.

## 5.2    Experiment Results and Discussion

In this section we will discuss our experiments using OptorSim and the results we obtained. We have done simulations for three kinds of mapping and schedulers.

The first one, *SRA*, corresponds to our algorithm. Scheduling and mapping that are used for the simulation are those that match the solution of our linear program.

In the *MCT* (for Minimum Completion Time) simulation, only the mapping has been done using the results of the linear program. The scheduling is on-line: at each request submission, it tries to find the computation server that should be able to finish this task first (considering time to retrieve data if needed and computation time of all jobs already scheduled on the CE).

Finally, the *greedy* simulation is done using the mapping of the greedy algorithm. The scheduling is done with the previous on-line scheduler. Simulations have been done for a pool a 40000 requests.

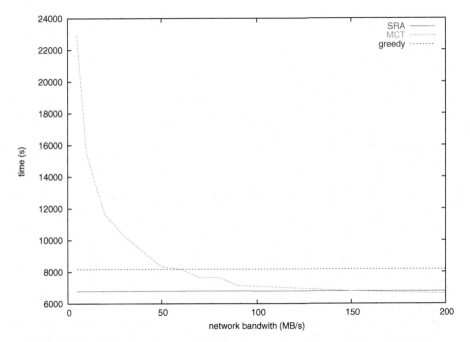

**Fig. 3.** Execution time for 40000 jobs as function of the network bandwidth

Figure 3 shows the execution time of whole set of requests depending on the network bandwidth. In this simulation, the bandwidth between nodes is chosen to be homogeneous to see more easily its impact on execution time. On Figure 3, we can see that for SRA and greedy, the time of execution is totally constant and independent of network bandwidth. It is due to the fact that there are no data movement with these two methods.

But reasons for which there are no movement are not the same in both cases. With SRA algorithm, the scheduling is computed at the same time as the placement. So the scheduler always schedules a job on a server that has needed data for this request. In the greedy case, the scheduling uses an on-line MCT

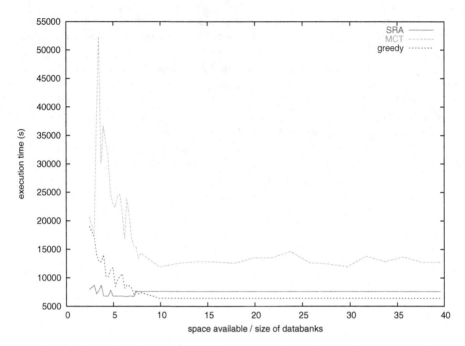

**Fig. 4.** Execution time for 40000 jobs as a function of available space on SE

method but there is still no data movement because the algorithm totally fills the available space in the platform. So the scheduler always schedules requests where a data is available. As MCT favours the execution time of current request to schedule, it does a lot of transfers. But its lack of knowledge on request usage scheme leads him to perform a lot of errors and useless data transfers. Then, it becomes efficient only when transfers costs are negligible in front of computation costs.

Figure 4 shows the execution time of same set of 40000 requests depending on the storage space available on the platform. The space is expressed as the ratio between the total volume of databanks and the global space available. For this simulation the network bandwidth is equal to 10MB/s. We can notice that for all kind of mapping and scheduling algorithms, the execution time decreases with the increase of available space. It can be easily explained by the fact that the more space is available, the more replicas can be placed on different servers. As we can expect, when storage space is small, less than 8 times the size of databanks, our solution gives better results than greedy and MCT. The linear program makes a better use of restricted resources. With the increase of available space, the results of the greedy algorithm improves regularly to become better than the SRA algorithm. This appears when next to all databanks can be stored on each server.

When space storage is very limited, the results of our algorithm are not regular. That comes from our heuristic that constructs an integer solution of the linear program from the solution over rational numbers. With a small storage space, the impact of a bad mapping choice has a high impact on the objective function. In this case, we notice very high differences between value of the objective function of the approximation integer solution and the solution in rational numbers. When available space becomes large enough, our integer approximation gives the same result of the objective value than resolution in rational number.

# 6    Conclusion and Future Work

In this paper, we have presented an algorithm that computes at the same time the mapping of data and computational requests on these data.

Our approach uses a good knowledge of databank usage scheme and of the target platform. Starting with these information, we have designed a linear program and a method to obtain a mixed solution, *i.e.*, integer and rational numbers, of this program. With the OptorSim simulator, we have been able to compare the results of our algorithm to other approaches: a greedy algorithm for data mapping, and an on-line algorithm for the scheduling of requests.

We came to the conclusion that when the storage space available on the grid is not large enough to store all databanks that lead to very time consuming requests on all computation servers, then our approach improves the throughput of the platform. But our heuristic for approximating an integer solution of the linear program does not always give the best mapping of data and can give results that are very far from the value of the objective function in the solution over rational number.

Our future work will consist on adding communication costs for the requests in the model to be able to consider other kind of applications. We are also working on an implementation of these algorithm in the DIET [20] environment to deploy efficiently the GriPPS [22] application. A replica manager will be designed and developed in this environment.

## Acknowledgements

The authors would like to thanks A. Legrand, L. Marchal and Y. Robert for their work on steady-state scheduling and discussion about the model used in this article. They also would like to thank C. Combet for his work and the access on the NPS@ bioinformatic portal.

## References

1. EBI. http://www.ebi.ac.uk.
2. Inst. de Biologie et Chime des Protéines. http://www.ibcp.fr.
3. NCBI. http://www.ncbi.nlm.nih.gov.
4. NPS@. http://npsa-pbil.ibcp.fr.
5. SIB. http://www.isb-sib.ch/.

6. The European DataGrid Project. http://www.eu-datagrid.org.

7. R. Apweiler, A. Bairoch, and C. H. Wu. Protein sequence databases. *Current Opinion in Chem. Bio.*, 8:76–80, 2004.

8. W. Bell, D. Cameron, L. Capozza, A. Millar, K. Stockinger, and F. Zini. Simulation of Dynamic Grid Replication Strategies in OptorSim. In *Proc. of the 3rd Int. Workshop on Grid Comput. (Grid'2002)*. Springer Verlag, Nov. 2002.

9. W. Bell, D. Cameron, L. Capozza, A. Millar, K. Stockinger, and F. Zini. OptorSim - A Grid Simulator for Studying Dynamic Data Replication Strategies. *International Journal of High Performance Computing Applications*, 17(4), 2003.

10. M. Berkelaar. LP_SOLVE. http://www.cs.sunysb.edu/~algorith/implement/lpsolve/implement.shtml.

11. F. Berman, G. Fox, and A. Hey, editors. *Grid Computing: Making the Global Infrastructure a Reality.* Wiley, 2003.

12. A. Bernal, U. Ear, and N. Kyrpides. Genomes OnLine Database (GOLD): A Monitor of Genome Projects World-Wide. *NAR*, 29:126–127, 2001.

13. B. Boeckmann, A. Bairoch, R. Apweiler, M.-C. Blatter, A. Estreicher, E. Gasteiger, M. Martin, K. Michoud, C. O'Donovan, I. Phan, S. Pilbout, and M. Schneider. The SWISS-PROT Protein Knowledgebase and its Supplement TrEMBL in 2003. *Nucleic Acids Res.*, 31:365–370, 2003.

14. P. Bucher and A. Bairoch. A Generalized Profile Syntax for Biomolecular Sequences Motifs and Its Function in Automatic Sequence Interpretation. In *Proceedings 2nd International Conference on Intelligent Systems for Molecular Biology*, volume 2, pages 53–61. AAAIPress, 1994.

15. V. Cardellini, E. Casalicchio, M. Colajanni, and P. Su. The State of the Art in Locally Distributed Web-Server Systems. *ACM Computing Surveys*, 34(2):263–311, June 2002.

16. A. Chakrabarti, R. Dheepak, and S. Sengupta. Integration of Scheduling and Replication in Data Grids. In *Proceedings 11th International Conference on High Performance Computing (HiPC 2004)*, pages 375–385. Springer, Dec. 2004.

17. A. Chervenak, I. Foster, C. Kesselman, C. Salisbury, and S. Tuecke. The Data Grid: Towards an Architecture for the Distributed Management and Analysis of Large Scientific Datasets. *J. of Netw. and Comp. Appl.*, 23:187–200, 2001.

18. C. Combet, C. Blanchet, C. Gourgeon, and G. Deléage. Nps@: Network protein sequence analysis. *TIBS*, 25, No 3:[291]:147–150, Mar. 2000.

19. R. C.-S. D.G. Cameron, A. Millar, C. Nicholson, K. Stockinger, and F. Zini. Evaluating Scheduling and Replica Optimisation Strategies in OptorSim. In *4th International Workshop on Grid Computing (Grid2003)*. IEEE Computer Society Press, Nov. 2003.

20. DIET. http://graal.ens-lyon.fr/DIET/.

21. I. Foster and C. Kesselman, editors. *The Grid 2: Blueprint for a New Computing Infrastructure.* Morgan Kaufmann, 2004.

22. GriPPS. http://gripps.ibcp.fr.

23. W. Hoscheck, J. Jaen-Martinez, A. Samar, H. Stockinger, and K. Stockinger. Data Management in an International Data Grid Project. In *First IEEE/ACM Int'l Workshop on Grid Computing (Grid 2000)*, Dec. 2000.

24. N. Jacq, C. Blanchet, C. Combet, E. Cornillot, L. Duret, K. Kurata, H. Nakamura, T. Sil-vestre, and V. Breton. Grid as a Bioinformatic Tool. *Parallel Comp.. Special issue: High-performance parallel bio-comp.*, 30(9-10):1093–1107, 2004.

25. T. Kosar and M. Livny. Stork: Making Data Placement a First Class Citizen in the Grid. In *Proceedings of 24th IEEE Int. Conference on Distributed Computing Systems (ICDCS2004)*, Mar. 2004.

26. A. Krishnan. A Survey of Life Sciences Applications on the Grid. *New Generation Computing*, 22:111–126, 2004.

27. P. Kunszt and L. Guy. *Grid Computing: Making the Global Infrastructure a Reality*, chapter The Open Grid Services Architecture for Data Grids, pages 385–435. Wiley, 2003.

28. H. Lamehamedi, B. Szymanski, Z. Shentu, and E. Deelman. Data Replication Strategies in Grid Environments. In *Proc. 5th International Conference on Algorithms and Architecture for Parallel Processing, ICA3PP'2002*, pages 378–383. IEEE Computer Science Press, Oct. 2002.

29. G. Perriere, C. Combet, S. Penel, C. Blanchet, J. Thioulouse, C. Geourjon, J. Grassot, C. Charavay, M. Gouy, L. Duret, and G. Deleage. Integrated Databanks Access and Sequence/Structure Analysis Services at the PBIL. *Nucleic Acids Res.*, 31:3393–3399, 2003.

30. S. Podlipding and L. Böszörmenyi. A Survey of Web Cache Replacement Strategies. *ACM Computing Surveys*, 35(4):374–398, Dec. 2003.

31. X. Qin and H. Jiang. Data Grid: Supporting Data-Intensive Applications in Wide-Area Networks. Technical Report TR-03-05-01, Univ. of Nebraska-Lincoln, May 2003.

32. K. Ranganathan and I. Foster. Decoupling Computation and Data Scheduling in Distributed Data Intensive Applications. In *Proc. of the 11th Int. Symp. for High Performance Distributed Computing (HPDC-11)*, July 2002.

33. K. Ranganathan and I. Foster. Simulation Studies of Computation and Data Scheduling Algorithms for Data Grids. *Journal of Grid Computing*, 1(1):53–62, 2003.

34. C. Wu, L. Yeh, H. Huang, L. Arminski, J. Castro-Alvear, Y. Chen, Z. Hu, P. Kourtesis, R. Ledley, and B. e. a. Suzek. The Protein Information Resource. *Nucleic Acids Res.*, 31:345–347, 2003.

35. C. Xu, H. Jin, and P. Srimani. Special Issue on Scalable Web Services and Architecture. *Journal on Parallel and Distributed Computing*, 63, 2003.

# The INFOBIOMED Network of Excellence: Developments for Facilitating Training and Mobility

Guillermo de la Calle[1], Mario Benito[1], Juan Luis Moreno[1], and Eva Molero[2]

[1] Biomedical Informatics Group, Artificial Intelligence Lab./Facultad de Informática,
Universidad Politécnica de Madrid, Madrid, Spain
{gcalle,mariobr,jlmoreno}@infomed.dia.fi.upm.es
[2] Biomedical Informatics Research Group,
Municipal Institute of Medical Research - IMIM, Barcelona, Spain
emolero@imim.es

**Abstract.** Enhancing training and mobility in the area of Biomedical Informatics (BMI) is one of the most important objectives of the European Network of Excellence INFOBIOMED. Based on the lessons learned from previous decades of experiences in teaching Medical Informatics and Bioinformatics, an action plan has been elaborated. This plan is structured into three actions: (a) a survey to analyze and evaluate the situation, needs and expectations in BMI. (b) A Biomedical Informatics course database (ICD) containing the relevant keywords in the area, and (c) the design and implementation of a Mobility Brokerage Service (MBS) to enhance mobility and exchanges in the area. This paper describes the overall approach and technical characteristics of the MBS. It follows an innovative service-oriented architecture based on Web Services, providing distributed access to on-line information sources. This approach is being evaluated and reused for different research applications within the Network.

## 1 Introduction

Medical Informatics (MI) and Bioinformatics (BI) have usually kept independent ways along their history. Experts on each discipline have rarely interacted between them to exchange their knowledge and expertise. Studies –such as the BIOINFOMED Study (EC-IST 2001-35024) funded by the European Commission– have stated the opportunities and challenges that might arise with the synergy generated between both disciplines. As a result of that collaborative approach, Biomedical Informatics (BMI) has been created as a novel scientific discipline.

From the perspective of the BMI, new methods, technologies, standards and tools are being developed to manage the enormous amount of data produced in hospitals or laboratories. These techniques would allow experts to extract, analyse and discover new knowledge that could be applied in many diseases [1]. Integrative approaches are needed to combine all the information coming from different studies, trials and experiments. Data from the sources will be completely heterogeneous, such as information from traditional health records and genomic or proteomic data [2].

The European Network of Excellence (NoE) INFOBIOMED [4] was born with the main objective of strengthening the BMI community from a European perspective, oriented to support individualised healthcare. It has been funded within the VI Framework Programme for Research and Technological Development of the Information Society Directorate-General of the European Commission (EC). The NoE consor-

J.L. Oliveira et al. (Eds.): ISBMDA 2005, LNBI 3745, pp. 274–282, 2005.

tium is composed by partners from ten different European countries. Benefits from the synergy between MI and BI have been widely studied by experts in the previous BIOINFOMED Study [5] which was the antecedent of the present INFOBIOMED network. Experts working on the BIOINFOMED study pointed out some inconvenients and challenges (cultural, legal, scientific and ethical) that had to be addressed to consolidate BMI as a new discipline.

Within the scope of the INFOBIOMED network, other goals have been established to demonstrate the benefits of the BMI approach for supporting individualized healthcare. From a pragmatic point of view, four pilot applications have been defined to solve several problems in different areas. New experts have to be trained on novel BMI techniques and therefore it is crucial to establish of a strong and robust community. People exchanges among different organizations should be promoted to eliminate barriers and obtain the maximum benefits of collaborative work. For such reason, mobility is an important objective of the INFOBIOMED network and efforts are being done to foster exchanges among partners. In this paper, several initiatives and informatics solutions, related to training and mobility, are presented.

## 2  Background

Both MI and BI have established their own programs, as separated disciplines. On the one hand, MI programs have been created since the 70s until present. Several postdoctoral programs were funded by the US National Library of Medicine. National hospitals and academic institutions offered the first proposals for MI in Europe and later the EC provided funding for some initiatives related with distance learning over the Internet and multimedia applications.

On the other hand, the first initiatives on BI arose along the 1990s. Currently, there exist many academic programs that cover all levels, from undergraduate to PhDs. Some of them are on-line courses or degrees (over the Internet) that were created to facilitate the training process of the candidates. These courses address the large demand of research groups and industry in the field. Both the EC and US National Institute of Health support these kinds of initiatives.

Regarding BMI, it is still a novel discipline. Since the conclusion of the Human Genome Project, new expectations have arisen within the scientific community. New drugs and novel diagnostic tests are expected to be found in order to foster and improve individualised healthcare, oriented to specific groups of people. Future informatics applications should be able to integrate different kinds of data (genomic, genetic, proteomic, health records...) and facilitate their management.

Recently, new education programs have been developed by several institutions, such as Stanford University which offers an integrated program, named 'Biomedical Informatics Training Program' [7]. Within the European framework, the Karolinska Institute in Sweden, has launched a similar PhD program in Medical Bioinformatics "to build up competence in bioinformatics with special emphasis on biomedical and clinical applications".

Within the INFOBIOMED analysis, a lack of efforts on BMI education and training has been detected. This is an objective that INFOBIOMED tries to address through its Training and Mobility program. In the following sections, we explain the activities carried out by the network to foster of researchers' training and mobility.

# 3  Methods

Three different actions have been adopted to analyse and address training and mobility problems in BMI: (1) a mobility survey to determine the actual needs, lacks and expectations of the professional on this field, at the European level; (2) the construction of a BMI course database, containing the most relevant educational programs, and (3) an on-line Mobility Brokerage Service (MBS), accessible through the Internet, to foster the exchanges of researchers among different organizations. These three activities are explained in more detail below.

## 3.1  A Survey on Mobility

The Survey on Mobility consisted on a set of questions, presented in HTML-based format and accessible over the Internet. Access was restricted to partners of the NoE. When a survey was completed, the information was stored automatically into a database for subsequent analysis. Finally, 90 questionnaires were collected and an statistical analysis (including average, frequency, maximum, minimum and standard deviation) was carried out to analyze the information. The objective was to identify the main obstacles and gaps that may hamper the deployment of BMI. Another additional goal of the survey was to use the results as an additional information, aimed to design and build the Mobility Brokerage Service and the Course Database.

## 3.2  INFOBIOMED Course Database (ICD)

Several courses and programs related with BMI were gathered into a database. Other MI and BI educational activities were also included. Some criteria were imposed before including a course into the database. For instance, it was preferred to include English-based programs or courses that were given in any European country.

In the first stages, the ICD was published in the private zone of the official web site of the NoE. The INFOBIOMED partners were allowed to access data in order to refine the contents before giving public access to the database to the outside scientific community. The main goal for collecting such kind of information was to analyze and establish the ideal profile expected for experts on BMI. Depending on the topic managed in those courses, that profile might be determined.

## 3.3  Mobility Brokerage Service (MBS)

The MBS is the principal activity developed by the NoE to promote and facilitate mobility of researchers among institutions, in order to improve their training and expertise. Most design requirements were obtained from the results of the mobility survey. The design followed an innovative service-oriented architecture based on Web Services technology. The MBS is accessible over the Internet and presents a traditional web interface, based on HTML pages and forms. More details about the implementation and visual aspects are given in the next sections.

# 4 Results

Once the analysis of the mobility survey was finished, some relevant results can be reported. As mentioned before, the goal for carrying out the survey was to determine the main barriers and difficulties that BMI professionals and researchers have to overcome to improve their training and expertise in this field. Table 1 presents a summary of these findings.

**Table 1.** Main barriers identified on the mobility survey, for participating in BMI activities. These obstacles hinder the definitive take-off of BMI, according with the opinion of the participants

| Description | % |
| --- | --- |
| Lack of funding | 73,33 % |
| Lack of information about host offers | 50,00 % |
| Lack of time | 46,67 % |
| Personal issues (family, etc) | 20,00 % |
| Lack of practical information about the host site | 15,00 % |
| Lack of permission from your organization | 10,00 % |
| Not interested in mobility activities | 5,00 % |
| No relevant benefits perceived | 3,33 % |

Results on Table 1 are shown on numerical descending order. As shown, there are three principal lacks: lack of time, lack of information about host offers and above all, lack of funding. Almost 75% of respondents considered these three reasons the main obstacles for participating on mobility activities. After analysing the results, the INFOBIOMED consortium proposed and implemented a series of methods to alleviate the two principal problems. In order to supply the lack of information, a Mobility Brokerage Services (MBS) has been designed and implemented. Host organizations can use it to publish their job offers and candidates can search for them, or express their demands or wishes. The MBS will be detailed later.

On the other hand, to deal with the lack of funding, the INFOBIOMED Training and Mobility Committee (TMC) proposed to the consortium the definition of a funding mechanism. A Mobility Funding System (MFM) was approved by all the partners. It consists on a grant available for candidates who wish to work for some time at another organization. Each candidate must complete an application form justifying the purpose of the mobility activity and send it to the TMC. Later, the TMC approves or denies each exchange. At this moment, those incentives are only available for members of the NoE and within partner organizations. In the future, if the system is a success, it is thought to be open for external applicants.

The INFOBIOMED Course Database (ICD) was designed with two major purposes; first, to have a comprehensive repository with information on already existing academic programs and training activities on BMI. The second purpose is closely related with the first one. Several professionals and experts have established an ideal curriculum on BMI. 'Actual' and 'ideal' curricula can be compared using the ICD and from that comparison, different lacks and gaps on BMI training might be identified. Besides having a complete and update repository of educational offers, another objective sought with the construction of the ICD is to promote BMI among young students.

In the first stages, the ICD only contained academic activities within the network. Later, it was enlarged with information from other European programs. To gather and manage the ICD, state of the art technologies on Information Retrieval (IR) were applied. A training topic thesaurus [8][9] was built in order to get better and more precise results for each search. Having a common terminology is useful for detecting gaps and comparing profiles. The reason for building a novel thesaurus is that, although there already exist other ones –e.g., the MeSH (Medical Subject Headings) from the US National Library of Medicine (NLM) – none of them covers completely the BMI field. To build the ICD, a software tool called Collexis© [10] has been used to retrieve and classify the data. Several scientific areas have been established to index the thesaurus which first prototype was released at the end of 2004.

## 4.1  Mobility Brokerage Service (MBS)

The INFOBIOMED Mobility Brokerage Service can be defined as a secure on-line marketplace accessible over the Internet. Its main objective is to facilitate the communication between host organizations and candidates. As mentioned above, according with the results of the analysis of the survey on mobility, the lack of information about offers on BMI was one of the most important drawbacks that may hinder the definitive consolidation of this novel discipline. The MBS is the Consortium approach to bridge the existing gaps. Both researchers and host organizations would have a common meeting place; institutions might publish offers and look for candidates to cover their needs and researches or professionals might use the system to find offers or publish their own demands or expectations. Benefits would be obtained on both sides. Besides traditional manual searches, the MBS has been provided with some advanced computing features to perform searches automatically without human intervention. Such "matching process" will be explained later.

During the development process, several prototypes have been deployed and tested by the members of the TMC in order to detect and polish different details. After multiple discussions and agreements, the first final release has been deployed on June 2005. From a technical point of view, the design of the MBS follows an innovative Service-oriented architecture [11] based on Web Services and standards like XML. Before to explain in more detail the features and possibilities of the system, some concepts about Web Services technology are introduced below.

## 4.2  Web Services Concepts

Numerous technologies have arisen to implement distributed systems and deal with their associated problems. Over the last 30 years many models and different architectures have appeared, with more or less success. For instance, Remote Procedure Call (RPC), Java Remote Method Invocation (RMI), Distributed Component Object Model (DCOM) or Common Object Request Broker Architecture (CORBA) are examples of technologies created for building distributed systems. They are completely independent technologies but also incompatible. At this time, many applications follow these approaches, but actually none of them has been adopted as a standard.

Web Services (WS) are a new technology aimed to integrate systems in a transparent way. WS are supported by consolidated protocols and languages such as HTTP and XML. Data are represented into a XML format and transported using a HTTP protocol. Usually, the HTTP traffic is not filtered or blocked by firewalls in the Internet. It is an advantage over traditional approaches.

Besides the standards mentioned above, another three new protocols or languages have been defined, closely related on Web Services:

- **WSDL (*Web Services Definition Language*)** [12] is a W3C proposal to define service interfaces in XML. The WSDL documents are expressed on a human/machine readable format. Thus, the applications can understand automatically how to use the offered service.
- **SOAP (*Simple Object Access Protocol*)** [13] is a protocol, also based on the XML standard, for the communication among applications over the Internet. SOAP defines the format of the messages too.
- **UDDI (*Universal Description, Discovery, and Integration*)** [14] is a protocol that allows the applications to register, search and use Web Services on Internet.

Several institutions and companies have developed their own tools and frameworks for the implementation of the Web Services and protocols and languages mentioned above. WS provide an enormous advantage: since they are based on XML standards, they are platform-independent. This means that heterogeneous systems might be integrated more easily. In the other hand, a drawback of WS is that their configuration and deployment might require important programming efforts.

### 4.3  Implementation

Several technologies have been used to implement the MBS. From the beginning, the system has been envisaged as a web application. *Java* was selected as the development platform and thus, all the web pages are dynamically generated using *JavaServer Pages (JSPs)* and *Servlets*, the Java technology created for the development of that kind of web-based systems. The INFOBIOMED official web site style, where the system will be integrated, has been applied for the visual design of the pages. A screenshot of the final appearance is shown on Fig. 1. *Apache Jakarta Tomcat 5.0* has been used as the web server and servlet container and *MySQL 4.0.20* has been the Database Management System (DBMS) selected to store data by the web application. Both *Tomcat* and *MySQL* have become very popular since they are free software.

The MBS can be reached at the INFOBIOMED web site. There exist public functionalities accessible to every external user but other functionalities demand users to get registered. During the registration process, users are asked to fill some forms with their personal information, depending on the type of user selected. Three different kinds of users have been defined. Basically, we call *Host Organizations* to the type of users who publish mobility or job offers and *Candidates* are the users who access the system to search mobility or job offers. Also, host organizations can look for candidates and candidates can publish their own job or mobility demands. Finally, to manage on-line the most important features of the MBS, an administrator role has been defined.

Besides typical manage functionalities, the MBS offers several ways to find mobility opportunities or job offers. As mentioned above, that feature is only available to registered users. An advanced automatic search process called *matching process* has been implemented. This process is explained in the following section. Additionally, searches can be manually performed by users using the different existing facilities; from simple textual Google-liked searches optionally filtered by countries to graphical searches by means of visual maps.

**Fig. 1.** The INFOBIOMED Mobility Brokerage Service main page. It has been integrated into the official web site of the NoE (www.infobiomed.org) following the same style. Latest offers and demands are displayed below a short description of the service. Control access to the private zone is also accessible from this page, just like additional sections (News, FAQs...)

From the beginning, the MBS has been envisaged as a distributed system. Design has been elaborated following an open service-oriented approach. The system architecture is shown on Fig. 2. Web Services technology has been used for implementing all the internal functionalities of the system.

WS are a platform-independent technology especially suitable for integrating heterogeneous systems. For the MBS, they have been used to perform several actions such as validation or identification of the users, store and retrieve information from databases. Each WS can be placed in the same computer or widespread over the Internet without limitations. That issue is one of their most important advantages. Also, databases containing the information managed by the WS, might be distributed in different physical places whenever a secure communication channel is established. The MBS has been designed as a web application composed by several Web services. These services are controlled and coordinated by the application and they are transparently accessed by means of common interfaces.

Security on distributed systems is especially critical. Sensitive data are exchanged between different computers or applications using communication networks. Inside Local Area Networks (LAN) or intranets, various security measures can be adopted to prevent undesirable accesses or attacks. But the situation is completely different when the Internet is used as communication channel. Data travel through the network 'without control' or supervision. Therefore, additional security mechanisms have to be deployed to guarantee data integrity and confidentiality. For the MBS, security measures have been defined at different levels. *Secure Socket Layer* (SSL) protocol assures the communication between the server and the clients. This protocol is based on cryptography techniques and X.509 Certificates emitted by a Certification Authority (CA). Additionally, WS-Security has been applied for supplying message authentication, integrity and confidentiality.

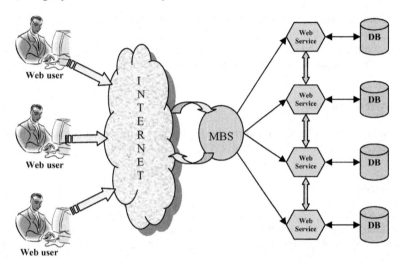

**Fig. 2.** Mobility Brokerage Service architecture

### 4.4 The Matching System

As exposed above, the *matching system* is a facility offered by the MBS in order to automatically find the opportunity or candidate wished. Those searches are based on a previously exposed concept. The keywords defined by the training topic thesaurus are used to improve the effectiveness and accuracy of the automatic searches. Keywords are classified in various main categories and subcategories. When a new offer or demand is introduced in the system, users are asked to indicate a set of keywords which define the area that fits better to their desires.

Each time an offer is stored in the database, the *matching process* is automatically performed. The system tries to find matches between the new offer and the rest of existing demands. Additionally, searches are filtered using other criteria such as the kind of collaboration sought, the starting and ending dates or duration. Analogous situation happens when a demand is introduced. The system looks for the offers which satisfy the demand. Anytime a match is found, the interested parts are notified by email. An email will invite each user to visit the MBS and check the news. The new matches are displayed using an animated icon, to attract the attention of the users.

# 5  Conclusions

Within the European framework, the INFOBIOMED initiatives aim to address the existing gaps on Biomedical Informatics education and training [15]. Several initiatives have been presented in this paper. A survey on mobility, the INFOBIOMED Course Database and the Mobility Brokerage Service are some of the actions adopted by the Network of Excellence. The MBS approach, based on a Service-oriented architecture of Web Services, is suitable to integrate heterogeneous data sources available in different physical places. Some of the components of the MBS might be reused for different tasks needed within the four NoE pilot projects; phamainformatics or genomic to microbiology, chronic inflammation and colon cancer. The MBS first release is active since the end of June 2005. Its main goal is to promote mobility exchanges within the INFOBIOMED Consortium but other European Networks of Excellence might use in the future the MBS to support their own mobility programs.

# Acknowledgments

The present work has been funded by the European Commission (FP6, IST thematic area) through the INFOBIOMED NoE (IST-507585).

# References

1. Martin-Sanchez, F., Maojo, V., Lopez-Campos, G.: Integrating genomics into health information systems. Methods Inf Med 2002; 41(1):25-30.
2. Liebman, M.: From Bioinformatics to Biomedical Informatics. Genome Technology 2001; N11: 64.
3. Guttmacher, A.E., Collins, F.S.: Genomic Medicine – A Primer. The New England Journal of Medicine 2002; 347 (19): 1512-1520.
4. http://www.infobiomed.org
5. http://bioinfomed.isciii.es
6. Maojo, V., Iakovidis, I., Martin-Sanchez, F., Crespo, J., Kulikowski, C.: Medical informatics and bioinformatics: European efforts to facilitate synergy. J Biomed Inform 2001; 34(6):423-7.
7. http://stanford.sanbi.ac.za/
8. Van Mulligen, E.M., Diwersy, M., Schmidt, M., Buurman, H., Mons, B.: Facilitating networks of information. Proc AMIA Symp. 2000: 868-72.
9. Van Mulligen, E.M., Diwersy, M., Schijvenaars, B., Weeber, M., Van Der Eijk, C., Jelier, R., Schuemie, M., Kors, J., Mons, B.: Contextual annotation of web pages for interactive browsing. *Medinfo 2004*: 94-8.
10. http://www.collexis.com
11. Erl, T.: Service-Oriented Architecture. Prentice Hall, 2004.
12. http://www.w3.org/TR/wsdl
13. http://www.w3.org/TR/soap/
14. http://www.uddi.org/
15. Maojo, V., Kulikowski C.A.: Bioinformatics and medical informatics: collaborations on the road to genomic medicine. Journal of the American Medical Informatics Association. 2003 Nov-Dec; 10(6):515-522.

# Using Treemaps
# to Visualize Phylogenetic Trees[*,**]

Adam Arvelakis[1,2], Martin Reczko[1], Alexandros Stamatakis[1],
Alkiviadis Symeonidis[1,2], and Ioannis G. Tollis[1,2]

[1] Foundation for Research and Technology-Hellas, Institute of Computer Science
P.O. Box 1385, Heraklion, Crete, GR-71110 Greece
[2] Department of Computer Science, University of Crete
University of Crete, P.O. Box 2208, Heraklion, Crete, Greece

**Abstract.** Over recent years the field of phylogenetics has witnessed significant algorithmic and technical progress. A new class of efficient phylogeny programs allows for computation of large evolutionary trees comprising 500–1.000 organisms within a couple of hours on a single CPU under elaborate optimization criteria. However, it is difficult to extract the valuable information contained in those large trees without appropriate visualization tools. As potential solution we propose the application of treemaps to visualize large phylogenies (evolutionary trees) and improve knowledge-retrieval. In addition, we propose a hybrid tree/treemap representation which provides a detailed view of subtrees via treemaps while maintaining a contextual view of the entire topology at the same time. Moreover, we demonstrate how it can be deployed to visualize an evolutionary tree comprising 2.415 mammals. The respective software package is available on-line at www.ics.forth.gr/~stamatak.

## 1 Introduction

Phylogenetic (evolutionary) trees are used to represent the evolutionary history of a set of $n$ organisms which are often also called taxa within this context. Usually, a multiple alignment of a—in a biological context—suitable small region of their DNA or protein sequences can be used as input for the computation of phylogenetic trees. Other computational approaches to phylogenetics also use gene order data [28].

In a computational context phylogenetic trees are usually strictly bifurcating (binary) unrooted trees. The organisms of the alignment are located at the tips (leaves) of such a tree whereas the inner nodes represent extinct common ancestors. The branches of the tree represent the time which was required for the mutation of one species into another—new—one. An example for the evolutionary tree of the monkeys and the homo sapiens is provided in Figure 1.

---

[*] Part of this work is funded by a Postdoc-fellowship granted by the German Academic Exchange Service (DAAD)

[**] This work was supported in part by INFOBIOMED code: IST-2002-507585 and the Greek General Secretariat for Research and Technology under Program "ARISTEIA", Code 1308/B1/3.3.1/317/12.04.2002

J.L. Oliveira et al. (Eds.): ISBMDA 2005, LNBI 3745, pp. 283–293, 2005.

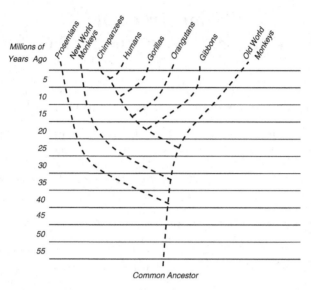

**Fig. 1.** Phylogenetic tree representing the evolutionary relationship between monkeys and the homo sapiens

The inference of phylogenies with computational methods has many important applications in medical and biological research, such as e.g. drug discovery and conservation biology. A paper by Bader *et al* [1] addresses potential industrial applications of evolutionary tree inference and contains numerous useful references to important biological results obtained by phylogenetic analysis.

Due to the rapid growth of available sequence data over the last years and the constant improvement of multiple alignment methods it has now become feasible to compute very large trees for datasets which comprise more than 500–1.000 organisms. The computation of the tree-of-life containing representatives of all living beings on earth is considered to be one of the *grand challenges* in Bioinformatics.

Unfortunately, phylogenetic inference under elaborate criteria such as Maximum Parsimony [10] (MP) or Maximum Likelihood [9] (ML) is an NP-complete[1] problem [7] [4]. However, the field has witnessed significant algorithmic progress over the last 2–3 years. Novel phylogeny programs and methods such as PHYML [11], MetaPigA [15], RAxML [25], [23], [26], or Rec-I-DCM3 [19] allow for inference of large evolutionary trees of up to 1.000 taxa with MP or ML within less than 24 hours on a single CPU. The largest ML-based tree computed to date with the parallel version of RAxML contains 10.000 taxa [24] and the largest MP-based phylogeny with Rec-I-DCM3 comprises more than 13.000 organisms [19].

Despite the algorithmic advances in the field only few adequate visualization tools are available for the analysis of such large trees. Thus, the design of novel

---

[1] Note that, this has not yet been demonstrated for ML due to the high mathematical complexity

tree viewing tools is crucial [22] in order to accelerate the analysis process as well as to extract useful information from the data and expedite the cognitive process. In this paper we describe the deployment of treemaps for visualization of phylogenies and present the respective software tool. Furthermore, we show how it can be used to visualize a phylogeny of 2415 mammalian mitochondrial DNA sequences which has been computed with RAxML.

The remainder of this paper is organized as follows: In Section 2 we survey the most common phylogenetic tree display tools and describe the basic concepts of treemaps. Thereafter, we describe the implementation, algorithms, and basic features of the visualization tool (Section 3). The advantages of displaying phylogenies with treemaps are outlined in Section 4 by example of a 2.415 taxon tree. Finally, we conclude in Section 5 and indicate directions of current and future research.

## 2   Related Work

### 2.1   Phylogenetic Tree Viewers

We review some popular tree viewing concepts and programs with respect to their ability for visualization of large evolutionary trees. Among the most popular representations are phylogram, radial, and slanted cladogram drawings [16]. Those representation are provided by common tree-viewing programs such as Treeview [17] and ATV [29]. However, these layouts and programs are targeted at medium-sized trees comprising a maximum of 300–400 taxa. Thus, they are not well-suited to visualize large trees with thousands of taxa (see Figure 3).

Approaches for larger trees make use of two-dimensional [3] and three-dimensional [12] hyperbolic space in order to simultaneously provide a detailed and contextual view of the tree. The two-dimensional hyperbolic tree-viewer Hypertree [3] is able to reasonably display tree with up to 1.000 taxa.

Other approaches such as SpaceTree [18] or TreeWiz [20] only display representative parts of very large trees. However, biologists usually prefer a simultaneous detailed display and contextual view of phylogenies. There also exist some approaches based on virtual reality [27],[21] which are however not accessible to most researchers due to the sheer cost of the respective infrastructure.

Carrizo [6] provides a readable and comprehensive review of efforts to appropriately display phylogenetic trees from an information visualization perspective.

To the best of our knowledge our implementation represents the first dedicated adaptation of treemaps to display phylogenetic trees.

### 2.2   Treemaps

The concept of treemaps for visualization purposes was initially proposed by Johnson and Shneiderman in 1991 [13]. It is particularly aimed at displaying tree structures. The standard treemap algorithm starts with a given area and positions a number of siblings within this area from left to right or top to bottom respectively in the same way as standard tree algorithms. The essential difference is that treemaps fill the entire available space and make use of rectangles to display nodes and hierarchies. In contrast, standard tree-representation such as e.g. cladograms draw a symbol for nodes at the center of the provided area and then connect the nodes via edges.

In particular, treemaps can—apart from simply dividing the space into sub-areas of equal size for each descendant—adjust the size of each sub-area depending on additional parameters/information located at the nodes. The color intensity of each treemap rectangle can be used to visualize branch length (weight) values. The method described so far represents the standard treemap-algorithm which is also known as slice & dice because the produced rectangles are relatively thin.

The squarified algorithm [5] is a variation of the slice & dice algorithm which seeks to solve the visibility problem of thin rectangles by drawing them in an as *square* as possible. Since fitting $n$ squares of a given area into a rectangle of pre-defined dimensions is an NP-complete problem [14], the optimal solution—if it exists—takes exponential time to compute. To solve this problem the restrictions are relaxed in [5] such that *perfect* squares are not required anymore. Those two algorithms (slice & dice, squarified) cover the two opposite ends of the aspect-ratio versus preservation-of-order spectrum.

Another variation for treemap drawing which intends to attain a compromise between the two previously mentioned approaches is the strip & ordered algorithm [2]. Apart from the different algorithms to draw treemaps with distinct rectangular shapes there exists a number of additional measures to enhance the visualization and interpretability of treemaps. The addition of frames (or borders) draws a border around each internal node such that the internal structure becomes more clearly visible [5]. When using a treemap without borders the risk of hiding color information of an internal node is high. The addition of cushions [5] can be used to create a pseudo-3D effect instead of displaying flat two–dimensional areas. At each hierarchy level of the tree a bump is added to the respective nodes to generate cushions.

# 3   Implementation

Now we describe how the treemap visualization mechanism has been adapted to display phylogenetic trees. One important property of the tree topologies—obtained e.g. by maximum likelihood analysis—is that they are strictly bifurcating unrooted trees, i.e., describe the *relative* evolutionary history of the organisms (see Figure 1). Moreover, the visualization of the branch lengths is very important since they denote evolutionary distances between organisms. Thus, mechanisms to sufficiently highlight branch lengths and to root the tree at arbitrary branches are required.

In general, branch lengths can be displayed by the addition of a see-through border to each node. The width of this border is proportional to the distance from the parent. However, this approach consumes valuable pixels and hence

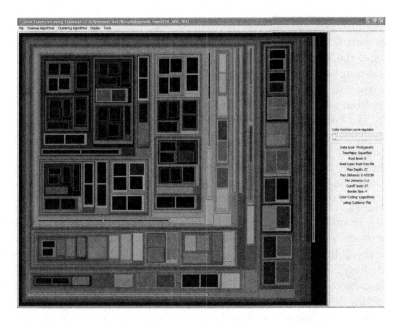

**Fig. 2.** The color of a node indicates the distance from the parent

limits the size of displayable trees. Therefore, we use the node color to indicate the distance from the parent (see Figure 2).

To visualize additional information for the tree, it is possible to load an annotation file that contains labels assigned to each leaf of the tree. A unique color is automatically assigned to each label and all leaves with the same label are shown in the treemap with the same color. We use this feature to indicate the taxonomic order of each species in the phylogenetic tree. The taxonomic orders can be assigned to a tree in the Newick file format using an external script and taxonomic information from the Integrated Taxonomic Information System on-line database available under http://www.itis.usda.gov. This kind of visualization can indicate very efficiently potential problems like outliers in the constructed tree.

The tree-viewer implements two of the aforementioned treemap algorithms (see Section 2.2): The standard slice & dice algorithm, and the squarified algorithm. The rationale for selecting those two is that they embrace the entire range of the aspect-ratio versus preservation-of-order spectrum.

The software has been entirely implemented in JAVA to ensure portability. The software uses JAVA swings and the JBCL library for the graphical interface.

In the following we list some of the main features of our software:

- Borders of up to 4 pixels can be added to each node to highlight the internal structure of the tree
- Nodes can be represented as cushions using a similar, but less compute-intensive approach, as described in [5]

- The tool offers a hybrid tree/treemap representation mode which draws the upper part (close to the root) of the phylogeny in standard tree representation and the lower part (subtrees containing tips) as treemaps.
- In addition to linear color coding w.r.t. branch lengths the coloring of the treemaps can be performed using exponential or logarithmic functions
- Apart from an equal rectangular area size which is usually assigned to each leaf of the treemap, the area of the leaves can be scaled according to their accumulated branch length distance from the root
- The initially unrooted phylogenetic tree is rooted at the center branch (the rooting that produces the tree with the smallest depth)
- The root can be moved with drag-and-drop to an arbitrary branch
- The user can zoom in into specific subtrees

Finally, in order to correlate the information of trees obtained by computational methods with phylogenetic information from other sources our tool provides the possibility to load taxonomic information about the tree species from a separate file. This additional information about families and subfamilies of organisms can be used to color groups (treemaps) of species accordingly. Moreover, this allows for detection of potentially misplaced organisms and identification of errors, at least with respect to expectations from non-computational approaches to phylogenetics.

## 4   Results

In order to demonstrate the features of our tree-viewer we used two large tree topologies which have been computed with RAxML and Phyml. The trees contain 2415 mammals and were computed based on a manually aligned mitochondrian DNA data set from Olaf Bininda–Emonds at the Technische Universität München.

In Figure 3 the tree produced by RAxML is displayed using a typical tree-drawing technique. In Figures 4 -7 the trees produced by both tools are colored according to the taxonomic order of each species using taxonomic information retrieved June 8, 2005, from the Integrated Taxonomic Information System on-line database, http://www.itis.usda.gov. Two problematic cases can be identified very easily.

1. The species *sturnaria-ilium* is of the taxonomic order *chiroptera*, but occurs in the tree on the branch of *carnivora* and
2. *thylamys-pusilla* belonging to order *didelphimorphia*, is on the branch of *rodentia*.

Finally, Figure 8 illustrates the hybrid tree/treemap viewing option.

## 5   Conclusion, Availability, and Future Work

Treemaps represent an effective visualization method to display and analyze hierarchical data. In this paper we have presented the—to the best of our

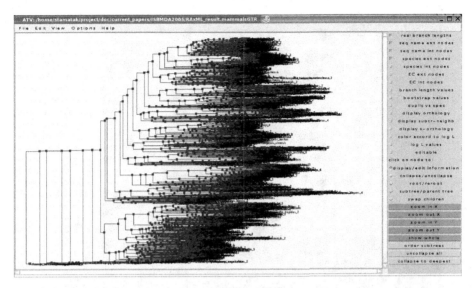

**Fig. 3.** Visualization of the 2415-taxon phylogeny of mammals with ATV

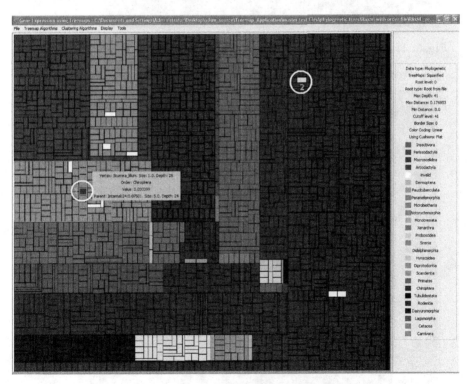

**Fig. 4.** The tree produced by RAxML, drawn with the *squarified* algorithm. The two problematic cases are indicated with circles

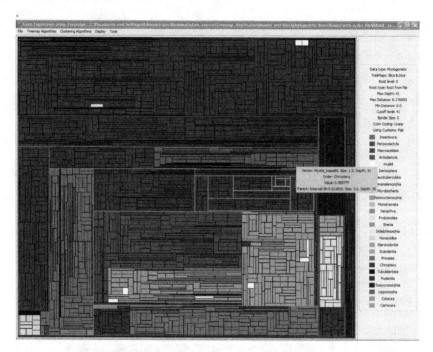

**Fig. 5.** The tree produced by RAxML, drawn with the *slice and dice* algorithm

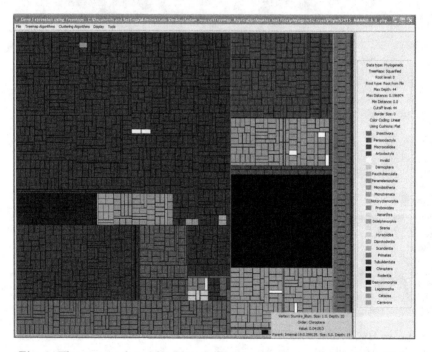

**Fig. 6.** The tree produced by Phyml, drawn with the *squarified* algorithm

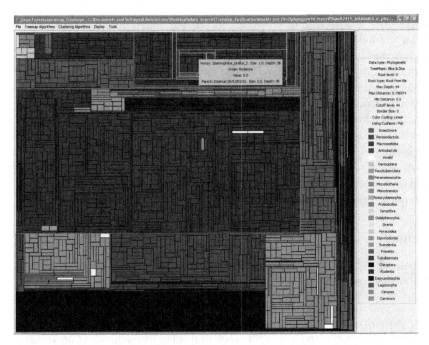

**Fig. 7.** The tree produced by Phyml, drawn with the *slice and dice* algorithm

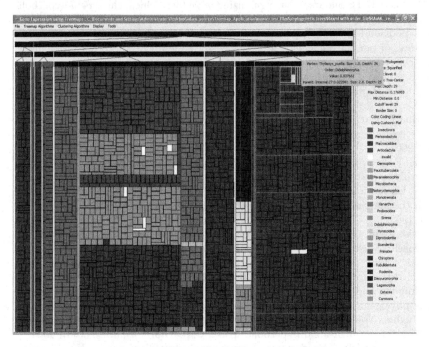

**Fig. 8.** The tree produced by RAxML, drawn as hybrid tree/treemap

knowledge—first adaptation of treemaps to the particular requirements of displaying large phylogenetic trees. Moreover, we have introduced a hybrid tree/ treemap representation of phylogenies which provides a detailed view of the subtrees containing the species (leaves) of the tree while maintaining a contextual view of the entire tree at the same time. Finally, we have demonstrated how taxonomical data from other sources can be used to easily detect errors originating either from the data assembly or the tree-building process.

The visualization tool is freely available for download at www.ics.forth.gr/~stamatak.

Future work will cover improved methods to visually emphasize that the displayed trees are unrooted. Finally, we intend to automate the retrieval process of taxonomical information about the species from public databases.

# References

1. D.A. Bader, B.M.E. Moret, and L. Vawter. Industrial applications of high-performance computing for phylogeny reconstruction. In *Proceedings of SPIE IT-Com: Commercial Applications for High-Performance Computing*, pages 159–168, 2001.
2. B. B. Berderson, B. Shneiderman, and M. Wattenberg. Ordered and quantum treemaps: Making effective use of 2d space to display hierarchies. *ACM Transactions on Computer Graphics*, 21(4):833–854, 2002.
3. J. Bingham and S Sudarsanam. Visualizing large hierarchical clusters in hyperbolic space. *Bioinformatics*, 16:660–661, 2000.
4. H. L. Bodlaender, M. R. Fellows, M. T. Hallett, T. Wareham, and T. Warnow. The hardness of perfect phylogeny, feasible register assignment and other problems on thin colored graphs. *Theoretical Computer Science*, 244:167–188, 2000.
5. D. M. Bruls, C. Huizing, and J.J. van Wijk. Squarified treemaps. In *Proceedings of the joint Eurographics and IEEE TVCG Symposium on Visualization*, pages 33–42, 2000.
6. Savrina F. Carrizo. Phylogenetic trees: an information visualisation perspective. In *Proceedings of the second conference on Asia-Pacific bioinformatics*, pages 315–320, 2004.
7. W. H. E. Day, D. S. Johnson, and D. Sankoff. The computational complexity of inferring rooted phylogenies by parsimony. *Math. Bios.*, 81:33–42, 1986.
8. M.W. Chase et al. Phylogenetics of seed plants: An analysis of nucleotide sequences from the plastid gene rbcl. *Annals of the Missouri Botanical Garden*, pages 528–580, 1993.
9. J. Felsenstein. Evolutionary trees from DNA sequences: A maximum likelihood approach. *Journal of Molecular Evolution*, 17:368–376, 1981.
10. W. M. Fitch. Toward defining the course of evolution: minimum change for a specified tree topology. *Syst. Zool.*, 20:406–416, 1971.
11. S. Guindon and O. Gascuel. A simple, fast, and accurate algorithm to estimate large phylogenies by maximum likelihood. *Syst. Biol.*, 52(5):696–704, 2003.
12. T. Hughes, Young Hyun, and D.A.Liberles. Visualizing very large phylogenetic trees in three dimensional hyperbolic space. *BMC Bioinformatics*, 5(48), 2004.
13. B. Johnson and B. Shneiderman. Treemaps: a space-filling approach to the visualization of hierarchical information structures. In *Proceedings of the 2nd International IEEE Visualization Conference*, pages 284–291, October 1991.

14. J. Y. T. Leung, T.W.Lam, C.S. Wong, G.H. Young, F.Y.L Chin. Packing squares into a square. *Journal on Parallel and Distributed Compouting*, 10:271-275,1990

15. A. R. Lemmon and M. C. Milinkovitch. The metapopulation genetic algorithm: An efficient solution for the problem of large phylogeny estimation. *Proceedings of the National Academy of Sciences*, 99:10516–10521, 2001.

16. T. Munzner, F. Guimbretiere, S. Tasiran, L. Zhang, and Y. Zhou. Treejuxtaposer: Scalable tree comparison using focus+context with guaranteed visibility. In *Proceedings of SIGGRAPH 2003*, 2003.

17. R.D.M. Page. Treeview: An application to display phylogenetic trees on personal computers. *CABIOS*, 12:357–358, 1996.

18. C. Plaisant, J. Grosjean, and B.B. Bederson. Spacetree: Supporting exploration in large node link tree, design evolution and empirical evaluation. In *Proceedings of the 2002 IEEE Symposium on Information Visualization*, pages 57–70, 2002.

19. U. Roshan, B. M. E. Moret, T. Warnow, and T. L. Williams. Rec-i-dcm3: a fast algorithmic technique for reconstructing large phylogenetic trees. In *Proceedings of the IEEE Computational Systems Bioinformatics conference (CSB)*, Stanford, California, USA, 2004.

20. U. Rost and E. Bornberg-Bauer. Treewiz: interactive exploration of huge trees. *Bioinformatics*, pages 109–114, 2002.

21. D.A. Ruths, E.S. Chen, and L. Ellis. Arbor3d: an interactive environment for examining phylogenetic and taxonomic trees in multiple dimensions. *Bioinformatics*, pages 1003–1009, 2000.

22. M.J. Sanderson and A.C. Driskell. The challenge of constructing large phylogenetic trees. *Trends in Plant Science*, 8(8):374–378, 2003.

23. A. Stamatakis. An efficient program for phylogenetic inference using simulated annealing. In *Proceedings of IPDPS2005*, Denver, Colorado, USA, 2005.

24. A. Stamatakis, T. Ludwig, and H. Meier. Parallel inference of a 10.000-taxon phylogeny with maximum likelihood. In *Proceedings of 10th International Euro-Par Conference*, pages 997–1004, 2004.

25. A. Stamatakis, T. Ludwig, and H. Meier. Raxml-iii: A fast program for maximum likelihood-based inference of large phylogenetic trees. *Bioinformatics*, 21(4):456–463, 2005.

26. A. Stamatakis, M. Ott, and T. Ludwig. Raxml-omp: An efficient program for phylogenetic inference on smps. In *Proceedings of 8th International Conference on Parallel Computing Technologies (PaCT)*, 2005. Preprint available on-line at WWW.ICS.FORTH.GR/~STAMATAK.

27. B. Stolk, F. Abdoelrahman, A. Koning, P. Wielinga, J.M. Neefs, A. Stubbs, A. de Bondt, P. Leemans, and P. van der Spek. Mining the human genome using virtual reality. In *Proceedings of the Fourth Eurographics Workshop on parallel Graphics and Visualization*, pages 17–21, 2002.

28. J. Tang, B.M.E. Moret, L. Cui, and C.W. dePamphilis. Phylogenetic reconstruction from arbitrary gene-order data. In *Proc. 4th IEEE Conf. on Bioinformatics and Bioengineering BIBE'04*, pages 592–599, 2004.

29. C. M. Zmasek and S. R. Eddy. Atv: Display and manipulation of annotated phylogenetic trees. *Bioinformatics*, 17:383–384, 2001.

# An Ontological Approach
# to Represent Molecular Structure Information

Eva Armengol and Enric Plaza

IIIA – Artificial Intelligence Research Institute,
CSIC – Spanish Council for Scientific Research,
Campus UAB, 08193 Bellaterra, Catalonia, Spain
{eva,enric}@iiia.csic.es

**Abstract.** Current approaches using Artificial Intelligence techniques applied to chemistry use representations inherited from existing tools. These tools describe chemical compounds with a set of structure-activity relationship (SAR) descriptors because they were developed mainly for the task of drug design. We propose an ontology based on the chemical nomenclature as a way to capture the concepts commonly used by chemists in describing molecular structure of the compounds. In this paper we formally specify the concepts and relationships of the chemical nomenclature in a comprehensive ontology using a form of relational representation called *feature terms*. We also provide several examples of describing chemical compounds using this ontology and compare our proposal with other SAR based approaches.

## 1 Introduction

The IUPAC (www.chem.qmul.ac.uk/iupac/) chemical nomenclature is a standard form to describe the (organic and inorganic) molecules from their chemical structure. In Artificial Intelligence some proposed representations of the molecules describe them atom by atom obtaining cumbersome descriptions that may be not easily understandable by chemists. From our point of view, a formal representation using the IUPAC nomenclature could be very useful since allows a direct description of the chemical structure, in a way very familiar to the chemist. For instance, chemists commonly describe the *anthracene* as a molecule formed by a group of three benzenes and they know some of its properties and the relative position of each atom. However, representations describing the molecules atom by atom do not take into account expert knowledge; therefore they need explicitly represent the 14 atoms of the *anthracene*, their bindings, interactions, etc. We propose and ontological approach to represent information about the molecular structure of a chemical compound. In this approach, a compound can be described as *anthracene* without any reference to individual atoms.

In the next section we briefly explain the chemical nomenclature and how the representation we propose capture this nomenclature. Then, in section 3 we explain a formal representation called *feature terms* and how the chemical compounds can be described using them. Finally, we discuss trade offs our proposal by comparing it with other SAR based approaches.

J.L. Oliveira et al. (Eds.): ISBMDA 2005, LNBI 3745, pp. 294–304, 2005.

## 2    Chemical Nomenclature Concepts

Following the recommendations of the IUPAC (1994) the organic compounds can be classified in four groups: 1) Based on Carbon, Hydrogen and Oxygen, 2) Based on other elements, 3) Natural products (antibiotics, lipids, nucleic acids, etc), and 4) Others.

We focus on compounds belonging to the groups 1 and 2 above, because we consider them as most elemental than the compounds included in the groups 3 and 4 in the sense that often compounds belonging to the two last groups are either extensions (lipids are chains of hydrocarbons) or particular cases (ions may be parts of functional groups) of compounds in groups 1 and 2.

Compounds included in the first group are the hydrocarbons, ring systems, alcohols, ethers, phenols and derivatives, aldehydes, ketones, quinones and derivatives, and carboxilic acids and derivatives. The second group includes compounds that are based on elements such as nitrogen, phosphorus, silicon, sulfur, halogens, metals, etc. Notice that these compounds could also be regrouped in two different classes taking into account whether they can be found alone (such as the hydrocarbons or the ring systems) or not (alcohols, ethers, aldehydes, etc). Therefore, we consider the following alternative classification of the first two groups of compounds above: a) Hydrocarbons, b) Ring systems, and c)Functional groups. In the following subsections we will analyze these groups separately.

### 2.1    Hydrocarbons

The *hydrocarbons* (also called *alkanes*) are chains of atoms that only contain carbon (C) and hydrogen (H). According to the number of C atoms of the chain they take different names (Fig. 1 shows some hydrocarbons). The left part of the figure shows hydrocarbons called *saturated* since all the bonds are single, i.e. all the C atoms (except those in the extremes) are bonded to two H and to two other C atoms. The right part of Fig. 1 shows *unsaturated* hydrocarbons (also called *alkenes* and *alkynes*) since some of the bonds are either double (*alkenes*) or triple (*alkynes*). For instance, the *1,3-butadiene* is an unsaturated hydrocarbon with two double bonds, one in position 1 and another in position 3. Notice that in the nomenclature of unsaturated hydrocarbons the position of the double and triple bonds is part of the compound name. Also, the suffix *-ene* (or *-en*) denotes hydrocarbons with double bonds whereas those with triple bonds have the suffix

| saturated HC | unsaturated HC |
|---|---|
| methane: $CH_4$ | ethyne: $CH-CH$ |
| ethane: $CH_3-CH_3$ | ethene: $CH_3=CH_3$ |
| propane: $CH_3-CH_2-CH_3$ | propyne: $CH\equiv C-CH_3$ |
| butane: $CH_3-CH_2-CH_2-CH_3$ | 1,3- butadiene: $CH=CH-CH=CH_2$ |
| pentane: $CH_3-CH_2-CH_2-CH_2-CH_3$ | 4-hexen, 1-yne: $CH\equiv C-CH_2-CH=CH-CH_3$ |

**Fig. 1.** Examples of acyclic saturated and unsaturated hydrocarbons

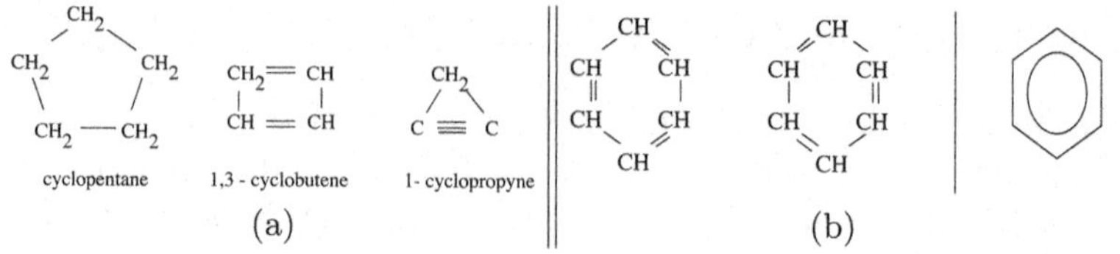

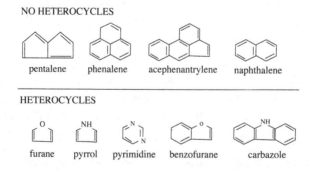

cyclopentane        1,3 - cyclobutene        1- cyclopropyne

(a)                                                                (b)

**Fig. 2.** (a) Three examples of cyclic saturated and unsaturated hydrocarbons. (b) The two resonant forms of the benzene are shown on the left, while the usual representation of the benzene is shown on the right

-*yne*. The name of compounds with double and triple bonds have both suffixes (as the *4-hexen, 1-yne* shown on the bottom right part of Fig. 1).

Moreover, both saturated and unsaturated hydrocarbons may be cyclic (i.e. *cycloalkanes*). Figure 2a shows some examples of saturated and unsaturated cyclic hydrocarbons. Concerning to the nomenclature, the name of cyclic hydrocarbons is the same that the name of the acyclic hydrocarbons preceded by the prefix *cyclo-* (i.e. cyclopentane; 1,3-cyclobutene).

## 2.2   Ring Systems

Ring systems include the cyclic hydrocarbons called *aromatic* or *arenes*. Aromatic rings are defined as those rings where the electrons are free to cycle around circular arrangements of atoms, which are alternately simply and doubly bonded. A typical example of aromatic rings is the *benzene*, that is a cyclohexane that can take the two forms show in the left part of Fig. 2b. In fact, since both forms of the benzene are equivalent (this is called *resonance*) and none of them accurately represents the benzene structure, the most common representation for the benzene is that shown at the right part of Fig. 2b. In this representation the double bonds have been replaced by a central circle meaning that the electrons have a free circulation among the atoms.

Ring systems could be *monocycles* or *polycycles*. Monocycles are those systems formed by only one ring; polycycles are those formed from the association of several cyclic hydrocarbons. The upper part of Fig. 3 shows a sample of mono-

NO HETEROCYCLES

pentalene        phenalene        acephenantrylene        naphthalene

HETEROCYCLES

furane        pyrrol        pyrimidine        benzofurane        carbazole

**Fig. 3.** Examples of both heterocyclic and no heterocyclic ring systems

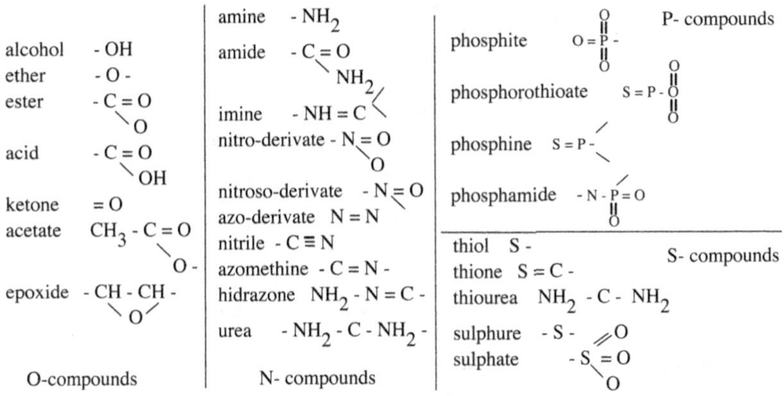

**O-compounds**

| | |
|---|---|
| alcohol | - OH |
| ether | - O - |
| ester | - C = O  \  O |
| acid | - C = O  \  OH |
| ketone | = O |
| acetate | $CH_3$ - C = O  \  O - |
| epoxide | - CH - CH -  \  O ∕ |

**N- compounds**

| | |
|---|---|
| amine | - $NH_2$ |
| amide | - C = O  \  $NH_2$ |
| imine | - NH = C < |
| nitro-derivate | - N = O  \  O |
| nitroso-derivate | - N = O |
| azo-derivate | N = N |
| nitrile | - C ≡ N |
| azomethine | - C = N - |
| hidrazone | $NH_2$ - N = C - |
| urea | - $NH_2$ - C - $NH_2$ - |

**P- compounds**

| | |
|---|---|
| phosphite | O = P -  (O, O) |
| phosphorothioate | S = P - O  (O, O) |
| phosphine | S = P -  \ |
| phosphamide | - N - P = O  ‖  O |

**S- compounds**

| | |
|---|---|
| thiol | S - |
| thione | S = C - |
| thiourea | $NH_2$ - C - $NH_2$ |
| sulphure | - S -  ∕∕ O |
| sulphate | - S = O  \  O |

**Fig. 4.** Functional groups

cycles and polycycles. The most common ring systems are based on the benzene. Both monocycles and polycycles can be classified as *heterocycles* (when all the atoms are C and H) or as *no heterocycles* (when there is some atom different of C). The bottom part of Fig. 3 shows some examples of heterocycles.

## 2.3   Functional Groups

A *functional group* is an atom or group of atoms that replaces one H atom in an organic compound and that defines the structure of a family of compounds and determines the chemical properties of that family. Based on the atoms they contain, we propose to classify the functional groups as follows: 1) *O-compounds* based on the oxygen, 2) *N-compounds* based on the nitrogen, 3) *P-compounds* based on the phosphorus, and 4) *S-compounds* based on the sulphur.

Some functional groups could be classified as belonging to more than one class since they may contain more than one atom different from H (for instance, oxygen and nitrogen). In such situations, we considered them as belonging to one class depending on which atom is considered as the most important of the functional group. Figure 4 shows some of the functional groups that we considered. Notice that the amide, the nitro-derivate and the nitroso-derivate could be considered both O-compounds or N-compounds.

## 3   Representation of the Chemical Compounds

In this section we introduce a formal specification of the chemical ontology based on feature terms. Several Artificial Intelligence approaches describe complex objects using a *relational* representation. In this kind of representation, an object is described by its parts and the relations among these parts. In particular, we use a relational representation called *feature terms*. Using feature terms, the concepts explained in the previous section have been specified by a hierarchy of *sorts* (Fig. 5). Moreover, a sort is described by a set of features where each

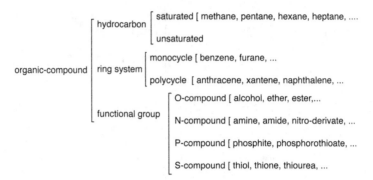

**Fig. 5.** The hierarchy of sorts for organic chemistry concepts

feature represents a relation of this sort with another sort. In the next section feature terms are briefly introduced and then, in section 3.2, we explain how the chemical compounds can be described using feature terms.

### 3.1   Feature Terms

*Feature Terms* (also called feature structures or $\psi$-terms) are a generalization of first order terms. The difference between feature terms and first order terms is the following: a first order term, e.g. $f(x, y, g(x, y))$ can be formally described as a tree and a fixed tree-traversal order. In other words, parameters are identified by position. The intuition behind a feature term is that it can be described as a labelled graph, i.e. parameters are identified by name. For instance, the definition of a particular object using feature terms is the following:

```
(define (sort object-name)
     (feature-1 obj-1)
     ....
     (feature-N obj-N))
```

where feature-1,..., feature-N are the names of the features that describe the object *object-name*. The object and also the values of the features belong to a *sort* and sorts are related among them by a hierarchy of sort/subsorts. Figure 5 shows the hierarchy of sorts we define to capture the chemical concepts that will be introduced later. The definition of a sort is as follows:

```
(define-sort sort
     (feature-1 sort-1)
     (.... )
     (feature-N sort-N))
```

where *sort* is the name of the sort that we are defining and feature-1,..., feature-N are the names of the features that describe the objects belonging to *sort*. When a *sort-i* is a subsort of another sort *sort-j* this is defined as follows: (define (*sort-j sort-i*) ... ), and *sort-i* inherits all the features of *sort-j*. For instance, Fig. 5 shows that *benzene* is a subsort of *monocycle* that, in turn, is a subsort of *ring system* that, in turn, is a subsort of *organic-compound*. The values of the features (e.g. feature-1) are restricted to the sort that is declared (e.g. sort-1) .

A more detailed explanation about the feature terms and the subsumption relation can be found in [1]. In the next section we explain how feature terms are used to represent chemical compounds. Also, we detail the sort hierarchy that represents the chemical concepts introduced in the previous section.

## 3.2   Chemical Compounds Described as Feature Terms

A chemical compound is described by a feature term of sort *chemical-compound* with features characterizing the compound. The definition of the sort *chemical-compound* is the following:

```
(define-sort chemical-compound
    (molecular-structure compound)
    (tests test-results))
```

Feature terms of sort chemical-compound are described by two features: the molecular-structure of the compound and the tests features that contains the results of some tests done on the compound. Notice that the value of molecular-structure has to be an object of sort *compound*. In this section we focus on the explanation of the representation of the molecular structure of the compounds.

Our ontology proposal is based on the chemical nomenclature but we also want to describe the molecular structure as accurately as possible. Nevertheless the nomenclature has some ambiguities since some compounds may have several synonym names. This means that in our ontology a compound can be described in several ways. To handle the synonyms of a compound we use the notion of *multi-instance* [2]. When a compound has synonym descriptions the only difference is that the feature molecular-structure from the sort *chemical-compound* is a set that contains all the possible synonym descriptions of that compound.

To describe the molecular structure of a compound, we defined the sort *compound* which has, in turn, two subsorts: *organic-compound* and *inorganic-compound*. The specification of the *organic-compound* sort is the following:

```
(define-sort (compound organic-compound)
    (main-group compound)
    (radical-set compound))
```

Organic compounds can be described as composed by two parts: the main group and the radical-set both with values of sort *compound*. The main group of a molecule is often the part of the molecule that is either the largest or the part located in a central position. Radicals are groups that are usually smaller than the main group (commonly they are functional groups). A main group can contain several radicals and a radical can, in turn, have a set of radicals. Both main group and radicals are the same kind of molecules, i.e. a molecule may appear as the main group in a compound and also as a radical in another compound.

Let us analyze now how to represent the different kinds of chemical compounds following the classification introduced in section 2.

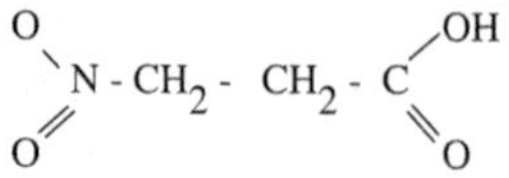

```
(define (saturated-hydrocarbon 3-nitropropionic-acid)
  (main-group propane)
  (radical-set acid nitro-derivate)
  (p-radicals (define (position-radical)
                (position one)
                (radicals acid))
              (define (position-radical)
                (position three)
                (radicals nitro-derivate))))
```

**Fig. 6.** Molecular structure of the *3-nitropropionic acid* and its representation using feature terms

**Hydrocarbons.** Although there are saturated and unsaturated hydrocarbons, their nomenclature follows the same idea: the basic name of the hydrocarbon is the number of C atoms. The name of unsaturated hydrocarbons has the suffix *-ene* when there are double bonds and *-yne* when there are triple bonds. When (saturated or unsaturated) hydrocarbons are cyclic the prefix (*cyclo* is added to the basic name. Using feature terms we define the sort *hydrocarbon* as a subsort of *compound* as follows:

```
(define-sort (organic-compound hydrocarbon)
    (cyclic? boolean)
    (p-radicals position-radicals))
```

Since *hydrocarbon* is a subsort of *organic-compound*, it inherits the features main-group and radical-set. When cyclic? is *true* means that the hydrocarbon is cyclic otherwise it is acyclic. The sorts *saturated-hydrocarbon* and *unsaturated-hydrocarbon* are subsorts of *hydrocarbon* so they inherit the features cyclic? and p-radicals.

The radicals of a compound are situated in a determined position with respect to the main group. This is represented with the sort *position-radical* as follows:

```
(define-sort position-radical
    (position numeric)
    (radicals compound))
```

Figure 6 shows the saturated hydrocarbon called *3-nitropropionic acid* that is a compound that has a propane (i.e. an hydrocarbon with three C atoms) and two radicals: a *nitro-derivate* in position 3 and an *acid* in position 1. The figure also shows the description of the *3-nitropropionic acid* using feature terms.

To represent unsaturated hydrocarbons the *unsaturated-hydrocarbon* sort is defined as follows:

```
(define-sort (hydrocarbon unsaturated-hydrocarbon)
    (main-group saturated-hydrocarbon)
    (p-bonds p-bond))
```

Notice that in this definition the sort of main-group is a *saturated-hydrocarbon*. Also, the *unsaturated-hydrocarbon* sort has a feature called p-bonds which values are of the sort *p-bond* defined as follows:

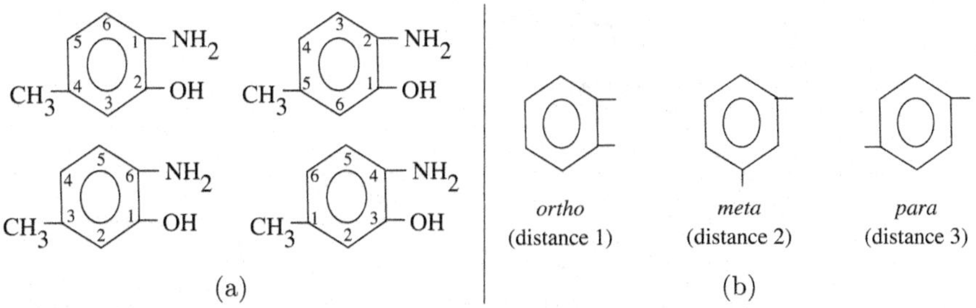

Fig. 7. a) Several possible numerings of the radicals in a molecule. b) Names used in chemistry for the relative positions of the radicals of a benzene ring

```
(define-sort p-bond
      (bond kind-of-bond)
      (position numeric))
```

i.e. by means of this sort we can define the kind of bonds of an unsaturated hydrocarbon and its position. For instance, the representation of the *4-hexen, 1-yne* shown in Fig. 1 is the following:

```
(define (unsaturated-hydrocarbon 4-hexen-1-yne)
      (cyclic? false))
      (main-group hexane))
      (p-bonds (define (p-bond)
                  (bond triple)
                  (position one))
             (define (p-bond)
                  (bond double)
                  (position four))))
```

This description comes directly from the chemical name that states that the double bond is in position 4 and the triple bond in position 1, that is to say, the C atoms are numerated from left to right.

**Ring systems.** are also defined as composed of a main group and radicals. When compounds have only one ring system then this ring system is the main-group. The problem, however, is how to determine the position of the radicals, since (without taking into account the nomenclature rules) the position 1 could be any of the radicals; once that position is fixed, the position 2 could be determined clockwise or counter clockwise. Figure 7a shows an example on how the positions of the radicals can change depending on which radical is considered to be in position 1.

In chemistry, when the main group is a benzene, some positions of the radicals (Fig. 7b) have particular names (*ortho, meta, para*). We take this idea for defining the positions of the radicals of a ring system. Thus, the *ring-system* sort is defined as a subsort of *organic-compound* as follows:

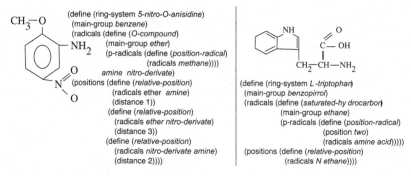

**Fig. 8.** Two examples of ring systems and their descriptions in our ontology

```
(define-sort (organic-compound ring-system)
    (radicals compound)
    (positions position))
```

where the sort *position* represents the positions of the radicals. We defined three subsorts of *Position*: 1) *absolute-position*, 2) *relative-position* and 3) *atom-position*. The sort *absolute-position* will be used in compounds where the positions of the radicals are straightforward (as in the hydrocarbons). The sort *relative-position* is used when the position of the radicals are defined by their distance (as in the positions *ortho*, *meta* and *para*). The sort *relative-position* is defined as follows:

```
(define-sort (position relative-position)
    (radicals compound)
    (distance number))
```

The sort *atom-position* is used when a radical is placed in a particular atom different of the C. The description of a ring system can contain the three kinds of positions. Figure 8 shows two examples of ring systems with radicals and their descriptions using feature terms.

**Functional groups.** The sort *functional-group* is a subsort of *organic-compound* defined as follows:

```
(define-sort (organic-compound functional-group)
    (radical-set compound)
    (p-radicals position-radical))
```

In turn, the sort *functional-group* has four subsorts: *O-compound*, *N-compound*, *S-compound* and *P-compound*. These sorts inherit the features of the *functional-group* sort, i.e. main-group and radical-set.

## 4   Discussion

Most Machine Learning (ML) tools used to build models of Toxicology are based on the *Structure-Activity Relationship (SAR)* descriptors. These descriptors represent the chemical compounds from several points of view (structural, physical

| | | |
|---|---|---|
| number of atoms | number of C atoms | number of 3-membered rings |
| number of non-H atoms | number of N atoms | number of 4-membered rings |
| number of bonds | number of O atoms | number of 5-membered rings |
| number of non-H bonds | number of P atoms | number of 6-membered rings |
| number of multiple bonds | number of S atoms | number of 7-membered rings |
| sum of conventional bond orders | number of F atoms | number of 8-membered rings |
| aromatic ratio | number of Cl atoms | number of 9-membered rings |
| number of rings | number of Br atoms | number of 10-membered rings |
| number of circuits | number of I atoms | number of 11-membered rings |
| number of rotatable bonds | number of B atoms | number of 12-membered rings |
| rotatable bond fraction | number of heavy atoms | number of benzene like rings |
| number of double bonds | number of halogen atoms | |
| number of triple bonds | | |
| number of aromatic bonds | | |
| number of H bonds | | |

**Fig. 9.** Constitutional descriptors used in representations based on SAR

properties, etc) and they are the basis to build equational models that relate the structure of a chemical compound with its physical-chemical properties. There is a number of commercial tools allowing the generation of these descriptors (CODESSA [5], TSAR (www.accelrys.com/products/tsar/), DRAGON [6], etc) and each one gives their own set of descriptors. Thus, methods that build toxicity models have to select a subset of these descriptors. As a consequence, the final model and, therefore its performance, will depend on which descriptors have been considered as the most important.

The main difference among the representations based on SAR and our ontological approach is that the former describe the molecular structure of the chemical compounds in an exhaustive way. SAR representations consist on a set of descriptors that can be grouped in several subsets according to the characteristics they describe. Thus, there are constitutional descriptions that capture structural features (Fig. 9 shows such descriptors), topological descriptors that capture 2D features, connectivity indices, WHIM descriptors, etc. Therefore, the description of a compound using SAR descriptors consists on giving a value for each descriptor. Notice that the descriptions of compounds based on SAR are vectors of attribute values, a very simple representation from which a comprehensive chemical ontology cannot be directly derived.

The representation we propose is more conceptual than SAR in the sense that directly uses the concepts understood by the chemists. We defined a chemical ontology with the chemical concepts in such a way that the molecular structure of a compound can be described using those concepts (Fig. 5). Thus, when our ontology describes a compound as formed by a benzene and an acid, chemists clearly understand the underlying structure of this compound, whereas using constitutional descriptors this compound should be described as composed of 22 atoms, 9 non-H atoms, 1 ring, 2 O atoms, etc. Therefore the molecular description using the ontology we propose is more understandable than descriptions based on SAR. Moreover, experimental evidence [3, 4] shows that the predictive performance of our approach is comparable to that of the approaches using representations based on SAR (although our ontology only incorporates structural information).

Some authors use approaches that are not centered on the representation of specific atoms but on molecular structures. For instance, González et al [7] and Deshpande and Karypis [8] represent chemical compounds as labeled graphs, using graph techniques to detect the set of molecular substructures (subgraphs) more frequently occurring in the chemical compounds of the data set. Conceptually, these two approaches are related to ours in that we describe a chemical compound in terms of its radicals (i.e. substructures of the main group).

Currently, we defined an ontology that only takes into account the structural aspects described by the constitutional descriptors of the SAR representations. In the future, we plan to extend this ontology with some other aspects of the chemical compounds that could be useful for predictive toxicology. Thus, our goal is not simply incorporating all SAR descriptors into the ontology, but rather developing a chemical ontology that captures the necessary molecular information.

## Acknowledgements

This work has been supported by the MCYT-FEDER Project SAMAP (TIC2002-04146-C05-01). The authors thank Dr. Lluis Bonamusa for his assistance in developing the representation of chemical compounds.

## References

1. Armengol, E., Plaza, E.: Bottom-up induction of feature terms. Machine Learning **41** (2000) 259–294
2. Dietterich, T., Lathrop, R., Lozano-Perez, T.: Solving the multiple instance problem with axis-parallel rectangles. Artificial Intelligence Journal **89** (1997) 31–71
3. Armengol, E., Plaza, E.: Relational case-based reasoning for carcinogenic activity prediction. Artificial Intelligence Review **20** (2003) 121–141
4. Armengol, E., Plaza, E.: Lazy learning for predictive toxicology based on a chemical ontology. In Dubitzky, W., Azuaje, F., eds.: Artificial Intelligence Methods and Tools for Systems Biology, In Press. Kluwer Academic Press (2004)
5. Katritzky, A., Petrukhin, R., Yang, H., Karelson, M.: CODESSA PRO. User's Manual. University of Florida (2002)
6. Todeschini, R., Consonni, V.: Handbook of Molecular Descriptors. Methods and principles in Medicinal Chemistry. Wiley-VCH, Weinheim (2000)
7. Gonzalez, J., Holder, L., Cook, D.: Application of graph-based concept learning to the predictive toxicology domain. In: Proceedings of the Predictive Toxicology Challenge Workshop, Freiburg, Germany. (2001)
8. Deshpande, M., Karypis, G.: Automated approaches for classifying structures. In: Proc. of the 2nd Workshop on Data Mining in Bioinformatics. (2002)

# Focal Activity in Simulated LQT2 Models at Rapid Ventricular Pacing: Analysis of Cardiac Electrical Activity Using Grid-Based Computation

Chong Wang[1], Antje Krause[1], Chris Nugent[2], and Werner Dubitzky[2]

[1] University of Applied Sciences Wildau, 15745 Wildau Germany
{cwang,akrause}@igw.tfh-wildau.de
[2] University of Ulster, BT52 1SA, Northern Ireland, UK
{cd.nugent,w.dubitzky}@ulster.ac.uk

**Abstract.** This study investigated the involvement of ventricular focal activity and dispersion of repolarization in LQT2 models at rapid rates. The Luo-Rudy dynamic model was used to simulate ventricular tissues. LQT2 syndrome due to genetic mutations was modeled by modifying the conductances of delayed rectifier potassium currents. Cellular automata was employed to generate virtual tissues coupled with midmyocardial (M) cell clusters. Simulations were conducted using grid-based computation. Under LQT2 conditions, early after-depolarizations (EADs) occurred first at the border of the M refractory zone in epicardium coupled with M clusters, but spiked off from endocardial cells in endocardium coupled with M clusters. The waveform of EADs was affected by the topological distribution of M clusters. Our results explain why subepicardial and subendocardial cells could exhibit surprisingly EADs when adjacent to M cells and suggest that phase 2 EADs are responsible for the onset of Torsade de Pointes at rapid ventricular pacing.

## 1 Introduction

Torsade de Points (TdP) is an atypical polymorphic ventricular tachyarrhythmia associated with long QT syndrome (LQTS). TdP is often self-terminated, but it can persist and deteriorate into ventricular fibrillation, leading to sudden cardiac death. TdP differs from other forms of polymorphic ventricular tachycardias, not only by its distinctive time-dependent changes in electrical axis and accompanied QT abnormality, but also by the characteristic pattern of onset.

Clinical reports [1, 2] have described a typical onset mode of TdP in patients with acquired LQTS as a short-long-short sequence (SLS), whereby TdP is preceded by a characteristic sequence of long RR intervals of the dominant cycle, followed by a short extrasystolic interval with premature depolarization interrupting the T wave[1]. Viskin *et al.* [3, 4] suggested that the SLS sequence,

---

[1] On an electrocardiogram, the QRS complex is associated with depolarization of the ventricles and the T wave is associated with repolarization of the ventricles. The RR interval is the time period between two consecutive R peaks

J.L. Oliveira et al. (Eds.): ISBMDA 2005, LNBI 3745, pp. 305–316, 2005.

which has been recognized as a hallmark of TdP in the acquired LQTS, plays a major role in the genesis of torsade in the congenital LQTS as well. Data from the Registry[2] of LQTS, however, shows that a SLS sequence occurrence accounts for only half of all patients with congenital LQTS, and implies the existence of other characteristic patterns of onset. Recently, Noda *et al.* [2] investigated the clinical data of patients with congenital LQTS, and classified TdPs into three predominant modes:

- a SLS pattern - defined as one or more short-long cardiac cycles followed by initiating short-coupled premature ventricular contractions (PVC),
- an increased sinus rate (ISR) pattern - defined as a gradual increase in sinus rate with or without T ave alternans,
- a changed depolarization (CD) pattern - defined as a sudden long-coupled PVC phase 2 or phase 3 EADs fusion beat followed by a short-coupled PVC.

Currently, a common understanding of TdP states that the initiating PVC of TdP is due to triggered activity arising from phase 2 or phase 3 EADs. TdP, if initiated, is at least maintained by a reentrant mechanism resulting from increased transmural dispersion of repolarization (DR) [5]. Whereas experimental studies and clinical observations indicate phase 3 EADs during the SLS sequence as a mechanism responsible for the initiating of the PVC of TdP, less evidence has been reported in support of the ISR and CD onset. Especially in the case of ISR, the underlying mechanism remains largely unknown. Burasnikov *et al.* [6] suggested that the phase 2 EAD is predominantly induced during a transient acceleration of the pacing rate from an initially slow rate. Based on this suggestion, Noda *et al.* [2] proposed that the initiating PVC in ISR is related to phase 2 EADs. This mechanism, however, is challenged by another opinion that rapid rates accompanying the onset of TdP abruptly shorten repolarization, thereby eradicating the prerequisite condition for EAD-mediated TdP [7, 8].

Molecular genetics has revealed that specific mutations in ion channels modulating cellular repolarization underlie the various congenital LQT and TdP arrhythmias. So far, seven genes with arrhythmia susceptibility have been identified for auto-dominant LQTS: KCNQ1, HERG, minK and MiRP1 encoding $\alpha$ or $\beta$ subunits of the delayed rectifier potassium channels; SCN5A, encoding cardiac sodium channel $\alpha$ subunit; KCNJ2 the cardiac inward rectifier potassium channel; ANK2 the sodium/calcium exchanger [5, 9]. Two genes, KCNQ1 (homozygous) and minK (homozygous), have been identified for Jervell and Lange-Nielsen syndrome. Many inward and outward potassium currents, including the rapidly and slowly activating delayed rectifier potassium currents ($I_{kr}$ and $I_{ks}$), and the inward rectifier potassium current ($I_{k1}$), are responsible for repolarization of the cardiac action potential [9]. A small disease-associated perturbation in plateau current can have a major impact on prolonging the action potential in the ventricular myocardium and the reflection of such a prolongation on the ECG is a prolonged QT interval. The HERG gene has been discovered to mediate $I_{kr}$ current in the heart. Mutations in multiple regions of the HERG gene

---

[2] International Long QT Syndrome Registry (ILQTR), PO Box 653, University of Rochester Medical Center, Rochester, NY 14642-8653

have been identified as the cause of LQT2 syndrome. Results from database searching in GeneCards shows that there has been a high relevance of LQT as well as TdP syndromes to HERG. Biolama's score[3] indicates that approximately 42% of TdPs stem from LQT2 syndrome.

The role of the M cell in the intrinsic heterogeneities of cardiac tissue has been supported in many reports [7, 10, 11]. However, discrepancy remains in the earlier studies regarding the functional expression of M cells in intact myocardium because the distribution of M cells varies considerably between experiments. Anyukovsky *et al.* [12] reported M cells are identified in all myocardial layers when studied in isolation, except for the most superficial endocardium (Endo) and epicardium (Epi), but are not functionally present *in vivo*. Akar et al. [7], used transmural optical imaging methods and demonstrated that the M cells exist in myocardial clusters that vary in spatial extent and location across the heart. They suggested that unique topographical distribution of M cells underlies the reentrant mechanisms of TdP in the long QT2 syndrome.

The purpose of our current study is to investigate the occurrence of EAD focal activity at rapid rates and the involvement of dispersion of repolarization in control and abnormal states. Our analysis focuses on the use of genetic and cellular information related to the cause of LQT2 syndrome. To meet the large memory and computational requirements, we perform computation on a cluster of workstations, by implementing the Sun Grid Engine (SGE)[4] with the integration of MPICH, the portable high-performance implementation of Message Passing Interface [13].

## 2  Methods

### 2.1  Computational Control and LQT2 Models

The dynamics in a 2D isotropic monodomain virtual tissue are described by the reaction-diffusion equation [11, 14]:

$$\frac{\partial V_m}{\partial t} = D\left(\frac{\partial^2 V_m}{\partial x^2} + \frac{\partial^2 V_m}{\partial y^2}\right) - \frac{1}{C_m}I_{ion} \tag{1}$$

where $V_m$ (mV) is membrane voltage, $C_m$ ($\mu Fcm^{-2}$) specific membrane capacitance, and D ($cm^2/s$) is a diffusion coefficient. To describe the electrical physiological properties of the ventricular cells, we used the Luo-Rudy phase II model which describes the mammalian cardiac ventricular action potential in a single cell in guinea pigs. This second version of the model [11, 15] is more comprehensive, describing a wide variety of ionic currents. Along with the sodium and

---

[3] Biolama's score refers to the relevance of the disease to individual genes based on their literature text-mining algorithms

[4] SGE is a form f resource management software that accepts jobs submitted by users and schedules them for execution on appropriate systems in the grid based upon resource management policies

potassium membrane currents, the model focuses on processes that regulate intracellular calcium and depend on its concentration. Thus, the phase II model is suitable for investigating channelopathies and their arrhythmic consequences. The detail of the Luo-Rudy model is contained in the description of the total ionic current $I_{ion}$ ($\mu Acm^{-2}$) flowing through the cardiac cell membrane.

Multiple types of voltage-gated $K$ channel currents have been distinguished physiologically and pharmacologically in cells from different regions of the heart. Interestingly, no transmural gradients in the expression of several other $K^+$ currents, including $I_{k1}$ and $I_{kr}$, are evident [16]. Therefore, to achieve the specific formulation for epi-, mid- and endocardial cells under control conditions, we set the scaling constant $gK_{s_{max}}$ of $I_{ks}$ to 0.433, 0.125 and 0.289 respectively. According to Viswanathan et al. [17], this models a $I_{ks}$ density ratio of about 23:7:15 in epi-/mid-/endocardial cells. These parameters represent the normal expression of $I_{ks}$ channel in epi-, mid- and endocardial cells.

The mutations in multiple regions of the HERG gene cause a dominant negative affect on the HERG, and the electrophysiological consequence is reduction or loss of the channel function of outward current $I_{kr}$ [1]. Thus, in the current study, we considered LQT2 states by integrating into the Luo-Rudy model the changes of current density of $I_{kr}$, caused by mutations of the HERG gene. We modified the $I_{kr}$ current by decreasing the scaling constant $gK_{r_{max}}$.

## 2.2 Preparation of Virtual Tissues

The ventricular myocardium is composed of 3 predominant types of cells, including epicardial, midmyocardial and endocardial cells [5]. We designed two groups of virtual tissues, i.e. the homogeneous tissue and the heterogeneous tissue.

In the first group, each of the virtual tissue was a 70x70 homogeneous ventricular sheet, consisting of only one type of cell. We generated each virtual tissue as a 2D domain using Matlab[5]. As the space step was set to 0.01cm in our design, this yields a tissue size of 7x7 $mm^2$. We prepared epicardial control tissues by setting the $gK_{s_{max}}$ to 0.433 and keeping the $gK_{r_{max}}$ as given in the Luo-Rudy model, and epicardial tissues with LQT2 states by keeping the $gK_{s_{max}}$ as the normal value in the model, but decreasing $gK_{r_{max}}$ to 95%, 85%, 75%, 65% and 50% of the default value. In a similar way, we built homogeneous endocardial control tissues as well as LQT2 endocardial tissues, but setting $gK_{s_{max}}$ to 0.289.

In the second group, each tissue was a 140x140 2D heterogeneous ventricular sheet, consisting of either the Epi coupled with M cell clusters or the Endo coupled with M cell clusters respectively. We built the M cell model by setting $gK_{s_{max}}$ to 0.125 as suggested by Viswanathan et al. [17], and simulated LQT2 by reducing the $I_{kr}$ conductance as conducted in the homogeneous tissue model. To represent the variability in location and distribution of M cells in the tissue, we used Cellular Automata (CA) [18], the scheme for computing using local rules and local communication. CA is defined on a grid, with each point on the grid

---

[5] Matlab is an interactive software system for numerical computations and graphics. http://www.mathworks.com

representing a cell with a finite number of states. A transition rule is applied to each cell simultaneously. Typical transition rules depend on the state of the cell and its (4 or 8) nearest neighbors, although other neighborhoods are used. In this study, we used the following rule.

```
sum := sum of the 8 nearest neighbors and the cell itself
IF sum < 4 OR sum = 5 THEN
    state := 0
ELSE
    state := 1
END IF
```

We implemented this rule in Matlab and generated epicardial or endocardial tissues coupled with different topological distributions of M cells randomly.

### 2.3   Pacing and Experimental Protocols

The ISR pattern, which is characterized by an increasing sinus rate, was simulated through a pacing acceleration. The pacing method was designed as follows:

1. stimulation at basic cycle length (BCL) of 500 ms,
2. subsequent pacing acceleration with a drive train of 500 ms, 420 ms and 360 ms.

Programmed stimulation was performed in all prepared virtual tissues. Action potential duration (APD) and EADs were investigated in control states as well as LQT2 states. To assess the gradient of DR, we computed APD90, the duration from the beginning of the action potential to 90% of repolarization. Concerning EADs, we investigated whether there were action potential disturbances during repolarization, the initial time of EADs and the amplitudes of EADs fluctuations. We evaluated the initial time of the first fluctuation if any disturbance was observed.

### 2.4   Parallel Simulation

A 70x70 2D model involves 4,900 systems of ordinary differential equations (ODEs). Accordingly, 19,600 systems of ODES have to be solved for a 140x140 2D model. The scale and the nature of the model induce a strong requirement for more memory and computational resources. A test with a 140x140 2D tissue was performed on a Platinum 1200 computer (2x 2.0 GHz CPUs. 2.0 GB memory). It took 567 hours to complete a whole computation with a simulation time of 750 ms on a single processor. Therefore, parallelism had to be used to increase the computing efficiency and to reduce the computational cost. By use of CardioWave [19] and the included kernel modules, the domain was decomposed across a 1-D processor grid. The MPICH library [13] was employed to handle communication. Furthermore, we used the Rung-Kutta modul of CardioWave instead of the Euler method to obtain a better balance between accuracy and time

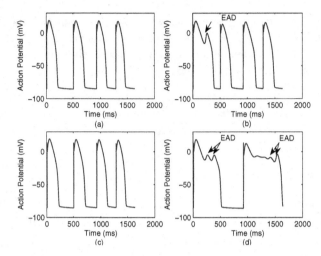

**Fig. 1.** Action potential in homogeneous tissues: (a) Epi under condition A, (b) Endo under condition A, (c) Epi under condition B, (d) Endo under condition B. A - a LQT2 condition with 75% of the normal $I_{kr}$ conductance, B - a LQT2 condition with 50% of the normal $I_{kr}$ conductance

consumed. To manage the computing resource dynamically, we implemented the SGE on a cluster of workstations with 34 CPUs. The SGE was configured with the integration of the parallel environment based on MPICH. The processing speed was enhanced by 30 fold on 34 CPUs.

## 3 Results

### 3.1 Electrical Characteristics in Homogeneous Tissue

Figure 1 illustrates the electrical characteristics observed in homogeneous tissue observed under LQT2 conditions. With $gK_{r_{max}}$ reduced to 50%, no EAD focal activity was observed in the homogeneous epicardial tissue, (Fig. 1a and Fig. 1c). In contrast to the Epi, the Endo was more prone to disturbances during the plateau phase, and phase 2 EAD was observed when $gK_{r_{max}}$ decreased to 75% of the normal value (Fig. 1b) and to 50% (Fig. 1d). Figure 2 shows APD90 of the homogeneous tissues with reference to simulated changing density of the $I_{kr}$ channel under LQT2 conditions. Again, the Endo showed a marked increase of APD90 when the $gK_{r_{max}}$ reached 75% of the normal value. The onset of the EADs was associated with a critical prolongation of APD.

### 3.2 Propagation of EADs in Heterogeneous Tissue

Compared to a homogeneous tissue, a heterogeneous tissue was double in size, i.e., 14x14 $mm^2$. Figure 3 shows where the EADs started and how they propagated in the tissues under a LQT2 condition with conductance of $I_{kr}$ reduced

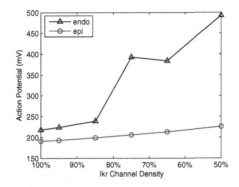

**Fig. 2.** Effect of change in $I_{kr}$ ion channel density on the action potential duration

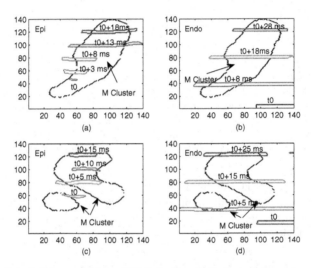

**Fig. 3.** Initiation and propagation of EAD focal activity in heterogeneous tissues under simulated LQT2 condition (gKr set to 50%): (a) Epi + Single M cluster, (b) Endo + Single M cell cluster, (c) Epi + separated M cell islands, (d) Endo + separated M cell islands. t0 is the initial time of EAD

to 50%. In an Epi coupled with M clusters (Fig. 3a and Fig. 3c), EADs occurred first at the border of the M cell formed refractory zone. But in the Endo coupled with M clusters, EADs were seen to initiate on the endocardial cells. Programmed stimulation was imposed on the bottom site in the present study. In each figure, the contour map with time intervals indicates how fast EADs propagated in the tissue. Notably, the topographical distribution of M clusters affected the onset position of EADs and the waveform of their propagation.

### 3.3   Dispersion of Repolarization in Heterogeneous Tissue

Figure 4 shows the DR across the 2D heterogeneous tissues in both the control situation and LQT2 states. In both the Epi and Endo coupled with either type of

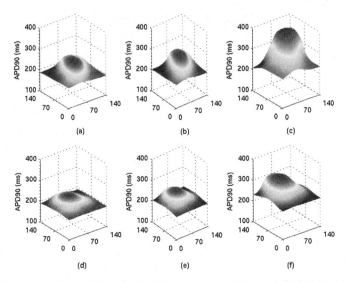

**Fig. 4.** Dispersion of repolarization in heterogeneous tissues (a) Epi + single M cell cluster in control state, (b) Epi + single M cell cluster in an abnormal state with 75% of normal $I_{kr}$, (c) Epi + single M cell cluster in an abnormal state with 50% of normal $I_{kr}$, (d) Endo + single M cell cluster in control state, (e) Endo + single M cell cluster in an abnormal state with 75% of normal $I_{kr}$, (f) Endo + single M cell cluster in an abnormal state with 50% of normal $I_{kr}$

M cell clusters, the APD increased with the enhancement of simulated severity of $I_{kr}$ dysfunction. The gradient of DR in the tissue coupled with a single M cell cluster (Fig. 4a, Fig. 4b and Fig. 4c) were steeper than in the tissue coupled with separated M cell islands (Fig. 4d, Fig. 4e and Fig. 4f).

### 3.4   Augmentation of EADs due to the Acceleration of Pacing

We compared the EAD amplitudes of two consecutive action potentials during the accelerating pacing in simulated LQT2 state with 50% $I_{kr}$ conductance. In the Endo coupled with a single M cell cluster, the amplitude of EADs increased from $11.41 \pm 0.76(mV)$ to $14.82 \pm 0.03(mV)$. In the Endo coupled with the separated M cell islands, EADs spiked off with amplitude $11.66 \pm 0.21(mV)$ after the last pacing, and no EADs occurred in the preceding beat. In any case, EADs were augmented by the accompanying pacing acceleration. Similar findings were obtained in the Epi coupled with M cell clusters. Interestingly, in Endo coupled with M cell clusters, spontaneous sarcoplasmic recticulum (SR) $Ca^{2+}$ release occurred during the pacing acceleration.

## 4   Discussion

### 4.1   Phase 2 EADs at Rapid Pacing Rates

EADs are perturbations occurring in membrane action potentials during the plateau phase (phase 2) or the repolarizing phase (phase 3). Usually, EADs are bradycardia or pause dependent. It is often believed that ventricular pacing at rapid rates will shorten repolarization and the QT interval and eliminate the pauses that precipitate torsades [7, 8]. ISR characterizes a type of onset mode of TdPs with increased sinus rate. Regarding the underlying mechanism, some questions arise: whether EADs could occur during pacing acceleration at rapid rates; what kind of EADs, phase 2 or phase 3 EADs, plays a critical role in the initiation of TdP if any EAD occurs.

Our data shows that at a BCL of 500 ms with subsequent acceleration, EAD was observed during the plateau phase (phase 2) in homogeneous endocardial tissues in the settings of LQT2 conditions. Although EADs were not observed in the homogeneous epicardial tissue, programmed stimulation-induced EADs were seen in both the Epi and Endo when coupled with M-cell clusters in simulated LQT2 situations. These results are consistent with the experimental evidence and findings of Brushnikov *et al.* using canine left ventricular myocytes exposed to a drug blocking $I_{kr}$[6], and the recent computational study of Huffaker *et al.* [20]. Both studies indicate that if SR $Ca^{2+}$ release occurs before repolarization is complete, activation of $Ca^{2+}$-sensitive inward currents such as Na/Ca exchange and the $Ca^{2+}$-activated nonselective cation channel cause EADs, leading secondarily to reactivation of the L-type Ca current and triggered activity.

The acceleration used by Burasnikov *et al.* is from a BCL range of 900 to 4000 ms to a range of 500 to 1500 ms. Our study used a shorter BCL. Thus our results may represent LQT2 arrhythmia caused by EAD focal activities observed in a subset of patients with fast heart beat (tachycardia). However, Our models have not yet taken into account the situation with increased sympathetic tone.

## 4.2   Role of Increased Dispersion of Repolarization

Differences in the response of the epi-, mid- and endocardial cells to pharmacological agents or pathological states often result in amplification of intrinsic electrical heterogeneities [5]. The increase on intrinsic transmural DR is thought to provide a substrate as well as a trigger for the development of TdP.

The prolongation of APD in the Endo under LQT2 condition was more pronounced than that in the Epi. The difference in the generation of EADs between the Epi and the endicardium was probably due to weaker expression of $I_{ks}$ channel in endocardial cells than in epicardial cells. Figure 3 shows EADs were firstly initiated at the border of M cells clusters in the Epi coupled with M cell clusters. Figure 4 demonstrates that there were steep gradients of repolarization along the border of the M cell refractory area. These results indicate that an increase of the DR provided a substrate for the initiation of EADs in neighboring epicardial cells. Using intracellular recordings in the intact left ventricle, Yan *et al.* [21] showed that EADs induced by the $I_{kr}$ blocker dl-sotalol usually are generated from rabbit subendocardial and endocardial layers in which APDs are prefentially prolonged. By comparing the tansmural DR induced by dl-sotalol and azimilide, they suggested that increased DR contributes to transmural propagation of EAD which produces an R-on-T extra systole initiating the onset of TdP. Noda *et al.* [3] showed that the initiating PVC occurs before the T-wave peak of the last beat before the onset of TdP in SLS mode. Our data provided computational evidence about the propagation of EADs and the effect of increased DR on the propagation, and imply that the propagation of EADs facilitated by increased DR contributed to the development of TdP.

Using optical imaging methods, Akar *et al.* [7] demonstrated that M cell clusters are not present uniformly at a given depth of myocardium, and they may extend to the epicardial and endocrinal surfaces. Due to eletrotonic interactions between cells, the M cell zone is necessarily blurred in space, possibly explaining why subepicardial cell could exhibit surprising long APD when near to M cells. Our data showed that in the neighboring epicardial and endocardial cells, not only was APD prolonged, but EADs could also be induced. Furthermore, the propagation demonstrated by the EADs contours indicates the topographical distribution of the M cells affected the topographical distribution and the propagation of EADs. These results are consistent with the experimental studies of Akar *et al.* [7] performed in the canine wedge preparation.

### 4.3   Genetic Implication of the LQT2 Models

Expression of HERG in heterogeneous systems led to the discovery that this gene encodes $\alpha$ subunits that form cardiac $I_{kr}$ potassium channels [1, 9, 22]. To date, there are 140 mutations found in HERG, including 108 substitutions, 19 deletions, 12 insertions and 1 complex rearrangement. These mutations account for 36% of the total number of LQTS mutations identified. Some mutations cause loss of function, whereas others cause dominant negative suppression of HERG function. These mutations are predicted to cause a spectrum of diminished $I_{kr}$ and delayed ventricular repolarization. Individuals with $\Delta$1261 and $\Delta$I500-F508 mutant proteins (intragenic deletions) would be predicted to express half the normal number of channels carrying $I_{kr}$ [22]. The A561V and G628S mutant channels (missence) do not express detectable currents in Xenopus oocytes injected with RNA encoding both wild-type and mutant channels. The N470D (missence) mutant proteins form functional channels with altered kinetic properties, and coexpression of normal and N470D channels reveal a dominant negative effect. The order of severity in reduction of HERG function for mutations characterized to date is G628S (most severe) > A561V > N470D > intragenic deletions (least severe) [1]. In our study, we changed the HERG channel ($I_{kr}$) to the spectrum of 95% to 50% of the control value. This implies that our models correlate only a LQT2 arrhythmia with a similar severity level caused by intragenic deletions. It is useful to extend such a study to investigate the effect of mutations having more severity than intragenic deletions and to integrate this spectrum of information into a cardiac simulation system.

## 5   Conclusion

This study has focused on the investigation of functional electrical instability arising from dysfunction at the molecular level, with the integration of genetic and cellular information, and with the help of grid-based computation. Our findings can be summarized as follows: (1) In the LQT2 conditions, phase 2 EADs could be observed in the homogeneous endocardial tissue as well as in the Epi or the Endo when coupled with M cell clusters. The amplitudes of EADs were augmented during the pacing acceleration. These findings suggest phase 2 EADs might play a critical role in the initiation of TdP at rapid heart rates. (2) EADs could not be induced in the homogeneous epicardial tissue in the simulated LQT2 conditions, but occurred in heterogeneous tissues at the border of the M cell refractory zone where steep spatial gradients of repolarization were observed. This implies that the increased DR provided a substrate for the spike off of EADs. (3) In heterogeneous tissues, the EADs were propagated mainly across the M cell clusters and the neighboring epicardial or endocardial cells. This would explain why subepicardial and subendocardial cells could exhibit surprisingly long APD and EADs when adjacent to M cells. (4) The waveform of EAD propagation depended not only on the topographical distribution of M cell clusters but also on the amplification of DR. These findings agree with the hypothesis that increased DR contributes to transmural propagation of EADs which produces an R-on-T extrasystole initiating the onset of TdP. In conclusion, our data suggests that the phase 2 EAD focal activity is the trigger responsible for the onset of TdP during rapid ventricular pacing. The whole computation has demonstrated that the SGE integrated with MPICH provided a useful means to solve the problem charaterized by lagre memory requirement and high computatioal load. This approach can increase the efficiency of dealing with not only multiple processors, but also multiple processors of jobs.

# References

1. Roden, D. M., Lazzara, R., et al., Multiple Mechanisms in the Long-QT Syndrome: Current Knowledge, Gaps, and Future Directions. Circulation, 1996. 94(8): p. 1996-2012.
2. Noda, T., Shimizu, W., et al., Classification and mechanism of Torsade de Pointes initiation in patients with congenital long QT syndrome. Eur Heart J, 2004. 25(23): p. 2149-2154.
3. Viskin, S., Long QT syndromes and torsade de pointes. Lancet, 1999. 354(9190): p. 1625-33.
4. Viskin, S., Alla, S. R., et al., Mode of onset of torsade de pointes in congenital long QT syndrome. J Am Coll Cardiol, 1996. 28(5): p. 1262-8.
5. Antzelevitch, C. and Shimizu, W., Cellular mechanisms underlying the long QT syndrome. Curr Opin Cardiol, 2002. 17(1): p. 43-51.
6. Burashnikov, A. and Antzelevitch, C., Acceleration-induced action potential prolongation and early afterdepolarizations. J Cardiovasc Electrophysiol, 1998. 9(9): p. 934-48.
7. Akar, F. G., Yan, G. X., et al., Unique topographical distribution of M cells underlies reentrant mechanism of torsade de pointes in the long-QT syndrome. Circulation, 2002. 105(10): p. 1247-53.
8. Viskin, S., Torsades de Pointes. Curr Treat Options Cardiovasc Med, 1999. 1(2): p. 187-195.

9. Keating, M. T. and Sanguinetti, M. C., Molecular and cellular mechanisms of cardiac arrhythmias. Cell, 2001. 104(4): p. 569-80.

10. Antzelevitch, C., Molecular biology and cellular mechanisms of Brugada and long QT syndromes in infants and young children. J Electrocardiol, 2001. 34 Suppl: p. 177-81.

11. Henry, H. and Rappel, W. J., The role of M cells and the long QT syndrome in cardiac arrhythmias: simulation studies of reentrant excitations using a detailed electrophysiological model. Chaos, 2004. 14(1): p. 172-182.

12. Anyukhovsky, E. P., Sosunov, E. A., and Rosen, M. R., Regional Differences in Electrophysiological Properties of Epicardium, Midmyocardium, and Endocardium: In Vitro and In Vivo Correlations. Circulation, 1996. 94(8): p. 1981-1988.

13. Gropp, W., Lusk, E., et al., A high-performance, portable implementation of the MPI Message-Passing Interface standard. Parallel Computing, 1996. 22(6): p. 789-828.

14. Clayton, R. H. and Holden, A. V., Dispersion of cardiac action potential duration and the initiation of re-entry: A computational study. Biomed Eng Online, 2005. 4(1): p. 11.

15. Luo, C. H. and Rudy, Y., A dynamic model of the cardiac ventricular action potential. I. Simulations of ionic currents and concentration changes. Circ Res, 1994. 74(6): p. 1071-96.

16. Nerbonne, J. M. and Guo, W., Heterogeneous expression of voltage-gated potassium channels in the heart: roles in normal excitation and arrhythmias. J Cardiovasc Electrophysiol, 2002. 13(4): p. 406-9.

17. Viswanathan, P. C., Shaw, R. M., and Rudy, Y., Effects of IKr and IKs heterogeneity on action potential duration and its rate dependence: a simulation study. Circulation, 1999. 99(18): p. 2466-74.

18. Wolfram, S., Cellular automata as models of complexity. Nature, 1984. 311: p. 419-424.

19. Pormann, J. B., Henriquez., C. S., et al. Computer Simulation of Cardiac Electrophysiology. in Proc. SC2000. 2000.

20. Huffaker, R., Lamp, S. T., et al., Intracellular calcium cycling, early afterdepolarizations, and reentry in simulated long QT syndrome. Heart Rhythm, 2004. 1(4): p. 441-448.

21. Yan, G. X., Wu, Y., et al., Phase 2 early afterdepolarization as a trigger of polymorphic ventricular tachycardia in acquired long-QT syndrome : direct evidence from intracellular recordings in the intact left ventricular wall. Circulation, 2001. 103(23): p. 2851-6.

22. Sanguinetti, M. C., Curran, M. E., et al., Spectrum of HERG K+-channel dysfunction in an inherited cardiac arrhythmia. PNAS, 1996. 93(5): p. 2208-2212.

# Extracting Molecular Diversity Between Populations Through Sequence Alignments

Steinar Thorvaldsen[1], Tor Flå[1], and Nils P. Willassen[2]

[1] Dept of Mathematics and Statistics, Faculty of Science
{steinart,tor}@math.uit.no
[2] Department of Molecular Biotechnology, Faculty of Medicine,
University of Tromsø, 9037 Tromsø, Norway
nilspw@fagmed.uit.no

**Abstract.** The use of sequence alignments for establishing protein homology relationships has an extensive tradition in the field of bioinformatics, and there is an increasing desire for more statistical methods in the data analysis. We present statistical methods and algorithms that are useful when the protein alignments can be divided into two or more populations based on known features or traits. The algorithms are considered valuable for discovering differences between populations at a molecular level. The approach is illustrated with examples from real biological data sets, and we present experimental results in applying our work on bacterial populations of *Vibrio*, where the populations are defined by optimal growth temperature, $T_{opt}$.

**Keywords:** sequence analysis; structural analysis; physicochemical properties; extremophiles; Fisher's exact test; Wilcoxon test.

## 1 Biological Motivation

Extreme environments are those that fall outside the limited range in which we, and most other eukaryotes can survive, and are inhabited by the *extremophiles*. Among extremophiles, which include thermophiles, psychrophiles, acidophiles, alkalophiles, halophiles, barophiles and xerophiles, those who live and prefer low temperatures are the largest and least studied group. Psycrophilic organisms are living at temperatures close to the freezing point of water. It is of great interest to understand how these organisms can function at "the limits of life" [1].

Living at extreme temperatures requires a multiplicity of crucial adaptations including preservation of membrane stability and maintenance of enzymatic activities at appropriate levels. At these temperatures a number of physiological factors are changed; the solubility of gases is not the same, the viscosity of water increases several folds as temperature is changed towards the extreme areas, for example.

The number of characterized cold or heat adapted proteins, reported sequences and high resolution structures is growing. The *Vibrios* are of the species with the greatest amount of published genomes, reaching five completed genomes this year, and seven ongoing whole genome sequencing projects including the cold adapted *Vibrio salmonicida*.

Alignment-free analysis has been used previously to compare amino acid compositions in whole genome and proteome datasets [2][3]. In this study, we focus on a set of homolog protein data from a relatively narrow range of closely related species

J.L. Oliveira et al. (Eds.): ISBMDA 2005, LNBI 3745, pp. 317–328, 2005.

belonging to the group *Vibrios* of gamma proteobacteria - a strategy also adopted by [4]. In our comparative study, we employ alignment based methods for examination of similarities and chemical differences at the molecular level by comparing amino acids and their *physicochemical properties* in the proteins. Different new methods of univariate analysis have been developed and used in this analysis.

The definition and analysis of chemical similarity has long been an active area of study in theoretical and computational chemistry [5-7]. Currently, there seems to be no generally agreed quantitative, or even qualitative, definition of chemical diversity. In formulating any description of quantitative chemical distance, one is obliged to make approximations and to use heuristically derived solutions. Many different ways have been used to represent chemical structures leading to many different approaches to assessing their similarity. These include methods based on three-dimensional representations. Two-dimensional approaches are, perhaps, even more numerous. The term 2D is a convention, as it is in general the properties of the molecular graph which are of interest, and not its pictorial representation in the plane. There have also been attempts to consider the measured biological properties of compounds as the basis for diversity analysis. The descriptors may take the form of measured or computed physical properties such as topological or constitutional indices. There are several approaches based on some count of shared features. Such features include atom or element types, bonds, topological torsions, etc.

## 2  The Algorithms

### 2.1  Residue Frequencies

We wanted to examine amino acid occurrences in relation to background distributions, and for this purpose we analysed the composition of amino acids in two different sequence populations. The distribution of the categorical variable (amino acid type) in the sequence samples can be modelled and compared statistically. The basic chain-like structure of proteins, allows an abstract view of these as strings, or sequences, over a finite alphabet. Protein sequences could in principle, as a first approximation, be considered as random samples taken from some distribution. Let $X$ be a discrete random variable with a finite set of possible values, or categories {A,R,N,D,Q,E,G,H,I,L,K,M,F,P,S,T,W,Y,V}. For convenience the amino acids are often encoded by ordinal numbers x = 1, 2,..., 20. If sites were independent, the *multinomial* distribution would follow. This is the multivariate generalisation of the common binomial distribution.

However, gene sequences are not completely random, but display various kinds of structure. E.g. a family of homologous proteins is likely to have similar amino acid residues in "equivalent" positions, and the amino acid frequencies are expected to be different from gene to gene. Let two populations be defined by some trait (like different temperature preference). For every gene family g = 1, 2,..., k; we may for each of the 20 amino acids compute the frequency change vector between the mean frequencies in the two populations:

$$\Delta f_g = (\Delta f_1, \Delta f_2, \dots, \Delta f_{20}), \quad g = 1, 2, \dots, k$$

This is a variable that can be studied statistically across the dataset concerning significant changes.

Still, the statistics above have serious limitations. They simply indicate the degree of evidence for an over- or under-representation of the variables, and are not adequate for answering other more interesting questions about the data. One should also study the nature and effects of these differences.

## 2.2  Residue Substitution Pairs

A statistical approach to molecular sequence analysis also involves the stochastic modelling of the substitution, insertion and deletion processes. We also present an analysis of amino acid substitution data matrices from an independent set of paired homolog protein sequences. The method is based on trusted alignments, where observed amino acid replacements are tallied in a raw residue replacement matrix. We also present an analysis of the amino acid substitution pattern.

The modelling of amino acid replacement by a Markov chain has been introduced by Dayhoff et al [8]. Our strategy is much of the same as the method Dayhoff' used to estimate the well known initial PAM1 (Percent Accepted Mutation) transition matrix based on just 1572 substitutions. But our technique is different and our aim is to study the internal distance between populations, not to extrapolate to higher PAM distances.

Brenner et al [9] were the first to point out the problem in Dayhoff's method of estimating *one* consistent model from an *inhomogeneous* pool of aligned sequence data. Bias in the sequence selection may influence the frequencies of substitutions. Though biases cannot be eliminated entirely when data are sparse, one has to minimize biases in the data selection. Without clustering, some closely related sequences may be over-represented, and special care must be taken to locate representative samples as a safe-guard against obtaining bias comparisons and ditto results in the investigations. Jones et al [10] presented an alternative method where the set of sequences are clustered at the 85 % identity level. The *closest* relating pairs of sequences are aligned, and observed amino acid exchanges tallied in a matrix.

We count by the Jones-method to minimize biases. Two pairs in an aligned site can be classified as invariant (where the same amino acid are conserved in the two populations) or variant (where there is a difference). Our main goal is not to measure conservation, but its opposite, deviation. A matched *Substitution Pair* (SP) is defined as the ordered combination of two amino acids observed in an alignment position, and a SP-matrix is the accumulation of all such pairs by summing over all positions. The accumulated array contains the frequency of all position specific pairing of residues. Thus, for any number of aligned amino acid sequences, the number of possible SP in each position is between 1 and 400, but only a small fraction of these SPs are observed, and the majority of sequence positions are covered by less than 10 SPs for closely related homologous proteins. This method of mapping and expressing alignment data by SP-matrices trim down problems caused by statistical dependences between the sequences and the uncertain phylogeny involved in the PAM procedure. Note that no counting is done when a residue is aligned to a gap. The accepted SPs are counted by the algorithm shown in Figure 1.

We may both calculate SP-matrices *between two* populations, and *within one* population. SPs may be used to address the following question: Which and how many SPs account for the major significant variations between the populations?

**Algorithm CountSubstitutionPairs**

| | |
|---|---|
| Input | k gene families each with m(k) aligned protein sequences $s_1, s_2, \ldots s_{m(k)}$ from two populations. |
| Process | find in all genes all closest sequence pairs between the populations, and count substitutions over all positions |
| 1 | for all genes |
| 2 |   for every sequence pair $(s_{Pop1}, s_{Pop2})$ with max similarity |
| 3 |     for every residue position $j = 1, \ldots, n$ |
| 4 |       find all residue pairs $SP_j = \{sp(x, x'): x \in Pop1 \text{ and } x' \in Pop2\}$ |
| 5 |       $SP := SP + SP_j$ |
| Output | substitution pair matrix SP between the two populations |

**Fig. 1.** An algorithm to compute the SP-matrix. The amino acids are denoted by x = 1,2,...,20

Instead of doing an overall test of the big SP-matrix, we partition the matrix in a meaningful manner and focus on more targeted tests. We want to analyse the over- or under representation of single SPs compared to a random model. The occurrence of SP might be expected to differ in some measure due solely to chance factors of sampling, and for other reasons which might be attributed to random causes. And what we shall need to find out is whether or not the observed differences are too large to be credited to such causes.

An enduring problem in statistics is the analysis of 2x2 contingency tables, and there has been a lot of research and debates [11, 12]. The main debate has at least two components. The first is to select either an exact test (e.g. Fisher's exact test) or an asymptotic test (e.g. Pearson's chisquared test). The second is which test procedure should be employed among many candidates in each group. There has been an effort to determine the best exact test among the Fisher's exact test, the exact chi-squared test and the exact likelihood ratio test in 2x2 tables in both large and small samples. Kang and Kim [13] compared the three conditional tests and showed that the Fisher's exact test turns out to be the best choice in most cases. Consequently, because of the practical values in our data, we decided to use *Fisher's exact test* to find statistically whether there is any non-random relation between any two categorical variables with two observed levels found from the SP-matrix.

## 2.3 Residue Properties

An alignment of homolog sequences is a set of matched *pairs* where there is a meaningful one-to-one correspondence between the data points in one group and those in the other. This gives us the possibility to investigate the mean property differences (like hydrophobicity) in the sequences by a probabilistic framework.

For two amino acids, $x$ and $x'$, we denote their linear chemical *difference measure*:

$$d(x,x') = q(x') - q(x)$$

This difference yields real values when we assume that we have a table of quantitative chemical values, $q$, for each amino acid. The measure is an expression of the diversity between the amino acids, and the choice of measures to be used depends on the test we want to perform.

Let $s_m$, $m = 1, 2, \ldots M$, be M aligned amino acid sequences, and let the amino acid at position $j$ in $s_m$ be denoted by $x_{m,j}$. We define the difference between the sequences from population $p_1$ and $p_2$ at position $j$ by averaging the measurements at position $j$ within each population:

$$d(p_{1,j}, p_{2,j}) = \overline{q}(x_{Pop2,j}) - \overline{q}(x_{Pop1,j})$$

By this we measure $n$ differential effects between population 1 and 2, where n is the length of the gapless alignment. We find the mean chemical difference, $D$, between the two populations by:

$$D(p_1, p_2) = \frac{1}{n} \sum_{j=1}^{n} d(p_{1j}, p_{2j})$$

Assuming that the distribution of the differences in each position, $j$, is identical, we obtain the expected value $\Delta$:

$$\delta = E(D(p_1, p_2)) = \frac{1}{n} \sum_{j=1}^{n} E(d(p_{1j}, p_{2j})) = E(d(p_{1j}, p_{2j})), j = 1,2,\ldots n$$

A standard parametric test would be to approximate $D$ with the normal distribution, and apply the paired $t$-test. However, since it is not always clear that this is appropriate with the protein properties; other statistical test should be considered. In the statistical analysis, it is also important that the significant difference found between means (or not found) be due to the different conditions of the populations, and not due to the organisation and conservation of the particular enzyme in the study.

A relevant alternative to the $t$-test is the *Wilcoxon signed-rank test* which is a nonparametric test that also can be used on continuous type of paired data, both when the underlying population is normal and when not [14]. This test automatically discards all differences equal to zero from the analysis (conserved sites). It can in some cases be better than the paired $t$-test for non-normal populations, although non-parametric procedures in general need larger sample size than $t$-tests.

It is not easy to compare the two test procedures in a general theoretical way. One widely used measure in the literature is *asymptotic relative efficiency* (ARE, [14]). The ARE of one test relative to another is the limiting ratio of the sample size necessary to obtain identical error probabilities for the two procedures. For normal populations the ARE of the Wilcoxon test relative to the $t$-test is $3/\pi \approx 0.95$, and for nonnormal populations the ARE is $\geq 0.86$ which means that it in some cases will exceed 1. Although these results are for large samples, and do not necessary tell us anything for small samples, one may generally conclude that the Wilcoxon signed-rank test will never be much worse than the $t$-test, and in many cases where the population is non-normal it may be better. Our experience with using the two tests on the protein data is that they do not make much difference, with the Wilcoxon test as the most conservative, except for properties with more than one peak distribution (like Kyte-Doolittle hydrophobicity).

This defines a useful and reliable statistical model when we are investigating a variable along the sequence in two population groups. The formal statement of the hypothesis of interest is

$$H_0 : \delta = 0, (\text{zero mean difference})$$

$$H_1 : \delta \neq 0$$

We may test the significance of property alterations, e.g. a decreased surface hydration free energy, in psychrophilic sequence populations. This gives an efficient and more reliable comparison of protein populations than earlier studies.

The differences between two sequence populations can be compared graphically as well as statistically, and we developed a smoothing technique to be able to recover and visualize underlying structure in the data set [15]. We use a rectangular box-filter where the vertical filter size is all the amino acids in the aligned sequence position of the population, and the horizontal window size can be varied. This filter can be used to plot smoothed lines of amino acid properties, such as comparative plots shown in Figure 5. All analyses reported in this work were implemented in Matlab.

### 2.4 False Discovery Rate

A common objection against the testing algorithms described above will be the multiple comparisons problem. Benjamini and Hochberg have suggested that the false discovery rate (FDR) may be the suitable error rate to control such multiple testing problems [16]. FDR is the expected proportion of false rejections among all rejections and is a new measure of error rate. A simple procedure was given by them as an FDR controlling procedure for independent test statistics and was shown to be much more powerful than comparable procedures like Bonferroni correction which may be much too conservative. The original formulation FDR presumes independence among the different amino acid properties, which is far from correct in our case. But in a recent paper [17], the FDR criterion has also been extended to multiple testing under dependency.

However, a more straightforward way to overcome this difficulty is just to analyse more than one dataset by the same procedure, and only report features that are significant across many independent protein groups.

## 3   Experimental Design and Results

### 3.1   Methodology and Datasets

Sequenced proteins from the *Vibrios* were downloaded from standard databases in order to identify homologues. With a minimum cut-off score of 70% sequence identity the corresponding amino acid sequences of 7-14 vibrio species (4-10 in the *mesophile* population and 2-4 in the *psychrophile*) were extracted and 25 alignments of intracellular sequences were made with T-coffee. Physicochemical, steric and other properties were downloaded from the database AAindex release 6 [18], which contain 494 quantitative properties of the amino acids, and collected from the literature [19].

We applied the methods of comparative analysis of protein sequences by focusing on discriminative features extracted from rationally selected parts of the data sets. In this approach we made extensive use of the sequence based predictors developed in the latest years, and our alignment-based data sets were decomposed and clustered in relevant subclasses. We used these tools: sub-cellular location (CELLO, Yu 2004), surface (SABLE, Adamczak 2004) and secondary structure (PSIPRED, McGuffin 2000). The secondary structure was predicted with default settings, and the solvent Accessible Surface Area (ASA) was predicted with thresholds 0-1: completely buried, 2-3: twilight zone, and 4-9: surface. All predictions were based on the sequences from

*Vibrio cholerae*. The alignments and annotations were automatically done with a program script, and made it possible to analyse the alignment data relative to its cellular, 2D and some of its 3D structural constraints:

- cellular location (intracellular, membrane, extracellular)
- 2D secondary structure region (alpha, beta, loop)
- 3D structure location (surface, twilight zone, core)

## 3.2 Experimental Results

**Compositional Analysis**

Different subsets of the protein data were used in the analysis; we examined the distribution in the compositional space. An example showing the variation in amino acid compositions are shown in Figure 2. In general it seems difficult to resolve general elements of cold adaptation at the basic compositional level. The standard deviations between the populations are overlapping. The greatest differences are found to be at the surface and the smallest in the interior of the molecule.

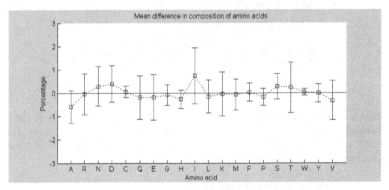

**Fig. 2.** Average amino acid difference in compositions from the mesophilic to the psychrophilic temperature domain of 25 cytoplasmic proteins from the bacteria *Vibro*. Bars are the empirical standard deviations

**Analysis of SP Patterns**

In the matrix shown in Figure 3, we present the SPs found between the mesophilic and psycrophilic populations in one of our data analysis. They were all tested for statistical significance by using Fisher's exact test (Figure 4).

The data may also be decomposed and studied in more detail, such as *secondary structure* and *external/internal* surface position. Analyses of residue properties reveal that many frequent SPs consist of the most hydrophilic/ hydrophobic residues, or residues with high propensities to form secondary structures. In general, residues forming the most significant SPs have extreme values of one or more essential Physicochemical property.

(V,I) is the most frequent SP. These residues have in common a maximal propensity to form beta-sheets, and the highest minimum width of the side chain. Among all pairs of residues, their steric similarity is the highest, as expressed by the partial specific volume and bulkiness. They are considered the most hydrophobic and consequently the most often buried residues which transfer with the highest free energy from the exterior to the core of the protein. The (L,I) has similar characteristics to (V,I) and is also frequent, but I is more similar to V than to the differently branching L.

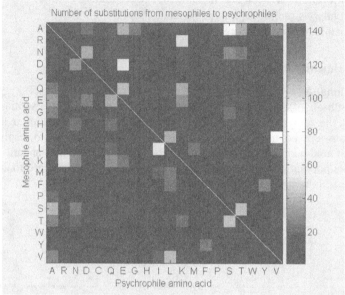

**Fig. 3.** Visualisation of all the pairwise substitutions from the mesophilic (left) to the psychrophilic (bottom) group. Two clear trends can be seen: Val (V) is replaced by Ile (I), P-value=0.007; and Glu (E) is replaced by Asp (D), P-value=0.21. Many of the other minor changes are found to be statistically significant (P<0.01)

Example: Substitution pair: V -> I

| Population | Yes | No |
|---|---|---|
| Mesophilic vs. Psychrophilic | 146 | 141 |
| Psychrophilic vs. Mesophilic | 83 | 132 |

**Fig. 4.** Example showing the use of Fisher's exacte test. P-value= 0.007

(E,D) is the second most frequent SP. These hydrophilic and accessible residues have high transfer free energies from water to organic solvents. They have the lowest propensity for bata-structures and the highest helix termination parameter.

(K,R) is also a frequent SP. These are very hydrophilic and accessible residues and have the most side chain heteroatoms. The gyration radius and the side chain interaction parameters are very high because of the long side chains of these amino acids.

In terms of involvement in replacement pairs, A is a very changeable residue. The high mutability of A is probably due to its role as a default residue, with positive contributions to alpha-helix propensity. Lack of gamma-carbon also allows substitutions with small steric obstructions. But replacements to the somewhat rigid A induce smaller changes in the fold than substitutions to the very flexible G.

(A,S) also frequently replace each other, mostly in surface and loop areas. Both residues have low free energy of hydration. (T,S) are also repeated substitutions.

In the literature, there are also published some paired indexes of amino acid *changes* [20]. This model is described by a 20x20 matrix $C$, and may also be used to define the chemical difference $d(x,x')$ between every pair $x$, $x'$ of amino acids in a similar manner as we did above.

## Physicochemical Properties

We analysed different sequence populations defined by origin and temperature, and conducted a large scale statistical test to identify systematic change and significant differences between the mesophilic and psychrophilic populations (described in part 2.3 of the paper). A summary of the results are shown in Table 1. Results are only reported in the table if the P-values are found to be significantly low (P<0.01). We observe that there are many interesting differences especially at the surface and in the alpha helices of the molecules.

The Gibbs free energy is a fundamental parameter that provides a measure of thermodynamic stability of the protein molecule. Most studies of the stability of proteins are concentrated on evaluation of the Gibbs free energy of unfolding. We found no significant difference for this parameter between the populations. However, the Gibbs energy, $\Delta G$, consists of two terms describing the enthalpic, $\Delta H$, and entropic, $-T\Delta S$, contribution, i.e. $\Delta G = \Delta H - T\Delta S$. The enthalpic and entropic contributions for a given system appear to have a close relationship, the so-called enthalpy/entropy compensation. In some cases the enthalpy/entropy compensation is significantly close to obscure the occurrence of the changes in a system, if the analysis is done only in terms of Gibbs energy. The differences between the separate changes in the enthalpy and entropy may be very significant, as we found it to be in our data analysis, where $\Delta H$ goes down and $-T\Delta S$ up (Table 1).

**Table 1.** List of main differences from mesophile to psychrophile populations. P-values are obtained by the Wilcoxon test of the total data set of 25 alignments from *Vibrio* bacteria. Particular 2D and 3D regions are also marked if they have significant hits (P<0.01): A=Alpha helix, B=Beta sheet, L=Loop, C=Core, T=Twilight zone and S=Surface. Details in the references to amino acid properties can be found in [18, 19]. Other properties gave no significant hits across the data set

| Property | Change | Entire seq. P-value | Significant 2D/3D region | Ref. property index |
|---|---|---|---|---|
| Hydrophobicity | ↓ | $<10^{-5}$ | A,L,C,T | Kyte-Doolittle, 1982 |
| Buriedness | ↓ | $<10^{-5}$ | A,L,C,T | Chothia, 1976 |
| Molecular weight | ↑ | $<10^{-5}$ | A,L,C,T | Fasman, 1976 |
| Volume | ↑ | $<10^{-5}$ | A,L,T. (C↓) | Zamyatin, 1972 |
| Metabolic costs | ↓ | $<10^{-5}$ | A,L,C,T | Akashi, 2002 |
| Alpha-helix, frequency of | ↓ | $4\ 10^{-3}$ | A | Chou-Fasman, 1978 |
| Beta-sheet, frequency of | ↓ | 0.6 | B | Chou-Fasman, 1978 |
| Side-chain contrib. to stab. | ↓ | $6\ 10^{-3}$ | B,C | Takano-Yutani, 2001 |
| Average flexibility | ↑ | $3\ 10^{-5}$ | A,B,C | Bhaskaran et al 1988 |
| $\Delta H$ (unfolding enthalpy) | ↓ | $<10^{-5}$ | A,B,L,C,T | Oobatake-Ooi, 1993 |
| $-T\Delta S$ (unfolding entropy) | ↑ | $5\ 10^{-5}$ | A,B,C,T | Oobatake-Ooi, 1993 |

Amino acids have different propensities to form helical structures, and the composition in helical regions may affect both the helix stability (Figure 5) and the overall stability of the protein. Our calculations of the alpha-helical sequences show a better stabilisation for mesophiles compared with psychrophiles (Table 1).

We observe increasing trends with cold adaptation for molecular weight, volume, and average flexibility. Hydrophobicity, side-chain contribution to stability and metabolic cost are decreasing properties.

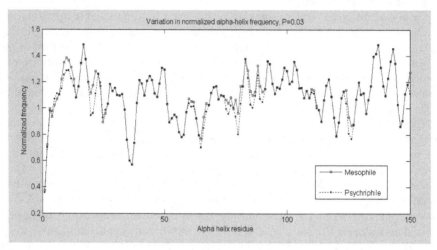

**Fig. 5.** Comparison of the mean helix formation parameter in the cytoplasmic protein *Isocitrate lyase* of 5 *Vibrio* gamma protobacteria (3 mesophilic and 2 psychrophilic). We used a box-filter of size m x 3 as smoothing technique to recover the underlying structure in the data, where m is the number of sequences in the population. For this particular scale (Chou-Fasman, 1978) the helix-favourable values are at the positive y-axis. In average the mesophilic sequences appear to have more favourable values than the psychrophile counterparts

## 4   Conclusion

We performed comparative analysis of genetic variability using protein sequences from bacterial populations of *Vibrio* with different temperature preferences. The use of data from the same taxonomic groups reduced problems associated with physiology and phylogenetic noise that have been a problem in other studies.

We have applied and expanded the methods of comparative analysis of proteins. The improved strategy is partly extensions of traditionally used statistics [19], e.g., residue frequencies, residue properties, but applied to *positions* of aligned sequence pairs rather than averaged over unaligned sequences. Statistics also include an amino acid replacement matrix approach to identify residue substitution pairs that differs between populations. The approach of using aligned sequence pairs yield better comparisons, and in this paper an appropriate probabilistic model of context-sensitive and property-dependent analysis of alignments is developed, including efficient algorithms for constructing them. We extracted compositional differences into several distinct physicochemical factors.

In the present *Vibrio* study we found that decreasing hydrophobicity and buriednes are generally (and especially for core residues) the most important properties for adaptation to cold in cytoplasmic proteins. Moreover, unfolding enthalpy and unfolding entropy are found to be different in a direction that compensates concerning the Gibbs free energy.

Furthermore, decreased stability parameters correlates both in alpha helices and in beta-strands. All these results suggest that the maintenance of proper balance between stability and flexibility is critical for proteins to function at their environmental temperatures.

Some of the features observed may be specific to intracellular proteins or to the *Vibrio* species, and more sequence families should be analyzed to detect both general and special determinants of cold adaptation.

## Acknowledgements

We thank Elinor Ytterstad for suggestions regarding the statistical analysis. Some of the data used in this work were sequenced at the University of Tromsø by Erik Hjerde.

Upon publication, our Matlab code can be downloaded from our web-site at: http://www.math.uit.no/bi/deltaprot/

## References

1. Russell, N.J.: Toward a molecular understanding of cold activity of enzymes from psychrophiles. Extremophiles 4: 83-90. 2000.
2. Karlin S, Brocchieri L, Trent J, Blaisdell BE, Mrazek J. Heterogeneity of genome and proteome content in bacteria, archaea, and eukaryotes. Theor Popul Biol. 61:367–390. 2002
3. Pe'er I, Felder CE, Man O, Silman I, Sussman JL, Beckmann JS: Proteomic signatures: Amino acid and oligopeptide compositions differentiate among phyla. Proteins-Structure Function and Genetics 54 (1): 20-40. 2004
4. Saunders NFW, Thomas T, Curmi PMG, et al.: Mechanisms of thermal adaptation revealed from the genomes of the Antarctic Archaea Methanogenium frigidum and Methanococcoides burtonii. Genome Res 13 (7): 1580-1588. 2003.
5. Nikolova N, Jaworska J.: Approaches to measure chemical similarity - A review. QSAR & Combinatorial Science (9-10): 1006-1026, 2004.
6. Kearsley, S. K.; Sallamack, S.; Fluder, E. M.; Andose, J. D.; Mosley, R. T.; Sheridan, R. P. Chemical Similarity Using Physicochemical Property Descriptors. J. Chem. Inf. Comput. Sci. 1996, 36, 118-127.
7. Basak, S. C.; Grunwald, G. D. Molecular Similarity And Estimation of Molecular-Properties. J. Chem. Inf. Comput. Sci. 1995, 35, 366-372.
8. Dayhoff, M. O., Schwartz, R. M. & Orcutt, B. C. (1978). A model of evolutionary change in proteins. Atlas of Protein Sequences and Structure, 5 suppl 3, 345–352.
9. Benner, S. A., Cohen, M. A. & Gonnet, G. H. (1994) Amino acid substitution during functionally constrained divergent evolution of protein sequences. Protein Eng., 7 (11), 1323–1332.
10. Jones DT, Taylor WR, Thornton JM (1992): The rapid generation of mutation data matrices from protein sequences. Computer Applications in the Biosciences 8 (3): 275-282.
11. Agresti A. (2001): Exact inference for categorical data: recent advances and continuing controversies. Statistics in Medicine. 20 (17-18): 2709-2722.
12. Agresti, A. (2002). Categorical Data Analysis. 2. ed. John Wiley & Sons.

13. Kang S.H., Kim S.J. (2004). A comparison of the three conditional exact tests in two-way contingency tables using the unconditional exact power. Biometrical Journal 46(3): 320-330.

14. Conover, W.J. (1999). Practical Nonparametric Statistics, 3.ed. John Wiley & Sons.

15. Oppenheim, A. V. & Schafer, R.W. (1999). Discrete-Time Signal Processing, 2.ed. Prentice-Hall.

16. Benjamini, Y & Hochberg, Y (1995). Controlling the false discovery rate: a practical and powerful approach to multiple testing. Journal of the Royal Statistical Society, Series B, 57:289-300.

17. Benjamini, Y. & Yekutieli, D. (2001). The control of the false discovery rate in multiple testing under dependency. Ann Stat 29 (4): 1165-1188.Rabus R, Ruepp A, Frickey T, et al.:

18. Kawashima, S., Ogata, H., and Kanehisa, M.; AAindex: amino acid index database. Nucleic Acids Res. 27, 368-369 (1999).

19. Gromiha, M.M., Oobatake, M. & Sarai, A. (1999). Important amino acid properties for enhanced thermostability from mesophilic to thermophilic proteins. Biophysical Chemsitry 82, 51-67.

20. Hua Tang, Gerald J. Wyckoff, Jian Lu, and Chung-I Wu A Universal Evolutionary Index for Amino Acid Changes. Mol Biol Evol 2004 21: 1548-1556

# Detection of Hydrophobic Clusters
# in Molecular Dynamics Protein Unfolding Simulations
# Using Association Rules

Paulo J. Azevedo[1], Cândida G. Silva[2], J. Rui Rodrigues[2],
Nuno Loureiro-Ferreira[2], and Rui M.M. Brito[2,3,*]

[1] Departamento de Informática, Universidade do Minho, 4710-057 Braga, Portugal
pja@di.uminho.pt
[2] Centro de Neurociências de Coimbra, Universidade de Coimbra, 3004-517 Coimbra, Portugal
[3] Departamento de Química, Faculdade de Ciências e Tecnologia, Universidade de Coimbra,
3004-535 Coimbra, Portugal
brito@ci.uc.pt

**Abstract.** One way of exploring protein unfolding events associated with the development of Amyloid diseases is through the use of multiple Molecular Dynamics Protein Unfolding Simulations. The analysis of the huge amount of data generated in these simulations is not a trivial task. In the present report, we demonstrate the use of Association Rules applied to the analysis of the variation profiles of the Solvent Accessible Surface Area of the 127 amino-acid residues of the protein Transthyretin, along multiple simulations. This allowed us to identify a set of 28 hydrophobic residues forming a hydrophobic cluster that might be essential in the unfolding and folding processes of Transthyretin.

## 1 Introduction

One of the most challenging problems in molecular biology today is the protein folding problem, *i.e.* the acquisition of the functional three-dimensional structure of a protein from its linear sequence of amino-acids. This sequence of amino-acids is encoded by the linear sequence of nucleotides in a gene, but protein function is mediated by its exquisite three-dimensional structure. Predicting the 3D structure of a protein from the linear sequence of amino-acids is as yet an unsolved problem today, and a challenge for those eager to harness the information content of the genomes.

In recent years, the issues of protein folding became also pivotal in the understanding of a series of human and animal diseases, generally know as conformational disorders or amyloid disorders, and ranging from Alzheimer´s to bovine spongiform encephalopathies (BSE). Although the proteins involved differ in sequence, structure and function, the amyloid pathologies share common molecular mechanisms. In particular, it seems that in all studied cases, due to proteolysis, mutation or unfolding

* The authors acknowledge the support of the "Fundação para a Ciência e Tecnologia" and the program FEDER, Portugal, through projects POSI/SRI/39630/2001/CLASS (to PJA), POCTI/BME/49583/2002 (to RMMB) and the Fellowships SFRH/BD/1354/2000 (to NLF), SFRH/BD/16888/2004 (to CGS) and AI/06/02 (to JRR). We thank the Center for Computational Physics, Departamento de Física, Universidade de Coimbra, for the computer resources provided for the MD simulations

J.L. Oliveira et al. (Eds.): ISBMDA 2005, LNBI 3745, pp. 329–337, 2005.

events, the normally soluble proteins are converted into molecular forms prone to aggregation, leading to cytotoxic oligomeric species and amyloid fibrils.

We have been particularly interested in the structural characterization of the molecular species present in the aggregation pathway of Transthyretin (TTR), a human plasma protein responsible for such amyloid diseases as Familial Amyloid Polyneuropathy (FAP), Familial Amyloid Cardiomyopathy (FAC) and Senil Systemic Amyloidosys (SSA), using both experimental and computational methodologies [1,2,3]. One way of exploring the unfolding events that may be responsible for TTR aggregation is through the use of Molecular Dynamics Protein Unfolding Simulations (MDPUS). However, we know today that in an ensemble of protein molecules not all of them follow the same folding or unfolding route. Thus, multiple simulations are required in order to have some idea of the conformational space available to a protein molecule in its unfolding process. These simulations are computationally expensive and generate a huge amount of data. In order to contrast, compare and characterize the molecular properties associated with each simulation, Data Mining techniques are required.

In the present paper, we report the use of Association Rules, a specific Data Mining technique to identify relations between atomic elements of the data, in order to detect potential coordinated movements of different amino-acid residues in unfolding simulations of the protein Transthyretin. In particular, through the analysis of the Solvent Accessible Surface Area (SASA) of each amino-acid residue along multiple unfolding simulations of TTR, we identify a group of hydrophobic residues moving in a coordinated fashion and most likely forming a hydrophobic cluster essential in the folding and unfolding processes of Transthyretin.

## 2  Molecular Dynamics Protein Unfolding Simulations

Molecular Dynamics (MD) simulations have recently become an important tool to explore folding and unfolding processes in proteins [4,5,6] and we have put forward some of the challenges facing the researcher when comparing and contrasting the results of multiple MD simulations in different proteins [7].

In Molecular Dynamics simulations, molecules are treated as spheres connected by springs and classical mechanics are used to calculate forces and velocities. Although this treatment is highly approximated and does not take into account quantum effects, the realism of the simulation depends on the ability of the potential energy function to reproduce the inter-atomic interactions characteristic of the molecular system under study. In fact, several decades of research in small molecules and biological macromolecules allowed the definition of a generally accepted set of empirical functions, and today MD is a well established method for studying equilibrium protein dynamics and non-equilibrium processes such as protein folding and unfolding.

### 2.1  Simulation Details

Initial coordinates for Transthyretin were obtained from the crystal structure (PDB entry 1tta) [8] and hydrogen atoms were added. All minimization and MD procedures were performed with the program NAMD [9], using version 27 of the CHARMM force field [10]. All atoms were explicitly represented. Internal waters were placed with the program Dowser [11] and the program Solvate (http://www.mpibpc.mpg.de/abteilungen/071/solvate/node8.html) was used to add solvent water molecules and Na$^+$Cl$^-$ ions around the protein. The complete system was comprised of 45,256 atoms.

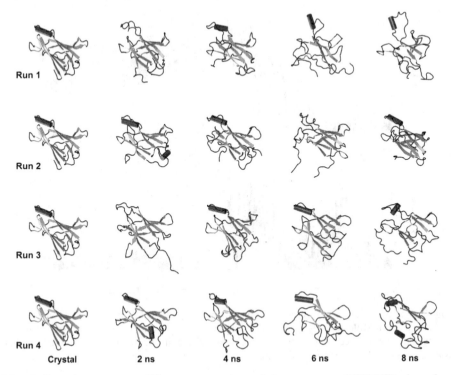

**Fig. 1.** Secondary structure ribbon representations of the monomer of WT-TTR along four different Molecular Dynamics unfolding simulations. Beta-strands, alpha-helices and turns and coil are represented by arrows, cylinders and tubes, respectively. The difference in the setup of each of the four runs resides in the assignment of initial atomic velocities

The system was minimized, equilibrated and heated to the target temperature. Several simulations were performed using *Centopeia*, a Linux computer cluster at UC. Control simulations, at 310 K, and several unfolding simulations, at 500 K, were performed for up to 10 ns. The simulations were carried out using periodic boundary conditions and a time step of 2 fs, with distances between hydrogen and heavy atoms constrained. Short range non-bonded interactions were calculated with a 12 Å cut-off, and long range electrostatic interactions were treated using the particle mesh Ewald summation (PME) algorithm. Figure 1 shows a set of representative trajectories for the thermally-induced unfolding of a TTR monomer.

## 2.2   Trajectory Analysis

Several global molecular properties, such as radius of gyration ($R_g$), root mean square deviation (RMSD), secondary structure and native contacts, among others, may be calculated along each trajectory in order to characterize and map the unfolding events. Here we calculated the Solvent Accessible Surface Area (SASA) of each individual amino-acid residue along the MD unfolding trajectories in order to study potentially correlated behavior among different residues.

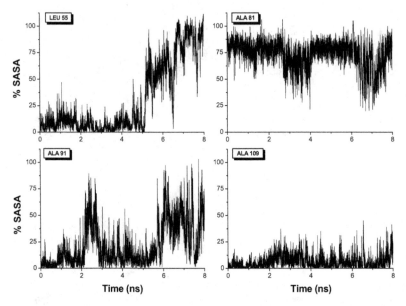

**Fig. 2.** Variation of the Solvent Accessible Surface Area (SASA) for individual amino-acid residues along a Molecular Dynamics unfolding simulation of the protein Transthyretin, at 500 K. 0% indicates an accessible surface of 0 $Å^2$ and 100% indicates a SASA for X equal to what is determined in the tripeptide Ala-X-Ala

The solvent accessible surface area (SASA) is the surface of the protein available to a spherical probe of 1.4 Å diameter, and was calculated using the program *naccess* [12]. The monomer of TTR has 127 amino-acid residues and each simulation trajectory analyzed here is constituted by 8,000 frames (one frame saved per ps simulated). Thus, for each simulation we have 127 plots (one per residue) of SASA *vs* time with 8,000 points (one point per frame). Figure 2 shows four examples of these plots. Leu55 is unexposed in the native structure and in the first half of the simulation, but it becomes highly exposed to the solvent late in the simulation. Ala109 is always unexposed to the solvent, even late in the simulation when the protein is already denatured. In general terms, some of the residues roughly follow the SASA patterns shown in Figure 2, but several other patterns are also observed. In order to find groups of residues that change solvent exposure in a coordinated fashion during the unfolding simulations, we have searched for Association Rules as detailed below.

## 3   Searching for Association Rules

Association Rules [13] represent a pattern language to describe relations among atomic elements (items) of the data. They hold simple and clear semantics and are of the form:

$$A_1 \ \& \ A_2 \ \& \ A_3 \ \& \ ... \ \& \ A_n \ \rightarrow \ C$$

A rule is derived from a co-occurring set of items (itemset). In the present case, the itemset could be $C, A_1, A_2, A_3, ..., A_n$. For the specific problem of SASA data analysis, items correspond to attribute/value pairs (residue / SASA value). The consequent C may be a set of items but here we only consider rules with a single item as consequent.

Quality and usability of the rules are measured through two types of metrics - predictability and incidence. Traditional metrics are *Support* (for incidence) and *Confidence*. *Support* is calculated by itemset counting among the transactions (records) contained in the data. *Confidence* corresponds to the strength of the rule and is obtained from the conditional probability of the consequent knowing the antecedent.

The aim of a rule generator algorithm is to derive high strength and interesting (surprising) rules. The user provides a minimal incidence (*Support*) value to avoid considering rare phenomena in the data. Thus, a minimal *Support* value filters out items (and itemsets) that occur in a low number of records. A minimal *Confidence* threshold is also supplied to select only high strength rules. The number of frequent items (which occurrence satisfies the minimal *Support*) is an important parameter for the computational complexity of the data analysis.

### 3.1 The Data

The studied data is composed of four different unfolding simulations of WT-TTR and L55P-TTR (with a Proline replacing a Leucine in position 55). The data describes SASA variations along 8000 frames (records) corresponding to the 8 ns of each run (simulation). 127 attributes (amino-acid residues that constitute the protein) are present. We removed the temporal label present in each frame so that only intratransactional relations would be extracted. These datasets turn out to be filled with very dense data, which makes rule generation into a computational hard task.

We exposed all datasets to a discretization process that, according to the analysis, reduced the number of values per attribute to 2 or at most 3. The values correspond to low ($[0,25[$ ), high ($[75,100[$ ) and medium SASA values. The latter was interpreted as a null value, which the system was programmed to ignore. Thus, the data was discretized to mainly consider unexposed or highly exposed amino-acid residues, along the MD simulations. Hence, we managed to reduce the number of frequent items being considered and consequently the complexity of the computational problem. In general, discretization reduces the complexity of the problem and leads to the derivation of higher quality rules.

### 3.2 Rule Generation

The standard algorithm to derive association rules is *Apriori* [13,16]. This algorithm is divided in two main steps: i) mining of frequent patterns (the extraction of itemsets that satisfy minimal support); and ii) rule generation (to derive rules using the frequent itemsets). The first step is the computational hard task and it has received considerable attention from the Data Mining community. Several proposals exist in the literature. We use CAREN (developed in [14]) which includes an algorithm for mining frequent patterns based on depth first expansion with bitwise representation.

CAREN also implements several features for rule derivation and selection, namely antecedent and consequent filtering (item or attribute specification), max/min number of items in a rule, different metrics, $\square^2$ test during itemset mining (which significantly reduces the number of relevant itemsets), improvement filtering on rules (to eliminate redundant rules), etc [15,16]. There are several metrics to evaluate association rules. We used the standard confidence metric. However, other metrics are available in the CAREN system.

Several extractions (queries) were performed on each discretized simulation. Each query was designed to answer a relevant biochemical question. For example, we were particularly interested in verifying which chemical classes of amino-acid residues behaved in a similar fashion and, among these classes, which particular residues behaved similarly. We designed queries to relate the following groups of amino-acid residues:

  i) hydrophobics *vs* hydrophobics
 ii) aromatics *vs* aromatics
iii) hydrophobics *vs* hydrophilics
 iv) positively charged *vs* negatively charged

For instance, to derive rules that represent the interaction between hydrophobic residues (i) we designed a script to invoke the CAREN system with specific parameters: consequent filter, list of hydrophobic residues as possible antecedents, specific minimal number of items as 4, among other less relevant parameters. The minimal number of items varied, depending on the query being addressed. For example, in query i) we used 4 and in query iv) we used 2 as the minimal number of items.

This difference is justifiable by chemical arguments. In the first case (i), we were looking for interactions between 4 or more residues - a hydrophobic cluster. In the last case (iv) we were looking for interactions between pairs of positively and negatively charged residues - a salt bridge.

### 3.3  Rule Selection

At this stage, we were mostly interested in finding amino-acid residues that had similar patterns of solvent exposure. Rules with high support are more significant since they represent a phenomenon that persists along a larger interval of data frames. As a first approach, we considered rules with *Support* above 30% and *Confidence* above 90%. It turns out that the only rules having very high *Support* are those relating hydrophobic residues with hydrophobic residues. All other queries generate rules with low *Support*. Figure 3 shows two examples of hydrophobic rules, one with relatively low *Support* (Fig. 3A) and another with very high *Support* (Fig. 3B). In both cases, the rules have 3 antecedents and 1 consequent, but there are rules involving 5, 6, 7 and more antecedents.

In order to have a deeper understanding of the processes being revealed by the Association Rules, we determined the inter-residue distances among all the residue pairs involved in each rule (Fig. 3, Panels on the right). While for the lower *Support* rule (Fig. 3A) the inter-residue distances vary more widely, for the high *Support* rule (Fig. 3B) inter-residue distances vary much less along the unfolding simulation. This metric clearly reveals amino-acid residues that not only have similar SASA behaviors but also maintain the same spatial relation among them along the unfolding simulation.

Using a similar treatment for all the rules obtained, we were able to define a set of 28 hydrophobic amino-acid residues that share two main characteristics: i) they remain unexposed to the solvent; and ii) they display a constant spatial relation during most of the unfolding simulation. As may be seen in Figure 4, the 28 identified hydrophobic residues are concentrated in the interior of the protein. Thus, we may conclude that this set of 28 residues out of 127 may constitute a hydrophobic cluster essential in TTR unfolding and folding.

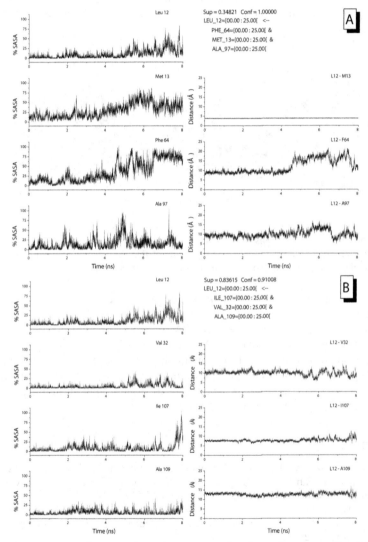

**Fig. 3.** Example of two Association Rules (at the top of each panel) generated by CAREN and involving hydrophobic residues. Panel **A** shows a low *Support* rule and Panel **B** a high *Support* rule. The plots on the left show the change in relative SASA along one MD unfolding simulation of WT-TTR for the residues involved in each derived rule. The plots on the right show inter-residue distances for each pair of residues involved in each derived rule

# 4  Conclusions

In this report we show that Data Mining techniques, such as searching of Association Rules, applied to the analysis of a massive quantity of data generated in Molecular Dynamics Protein Unfolding Simulations, are in fact very useful in the detection of hidden relations among the constituents of the molecular system under study. Association rules appear as a pattern language with high potential to express common behavior among amino-acid residues during the protein unfolding process. Rules are simple objects with clear reading. They are easy to interpret and useful for future prediction tasks.

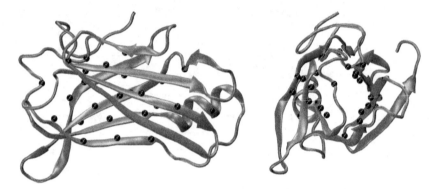

**Fig. 4.** Schematic representations of the backbone structure of the monomer of Transthyretin. Black spheres indicate the positions of the $C_\alpha$ atoms of the 28 hydrophobic residues identified by the Association Rules. The two views are related by a 90° rotation

In the case of the molecular system studied here - the unfolding behavior of the protein Transthyretin -, we searched for Association Rules among the variation profiles of Solvent Accessible Surface Area (SASA) of each one of the 127 residues of TTR in several unfolding simulations. This allowed us to define a group of 28 hydrophobic residues which appear to form a hydrophobic cluster essential in the unfolding and folding processes of TTR. Thus, the application of specific data mining techniques to an extremely large set of data generated from protein unfolding simulations helped us uncover new biochemically relevant knowledge.

# References

1. Quintas, A., Vaz, D.C., Cardoso, I., Saraiva, M.J.M., Brito, R.M.M.: Tetramer Dissociation and Monomer Partial Unfolding Precedes Protofibril Formation in Amyloidogenic Transthyretin Variants. J. Biol. Chem. 276 (2001) 27207-27213
2. Brito, R.M.M., Damas, A.M., Saraiva, M.J.S.: Amyloid Formation by Transthyretin: From Protein Stability to Protein Aggregation. Current Medicinal Chemistry - Immun. Endoc. & Metab. Agents 3 (2003) 349-360
3. Rodrigues, J.R., Brito, R.M.M.: How important is the role of compact denatured states on amyloid formation by transthyretin? In: Gilles Grateau, Robert A. Kyle and Martha Skinner (eds.): Amyloid and Amyloidosis. CRC Press (2004) 323-325
4. Daggett, V.: Molecular Dynamics Simulations of the Protein Unfolding/Folding Reaction. Acc. Chem. Res. 35 (2002) 422 - 429

5. Pande V.S., Baker I., Chapman J., Elmer S.P., Khaliq S., Larson S.M., Rhee Y.M., Shirts M.R., Snow C.D., Sorin E.J., Zagrovic B.: Atomistic protein folding simulations on the submillisecond time scale using worldwide distributed computing. Biopolymers 68 (2003) 91-109
6. Beck, D.A.C., Daggett, V.: Methods for molecular dynamics simulations of protein folding/unfolding in solution. Methods 34 (2004) 112-120
7. Brito, R.M.M., Dubitzky, W., Rodrigues, J.R.: Protein Folding and Unfolding Simulations: A New Challenge for Data Mining. OMICS A Journal of Integrative Biology 8 (2004) 153-166
8. Hamilton, J.A., Steinrauf, L.K., Braden, B.C., Liepnieks, J., Benson, M. D., Holmgren, G., Sandgren, O., Steen, L.: The X-ray crystal structure refinements of normal human transthyretin and the amyloidogenic Val-30-Met variant to 1.7 Å resolution. J. Biol. Chem. 268 (2003) 2416-2424
9. Kalé, L., Skeel, R., Bhandarkar, M., Brunner, R., Gursoy, A., Krawetz, N., Phillips, J., Shinozaki, A., Varadarajan, K., Schulten, K.: NAMD2: Greater scalability for parallel molecular dynamics. J. Comp. Physics 151 (1999) 283-312
10. 10.MacKerell, A.D., Bashford, D., Bellott, M., Dunbrack, R.L., Evanseck, J.D., Field, M.J., et al.: All-atom empirical potential for molecular modeling and dynamics studies of proteins. J. Phys. Chem. B 102 (1998) 3586-3616
11. 11.Zhang, L., Hermans, J.: Hydrophilicity of cavities in proteins. Proteins: Struct. Func. Gen. 24 (1996) 433-438
12. 12.Hubbard, S.J., Thornton, J.M.: NACCESS, Computer Program, Department of Biochemistry and Molecular Biology, University College London (1993)
13. 13.Agrawal, R., Srikant, R.: Fast Algorithms for Mining Association Rules. In: Proceedings of the 20th International Conference on Very Large Databases, Chile (1994)
14. Azevedo, P.J., Jorge, A.M.: The CLASS project: http://www.niaad.liacc.up.pt/~amjorge/Projectos/Class/
15. Azevedo, P.J.: The Caren System: http://www.di.uminho.pt/~pja/class/caren.html
16. Azevedo, P.J.: CAREN – A java based Apriori implementation for classification purposes. Research Report – Departamento de Informática, Universidade do Minho (2003)

# Protein Secondary Structure Classifiers Fusion Using OWA

Majid Kazemian[1], Behzad Moshiri[1], Hamid Nikbakht[2], and Caro Lucas[1]

[1] Control and Intelligent Processing Center of Excellence, Electrical and Computer Eng. Department, University of Tehran, Tehran, Iran
am.kazemian@ece.ut.ac.ir, moshiri@ut.ac.ir
[2] Laboratory of Biophysics and Molecular Biology, Institute of Biochemistry and Biophysics, University of Tehran, Tehran, Iran

**Abstract.** The combination of classifiers has been proposed as a method to improve the accuracy achieved by a single classifier. In this study, the performances of optimistic and pessimistic ordered weighted averaging[1] operators for protein secondary structure classifiers fusion have been investigated. Each secondary structure classifier outputs a unique structure for each input residue. We used confusion matrix of each secondary structure classifier as a general reusable pattern for converting this unique label to measurement level. The results of optimistic and pessimistic OWA operators have been compared with majority voting and five common classifiers used in the fusion process. Using a benchmark set from the EVA server, the results showed a significant improvement in the average Q3 prediction accuracy up to 1.69% toward the best classifier results.

## 1 Introduction

There are three main classes of secondary structural elements in proteins named as alpha helices (H), beta strands (E) and irregular structures (turns or coils that are shown as C), so prediction engines can be assumed as structural classifiers. In the process of predicting protein secondary structure classification, it is usual to use a variety of approaches that each of them has its own strengths and weaknesses [1, 2, and 3].

The reasons for combining the outputs of multiple classifiers are compelling, because different classifiers may implicitly represent different useful aspects of the problem, or of the input data, while none of them represents all useful aspects [4]. In this context, the idea of decisions fusion of several classifiers has been well explored. Information fusion techniques have been intensively investigated in recent years and their applications for classification domain have been widely tested [5]. Methods for fusing multiple classifiers can be classified according to the type of information produced by the individual classifiers. Classifiers can differ in both the nature of the measurements used in the classification process and in the type of classification produced. A Type I classification is a simple statement of the class, a Type II classification is a ranked list of probable classes, and a Type III classification assigns probabilities to classes [6]. These three types are known as *Abstract level* outputs, *Rank level* outputs and *Measurement level* outputs respectively. Most of protein secondary struc-

---

[1] Ordered Weighted Averaging (OWA)

J.L. Oliveira et al. (Eds.): ISBMDA 2005, LNBI 3745, pp. 338–345, 2005.

ture classifiers are type I classifier. The best known approach for consensus of type I classifiers is majority voting. In Majority voting approach there are some problems, for instance, where there is no majority winner, what the majority voter should do? There are some better techniques for classifier's outputs fusion but most of them need the results of type II or type III classifiers [7].

The results of five common protein secondary structure prediction engines of a benchmark dataset have been used for testing this fusion approach. The rest of the paper is organized as follows: the confusion matrix and the algorithm for converting type I classifier to type III classifier have been explained in section 2. Section 3 describes two simple OWA operators and demonstrates the application of these operators in the protein secondary structure classifiers fusion context. The classifiers and test dataset were introduced briefly in section 4. Section 5 presents the criteria of secondary structure prediction accuracy. Section 6 reveals the results of the fusion and finally the conclusion has been posed.

## 2 Measurement Level from Abstract Level Classifications

In this study, it is suggested that measurement level classifications could be created from the confusion matrix (a posteriori probabilities of each classification) of a Type I classifier. The assumptions are that, first, the behavior of the classifier is known and is characterized in a confusion matrix and, second, that the prior behavior or the classifier is representative of its future behavior. The larger the data set on which the classifier has been tested, the more thoroughly will the second assumption be true.

A confusion matrix is a matrix in which the actual class or a datum under test is represented by the matrix row, and the classification of that particular datum is represented by the confusion matrix column. The element $M[i][j]$ gives the number of times that a class i object was assigned to class j. The diagonal elements indicate correct classifications and, if the matrix is not normalized, the sum of row I is the total number of elements of class I that actually appeared in the data set [8]. The columns of such a matrix can be used to convert Type I to Type III classification, which can then be sorted to yield the ranks (Type II classification).

Consider a classifier that produces only a single output class (Type I) and has been trained and tested on many thousands of data elements. During this process, the following confusion matrix was generated (assume there are three classes): (Table 1)

Table 1. Confusion Matrix for a classifier

|                  | Classified as H | Classified as E | Classified as C |
| ---------------- | --------------- | --------------- | --------------- |
| Actual Class H   | 1600            | 100             | 300             |
| Actual Class E   | 200             | 1200            | 600             |
| Actual Class C   | 100             | 500             | 1400            |

Each row sums to 2000, which was the number of elements of each class in the data set. Note that if there was not the same number of elements in each row, it must be normalize by the number of its elements. Now as an example, presume that this classifier issues a classification of H for a given input datum. From the first column of the confusion matrix, it can be seen that the most likely actual class is H, the second most likely class is E, followed by C. This is a fair ranking of the possible classes based on the past of the classifier history. In other words, given a classification of H:

$$1600/1900 \text{ will be correct (class H)}$$
$$200/1900 \text{ will actually be class E}$$
$$100/1900 \text{ will actually be class C}$$

This is the scheme suggested for converting abstract level classifications into measurement level.

## 3  Ordered Weighted Averaging

The Ordered Weighted Averaging Operators (OWA) were originally introduced by Yager to provide a means for aggregating scores associated with the satisfaction of multiple criteria, which unifies in one operator the conjunctive and disjunctive behavior [9]. An OWA operator of dimension n is a mapping F: $R^n \rightarrow R$ and is given by:

$$OWA(x_1, x_2, ..., x_n) = \sum_{j=1}^{n} w_j x_{\sigma(j)} \tag{1}$$

Where $\sigma$ is a permutation that orders the elements: $x_{\sigma(1)} \le x_{\sigma(2)} \le .... x_{\sigma(n)}$. The weights are all non-negative ( $w_i \ge 0$ ) and their sum equals to one ( $\sum_{i=1}^{n} w_i = 1$ ).

This operator has been proved to be very useful, because of its versatility, The OWA operators provide a parameterized family of aggregation operators, which include many of the well-known operators such as the maximum, the minimum, the k-order statistics, the median and the arithmetic mean. In order to obtain these particular operators we should simply choose particular weights. The Ordered Weighted Averaging operators are commutative, monotone, idempotent, they are stable for positive linear transformations, and they have a compensatory behavior. This last property translates the fact that the aggregation done by an OWA operator always is between the maximum and the minimum. It can be seen as a parameterized way to go from the *min* to the *max*. In this context, a degree of maxness (initially called orness) was introduced in [9], defined by:

$$maxness(w_1, w_2, ..., w_n) = \frac{1}{n-1} \sum_{i=1}^{n} (n-i) w_i \tag{2}$$

We see that for the minimum, we have that *maxness*(1,0,...,0)=0 and for the maximum *maxness*(0, ...,0,1)=1.

A simple class of OWA operators as exponential class of OWA operators was introduced to generate the OWA weights satisfying a given degree of maxness. The optimistic and pessimistic exponential OWA operators were introduced as follows [9]:

- Optimistic weights:

$$w_1=a; \; w_2=a(1-a); \; w_3=a(1-a)^2; \; ...; \; w_{n-1}=a(1-a)^{n-2}; \; w_n=(1-a)^{n-1} \tag{3}$$

- Pessimistic weights:

$$w_1 = a^{n-1}; \; w_2 = (1-a)\,a^{n-2}; \; w_3 = (1-a)a^{n-3}; \; ...; \; w_{n-1} = (1-a)a; \; w_n = (1-a) \tag{4}$$

*Where parameter a (alpha) belongs to the unit interval, [0 1] and is related to orness value regarding the n.*

Each protein secondary structure classifier outputs a label (H or E or C). Meanwhile a list of measured level classification (MLC) is constituted by using confusion matrix as descried in previous section $MLC(i) = \{W_i(H), W_i(E), W_i(C)\}$. This list shows the confidences of a classifier to its possible outputs. For example, consider the MLC of two classifiers:

$$\begin{array}{ccc} \textbf{H} & \textbf{E} & \textbf{C} \end{array}$$

**Classifier 1: [ 0.7  0.2  0.1 ]**
**Classifier 2: [ 0.3  0.5  0.2 ]**

$W_i(H)$, $W_i(E)$ and $W_i(C)$ are fused by OWA operator separately for all of classifiers. After fusion process, the secondary structure of a certain amino acid is extracted from the Fused MLC, $\{W(H), W(E), W(C)\}$ as below:

$$PC = arg\ max\{W(H), W(E), W(C)\} \tag{5}$$

The general architecture of proposed Meta classifier is shown in Figure 1.

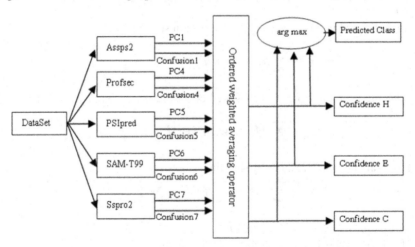

**Fig. 1.** Meta classifier schema-(PC: predicted class)

## 4   Experimental Evaluations

An experimental evaluation was carried out on EVA1 dataset. Novel test set which is provided by the datasets available from the real-time evaluation experiment [10], which compares a number of prediction servers on a regular basis using the sequences deposited in the PDB every week. In particular, we have used the dataset labeled "common1" published on 20/10/2002. Some information about the prediction method and location of five used prediction servers are shown in Table 2. For more information about these servers, see the references.

**Table 2.** Secondary structure prediction servers on Internet

| Secondary Structure Prediction Servers | | |
| --- | --- | --- |
| **Server** | **Location** | **Prediction method** |
| Apssp2 [11] | Institute of Microbial Technology, INDIA | EBL[*] + Neural network |
| Profsec [12] | Columbia University, USA | Profile-based Neural network |
| PSIPRED [13] | University College London, UK | Neural network |
| SAM-T99 [14] | University of California, Santa Cruz, USA | Hidden Markov Model |
| SSPro2 [15] | University of California, Irvine, USA | Recurrent Neural Network |

[*] Example Based Learning

To choose the parameter *alpha* in optimistic and pessimistic OWA, an iterative approach is used. In this purpose, a dataset is divided into three parts randomly; one of them is assigned for training and the rest for testing purpose. In training set, the *alpha* value was increased from zero to one by step 0.01 and consequently the prediction accuracy was calculated. The *alpha* value, in which the prediction accuracy was the maximum there, is selected as a needed parameter.

## 5   Accuracy of Predicting Secondary Structure Content

- *Prediction accuracy matrix:*
  $M_{ij}$ = number of residues observed in state I and predicted in state j, with i, j {H, E, C}
  Note: the total number of residues observed in state i is:

$$obs_i = \sum_{j=1}^{3} M_{ij} \ , \ with \ j \in \{H,E,C\} \tag{6}$$

Note: the total number of residues predicted in state i is (helix, strand, other)

$$prd_i = \sum_{j=1}^{3} M_{ij} \ , \ with \ j \in \{H,E,C\} \tag{7}$$

The total number of residues is simply:

$$N_{res} = \sum_{i} obs_i = \sum_{i} prd_i = \sum_{i,j} M_{ij} \tag{8}$$

- *Three-state prediction accuracy: $Q_3$*
  The three-state per residue accuracy $Q_3$ becomes:

$$Q_3 = 100 \times \frac{1}{N_{res}} \times \sum_{i=1}^{3} M_{ij} \tag{9}$$

- *Per-state percentages:*
  To define the accuracy for a particular state (helix, strand, other), two possible variants could be considered. As a result, the following questions could be raised up:
  How many observed helix residues (strand or coil) were correctly predicted? Given are the correctly predicted residues as percentage of all residues OBSERVED in a particular state (% obs).

$$Q_i^{\%obs} = 100 \times \frac{M_{ij}}{obs_i} \tag{10}$$

How many predicted helix (strand or coil) residues were correctly predicted? Given are the correctly predicted residues as percentage of all residues PREDICTED in a particular state (% prd)

$$Q_i^{\%prd} = 100 \times \frac{M_{ij}}{prd_i} \tag{11}$$

## 6  Results

Statistics of the predictions performed by the five selected servers (described in previous section) are presented in Table3. The results demonstrate that the best classifier for this dataset is PSIPRED. Although its predictions of the secondary structure are of the highest accuracy, it has been further improved by our meta-classifier. Improvements in terms of the accuracy of the OWA based meta-classifier are presented in Table 5. The results show that the OWA based meta-classifier has absolute improvement of 1.69% compared to PSIPRED. The most interesting results have been achieved for β strand prediction. PSIPRED predicts accurately 68.25% of the cases while OWA gets 73.08% giving an improvement of 4.83%. In addition, there is 5.78% improvement in helix structure prediction with -3.75% changes in coil structure classification.

Comparison between MV and OWA shows that the OWA based meta-classifier has improvement of 0.79% compared to MV, which is not very interesting in first look, but with deeper look into the results, we found that the OWA caused an improvement of 6.71% in β strand and 4.14% in helix structure. Recognition rate in helices and strands is more important than coils because due to general definition of protein secondary structures, each residue that is not in helix or strand structure will be posed in coil structure.

**Table 3.** The results of EVA1 dataset prediction by five common selected engines

|          | $Q_3$ | $Q_h^{\%obs}$ | $Q_e^{\%obs}$ | $Q_c^{\%obs}$ | $Q_h^{\%prd}$ | $Q_e^{\%prd}$ | $Q_c^{\%prd}$ |
|----------|-------|--------------|--------------|--------------|--------------|--------------|--------------|
| apssp2   | 74.49 | 78.00 | 65.65 | 77.01 | 79.4  | 76.38 | 70.08 |
| Profsec  | 74.71 | 75.38 | 74.48 | 74.05 | 82.95 | 71.29 | 70.76 |
| Psipred  | 74.78 | 78.53 | 68.25 | 75.67 | 79.21 | 75.32 | 70.99 |
| samt99_sec | 74.63 | 82.60 | 63.12 | 75.06 | 77.90 | 79.37 | 69.74 |
| sspro2   | 73.58 | 78.14 | 62.79 | 76.45 | 78.70 | 76.55 | 68.35 |

**Table 4.** The calculated maxness value and corresponding alpha value (achieved by the Fig.1 of [14]) of OWA

|                 | alpha | maxness |
|-----------------|-------|---------|
| Majority Voting | *     | *       |
| OWA-optimistic  | 0.3   | 0.6     |
| OWA-pessimistic | 0.7   | 0.4     |

Table 5. The results of Majority Voting and OWA operators

|  | $Q_3$ | $Q_h^{\%obs}$ | $Q_e^{\%obs}$ | $Q_c^{\%obs}$ | $Q_h^{\%prd}$ | $Q_e^{\%prd}$ | $Q_c^{\%prd}$ |
|---|---|---|---|---|---|---|---|
| MV | 75.68 | 80.17 | 66.37 | 77.68 | 80.28 | 78.94 | 70.58 |
| OWA-optimistic | 76.47 | 84.31 | 73.08 | 71.92 | 78.63 | 75.51 | 75.03 |
| OWA-pessimistic | 76.47 | 84.31 | 73.08 | 71.92 | 78.63 | 75.51 | 75.03 |

# 7 Conclusions and Future Research

Combining protein secondary structure classifiers requires a uniform representation of their decisions with respect to an observation. Confusion matrix is a well-known evaluator for each type of classifier which is used here as a general reusable pattern for fusion of protein secondary structure classifiers. Such a general assessor could be used in better weighting assignment in all fusion approaches. Moreover, a confusion matrix is used for converting Type I classifier to Type III classifier. In these types of classifiers, heuristic functions or theories for decision fusion may be more applicable.

The performance of a Meta classifier system can be better than each individual classifier; also, such systems can provide a unified access to data for users.

There are still open issues ahead:

- To obtain the number of classifiers those are required to achieve a desired accuracy.
- To obtain better identifiers to convert Type I classification to Type III classification.
- To publish the protein secondary structure meta-classifiers as an open-access web service.

## References

1. Argos P, Hanei M, Garavito RM. The Chou-Fasman secondary structure prediction method with an extended database. FEBS Lett. 1978 Sep 1;93(1):19-24.
2. Cai YD, Liu XJ, Chou KC. Prediction of protein secondary structure content by artificial neural network. J Comput Chem. 2003 Apr 30;24(6):727-31.
3. Kim S. Protein beta-turn prediction using nearest-neighbor method. Bioinformatics. 2004 Jan 1;20(1):40-4.
4. T.H. Ho, J. J. Hull, S.N. Srihari, "Decision Combination in Multiple Classifier System", IEEE Transactions on Pattern Analysis and Machine Intelligence, vol. 16, pt. 1, pp. 66-75, 1994.
5. Dymitr Ruta and Bogdan Gabrys, An Overview of Classifier Fusion Methods, computing and Information Systems, 7 (2000) p.1-10.
6. L. Xu, A. Krzyzak, and C.Y. Suen, "Methods of Combining Multiple Classifiers and their Applications to Handwriting Recognition", IEEE Trans. SMC, vol. 22, No. 3, 1992. pp 418-435.
7. Robles V, Larranaga P, Pena JM, Menasalvas E, Perez MS, Herves V, Wasilewska A., "Bayesian network multi-classifiers for protein secondary structure prediction.", Artif Intell Med. 2004 Jun;31(2):117-36.
8. J. R. Parker: Rank and response combination from confusion matrix data. Information Fusion 2(2): 113-120 (2001).

9. Yager, R. R., On ordered weighted averaging aggregation operators in multi-criteria decision making, IEEE transactions on Systems, Man and Cybernetics 18, 183-190, 1988.
10. B. Rost and V.A. Eyrich. EVA: large-scale analysis of secondary structure prediction. Proteins, 5:192–199, 2001.
11. G. P. S. Raghava, Protein secondary structure prediction using nearest neighbor and neural network approach. CASP4: 75-76, 2000.
12. B Rost, PROF: predicting one-dimensional protein structure by profile based neural networks. http://cubic.bioc.columbia.edu/predictprotein.
13. Jones DT. Protein secondary structure prediction based on position-specific scoring matrices. J. Mol. Biol. 292: 195-202.1999.
14. K Karplus, C Barrett, and R Hughey: Hidden Markov Models for Detecting Remote Protein Homologies, Bioinformatics, 14, 846-856, 1998.
15. G.Pollastri, D.Przybylski, B.Rost, P.Baldi, "Improving the Prediction of Protein Secondary Structure in Three and Eight Classes Using Recurrent Neural Networks and Profiles", Proteins, 47, 228-235, 2002.

# Efficient Computation of Fitness Function
# by Pruning in Hydrophobic-Hydrophilic Model

Md. Tamjidul Hoque, Madhu Chetty, and Laurence S. Dooley

Gippsland School of Information Technology
Monash University, Churchill VIC 3842, Australia
{Tamjidul.Hoque,Madhu.Chetty,Laurence.Dooley}
@infotech.monash.edu.au

**Abstract.** The use of Genetic Algorithms in a 2D Hydrophobic-Hydrophilic (HP) model in protein folding prediction application requires frequent fitness function computations. While the fitness computation is linear, the overhead incurred is significant with respect to the protein folding prediction problem. Any reduction in the computational cost will therefore assist in more efficiently searching the enormous solution space for protein folding prediction. This paper proposes a novel pruning strategy that exploits the inherent properties of the HP model and guarantee reduction of the computational complexity during an ordered traversal of the amino acid chain sequences for fitness computation, truncating the sequence by at least one residue.

## 1 Introduction

Proteins are made up of an alphabet set of 20 different amino acids [1]. Variations in protein conformation depend upon the different combination of amino acids in the sequence and their properties [2]. In addition to these variations, a number of chemical bonds, variations in side-chain, and a number of dihedral angles with a number of degrees of freedom within the amino acid chain make the search space for the optimum folding intractable [3]. This provides motivation to design an effective search algorithm.

Initially, the focus was upon computer-based protein folding prediction algorithms [4], with Molecular Dynamics (MD) and Monte Carlo (MC) technique being heavily employed, though these conformational search methods proved to be too slow. Subsequently, improvements in the speed and efficiency of the search methods became the primary concern [4], with Unger *et al.* [5] designing a Genetic Algorithm (GA) implementation that was much faster than the traditional MC technique. Other strategies including, Hydrophobic Zipper (HZ) [6], Contact Interaction methods (CI) [7], Constraint Programming [8], have developed statistical approaches successfully but only for sequences having limited length around 60 residues or less.

One basic, yet highly effective [8] representation of lattice models for protein folding investigation is the 2D Hydrophobic-Hydrophilic (HP) model proposed by Dill [9], which uses two letter alphabets, namely H and P. Based on dominating hydrophobic force this model has been designed which is well accepted, and used for evaluating search strategies. H indicates the hydrophobic amino acid, while P represents the polar or hydrophilic amino acids. The energy function for the HP-model is calculated as follows. If two residues are Topological Neighbours (TN) - indicated by

J.L. Oliveira et al. (Eds.): ISBMDA 2005, LNBI 3745, pp. 346–354, 2005.

the dotted lines in Figure 1, and they are both H then an $\varepsilon$ energy contribution is made where $\varepsilon$ is having a value -1. The sum of $\varepsilon$ in a conformation becomes the *fitness function* (F) of that particular conformation.

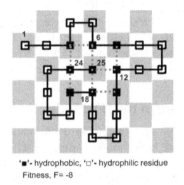

'■'- hydrophobic, '□'- hydrophilic residue
Fitness, F= -8

**Fig. 1.** HP coordinate model, presenting a sequence of amino acids connected by solid line

Searching for the optimum conformation using an HP model is an NP-complete [10] problem, which has motivated researchers to explore alternative solutions such as, the application of GAs [2, 11-14]. The search space however, is enormous and convergence takes a significant time even for short sequences [5]. Pruning strategies to reduce the search space [15] have therefore recently been developed. Caching techniques [16-17] within GA has also proved to be able to reduce the computational load, further. The paper shows pruning residues or truncated traversal while computing fitness. The pruning affords the potential to reduce the computational overhead; as such a traversal is a repetitive process during the search algorithm, irrespective of the dimension (2D or 3D) of structure prediction. Hoque *et al.* [11-12] have previously proposed an improved fitness computation, and this paper extends the work to show that it is not essential to traverse all the hydrophobic residues in computing the fitness computation, which will reduce the computational load further.

The remainder of the paper is organized as follows. Section 2 explains the nomenclature used in the paper, while section 3 describes the fitness computation in the HP model. Section 4 defines *lemma* for the identification of pruning residues and bounds while section 5 explores the searching strategies for pruning and presents the new pruning algorithm. Section 6 examines the impact of pruning and finally, section 7 provides key conclusions.

## 2  Nomenclature

An amino acid chain traversal for fitness function computation can either be from a higher-numbered residue to a lower numbered residue or vice versa. Throughout this paper, the amino acid chain traversal direction is indicated by *L2H* and *H2L* to respectively represent travel from a lower to a higher-numbered residue and vice versa. In a *L2H* traversal, after pruning, the highest numbered remaining H residue is represented by $LCR_{L2H}$, while $LCR_{H2L}$ indicates the last computable residue in a H2L traversal after pruning. $Prune_{L2H}$, $Prune_{H2L}$, $MaxPrune$ are respectively the number of pruned residues in a L2H traversal, the number of pruned residues in H2L traversal and the maximum of $(Prune_{L2H}, Prune_{H2L})$.

For clarity, the amino acid sequence is represented as a binary string, $S = [s_1, s_2, s_3, \cdots, s_m]$ where $m$ is the total number of residues. $s_i$ can either have a value of '1' indicating a hydrophobic residue, or '0' representing a hydrophilic residue. Let $E$ be an ordered number set holding the index $i$ of $s_i$ where $s_i$ is '1'; thus $E = [e_1, e_2, e_3, \cdots, e_n]$, where $n$ is the number of total hydrophobic residues in $S$, and $n \le m$.

## 3   Fitness Computation in the HP Model

In a 2D HP model, possible protein folding conformations are represented by the amino acid chain on a square lattice model forming a self-avoiding walk as shown in Figure 1. For a particular sequence, a number of valid conformations are possible, with the corresponding Fitness function F defined [2], as the negative of the sum of all the TN pairs possible in a particular conformation. Hence, the conformation with the highest number of TN pairs has the lowest energy.

In fitness computation, two possible directions of traversal (L2H and H2L) are considered. The amino acid sequence is numbered for ordered traversal, so for example, a L2H traversal starts from the first hydrophobic residue (Number 3 in the sequence in Figure 1) and searches for the TN amongst its four possible neighbours (in 2D and six in the 3D representation). A residue is identified if and only if, there is a TN from a lower numbered hydrophobic residue to a higher numbered residue. In the example in Figure 1, for a L2H traversal, a TN is encountered from residue number 3 to 6 (3, 6) but not (6, 3), while (12, 25) is a TN, while (25, 12) is not. Thus for a L2H traversal, 8 TNs; (3, 6), (3, 24), (6, 25), (7, 12), (12, 25), (13, 18), (18, 25) and (19, 24) are obtained so F = -8 for the conformation in Figure 1.

This fitness computation is performed after every crossover and mutation operations when a GA is applied to the HP model, which makes it an extremely time-consuming process. Any improvement in the fitness computational cost will therefore reduce the overall computational load significantly.

## 4   Identification of Pruning Residues and Bounds

During any ordered L2H or H2L traversal, it is clear that the final hydrophobic residue will not encounter a TN, so the last hydrophobic residue 25 for instance in Figure 1 will encounter no TN. Similar reasoning applies to a H2L traversal, so the hydrophobic residue that is traversing last can *always* be omitted from the residue list, which guarantees at least one fewer hydrophobic residues to be traversed.

### 4.1   Pruning Residues

The objective in this paper is to identify the number of hydrophobic residues that can be pruned in any arbitrary sequence during traversal from one end to the other. The following *lemmas* form the basis for this pruning strategy when searching for either LCR$_{L2H}$ or, LCR$_{H2L}$ in $n_H$, where $n_H$ is the total number of hydrophobic residues in a sequence.

**Lemma 1:** To have a TN, the minimum sequence distance between two hydrophobic residues must be greater than 2.

**Proof.** Let $x \in E$ and $y \in E$. If $|x-y|=1$ then residues are sequentially connected so no TN is possible (see in Figure 2(a)). If $|x-y|=2$, then $x$ and $y$ can at best be diagonally positioned so again no TN is possible (see Figure 2 (b)). However, if $|x-y|=3$ then placement of two residues at two non-diagonal lattice points of a unit square is feasible as shown in Figure 2(c).  □

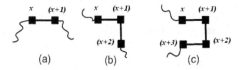

**Fig. 2.** Instances attempting a TN. The residues in (a) and (b) cannot have TN, but x and (x+3) in (c) can

**Lemma 2:** The distance between any two TN candidates must be odd.

**Proof.** Consider any two TN candidates, one hydrophobic residue must be odd while the other will be even in the sequence number. Since in a lattice presentation, even sequenced hydrophobic residues can only be surrounded by odd sequenced [18] residues and visa versa, the distance between two candidates of TN is therefore always odd.  □

To illustrate *lemma 2*, consider Figure 1. Residue 25 (odd indexed) is surrounded by residues 6, 12, 18 and 24 (all of which are even indexed). Since, shaded square (odd indexed) is always surrounded by white squares (even indexed) and visa versa. It shows that, opposite indexed residues are separated by odd distance.

**Lemma 3:** The minimum pruning for any sequence is always 1.

**Proof.** Based on traversal direction *lemma 1* is further extended. During *L2H* traversal, for $x$ the TNs ($y$) can be encountered provided $x < y$. Hence, if $x$ is the last hydrophobic residue then there will be no $y$, such that $x < y$. Therefore, visiting the last hydrophobic residue is not required and can be pruned. A similar conclusion applies to H2L traversal.  □

**Lemma 4:** The maximum pruning for traversal is equal to the total number of hydrophobic residue for a particular sequence.

**Proof.** If there exists a sequence such that all the hydrophobic residues are an even distant apart with respect to each other, then according to *lemma 2*, there will be no TN. Therefore, for such sequences there is no need to traverse any hydrophobic residue.  □

## 4.2 Defining the Pruning Bounds

For defining the lower bound of pruning, assume a sequence of $m$ residues. For a number, $r$ is such that $r \leq m$, and we have the $r^{th}$ residue as the last hydrophobic residue in a sequential traversal. Also, if the $r^{th}$ and $(r-3)^{th}$ residues are hydrophobic but $(r-1)^{th}$ and $(r-2)^{th}$ are hydrophilic, then from *lemma 1*, TN pair of $r^{th}$ or last residue with the shortest distance will be the $(r-3)^{th}$ residue. Using *lemma 3*, we can thus prune the $r^{th}$ single residue so the minimum pruning for any sequence is always 1.

To develop an upper-bound for this new pruning strategy, consider a sequence in which all the hydrophobic residues are either only even or odd indexed. From *lemma 4*, there will be no TN at all for these residues, so every hydrophobic residue is pruned.

Hence, the pruning bound is $[1, n)$, where $n$ is the number of hydrophobic residues. For short sequences, pruned traversal for fitness computations will always be significant, while for relatively larger sequences (having length around 100 residue or more) with trivial patterns, such as mostly odd indexed or even indexed residues at a location preferably at the start or at the end of the sequence as discussed for upper-bound, will be extremely significant for pruned traversal.

## 5  Pruning Algorithms

*Lemmas* 1 and 2 (section 4) allow us to define the following two functions, which are used in the pruning algorithm-1 (given below) to detect $LCR_{L2H}$ or $LCR_{H2L}$.

$$f_{LCR_{L2H}}(x) = \begin{cases} i & \text{where } 1 \le i < x \le n \text{, and } d \in (e_x - e_i) \text{ such that } d \text{ is minimum} \\ 0 & \text{provided } d > 2 \text{ and } d \text{ is odd. If no such } i \text{ exists then return } 0. \end{cases}$$

$$f_{LCR_{H2L}}(x) = \begin{cases} i & \text{where } 1 \le x < i \le n \text{, and } d \in (e_i - e_x) \text{ such that } d \text{ is minimum} \\ (n+1) & \text{provided } d > 2 \text{ and } d \text{ is odd. If no such } i \text{ exists then return } (n+1). \end{cases}$$

**Algorithm 1.** To find maximal truncated traversal sequence

---

**Input**: Sequence $S$ and Traversal Direction $TD$.

**Output**: Number of pruned residue.

*Step 1:* If $TD = L2H$ then

*Step 2:*        Compute $e_a$ and $e_b$, respectively maximum odd, maximum even in E.

*Step 3:*        $LCR_{L2H}$ = maximum of $\{f_{LCR_{L2H}}(a), f_{LCR_{L2H}}(b)\}$

*Setp 3:*        Return: $Prune_{L2H} = (n - LCR_{L2H})$

                else

*Step 4:*        Compute $e_u$ and $e_v$, respectively minimum odd, minimum even in E.

*Step 5:*        $LCR_{H2L}$ = minimum of $\{f_{LCR_{H2L}}(u), f_{LCR_{H2L}}(v)\}$

                Return: $Prune_{H2L} = (LCR_{H2L} - 1)$

                endif

---

The Algorithm-1 assumes that both the sequence and the traversal direction are given as input and it will return number of pruned residue. Depending on odd and even hydrophobic residue groups in a sequence, the function $f_{LCR_{L2H}}(x)$ is invoked twice (Step 3), with $x=a$ and $x=b$, indicating the index of the maximum odd and maximum even numbered hydrophobic residue (Step 2) respectively. With $x=a$, the function will return an even indexed candidate of $LCR_{L2H}$ and with $x=b$, the function will return an odd indexed candidate of $LCR_{L2H}$. The maximum return of the two is the $LCR_{L2H}$. Similarly we, invoked function $f_{LCR_{H2L}}(x)$ (Step 5) to detect the $LCR_{H2L}$, where $x$ is being assigned the indices of *first odd* and *first even* hydrophobic residues (Step 4) and in this case the minimum of the two return is taken. These two functions help choosing traversal direction to have maximum pruning from either direction.

For example, in Figure 1, $S = [0010011000011000011000011]$ and $m = 25$, that is, $E = [3, 6, 7, 12, 13, 18, 19, 24, 25]$ and $n = 9$. The value of $e_a$, $e_b$, $e_u$ and $e_v$ are respectively 25, 24, 3 and 6. According to Algorithm 1, $(LCR_{L2H} = 7)$ as maximum of $\{(f_{LCR_{L2H}}(9) = 6), (f_{LCR_{L2H}}(8) = 7)\}$. Similarly, $(LCR_{H2L} = 4)$ as minimum of $\{(f_{LCR_{H2L}}(1) = 4), (f_{LCR_{H2L}}(2) = 5)\}$. So, $Prune_{H2L} = 3$, $Prune_{L2H} = 2$, hence $MaxPrune = 3$. Therefore, the traversal direction is H2L will provide the *MaxPrune* and it saves fitness computation traversal by 33.33%.

## 6  Simulation Results

For each sequence length, the occurrence frequency of H (the total number in a sequence) is considered as a percentage of sequence length. So for a sequence length of 1000 and having 20 H residues means H% = 20. To identify the impact of H%, it is varied from between 10% to 90% in steps of 10. Since these residues are randomly distributed, for each value of H%, the average improvement is computed from 1000 simulation runs with the measure of pruning defined as $I = \dfrac{k}{n} \times 100\%$, where $n$ is the total number of hydrophobic residues in a sequence and $k$ the number of pruned hydrophobic residues from that sequence. To establish the significance and impact of pruning on computational throughput, simulations were undertaken using both randomly generated sequences for analyzing robustness and popular benchmark sequences [19] used by the broader research community for testing the practical impact.

a) Randomly generated sequences:

The studies are performed to investigate the impact upon both the sequence length as well as the occurrence of H in the sequence. For this purpose, the length is varied from 20 till 1000 with a step increase of 20. It can be seen from Figures 3 and 4(a) that the significance of pruning depends on the occurrence frequencies and pattern of H and P in the sequence. It is observed that the lower the frequency of H in a sequence, the higher the pruning improvement I. Also, it is observed that the pruning performance is higher for shorter sequences which are less than around 100 residues. Figure 4(b) shows the maximum, average and minimum percentage 'improvement', which also reveals that for a sequence with a relatively lower number of hydrophobic residues, the line showing maximum improvement (%) has higher value. For any sequence, the minimum number of pruned residue is at least 1, so the significance of the minimum pruning decreases uniformly with increasing numbers of hydrophobic residues in a sequence.

b) Benchmark sequences [19]:

The pruning algorithm was next applied to a selection of the 2D benchmark sequences given in Table 1. The Table also shows the corresponding improvement in results with respect to all hydrophobic residue traversal thus establishing the practical significance of the pruning technique presented here in this paper.

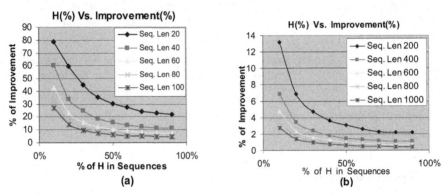

**Fig. 3.** Percentage of H and corresponding pruning improvement (%). (a) Sequence length 20 to 100, step 20. (b) Sequence length 200 to 1000, step 200

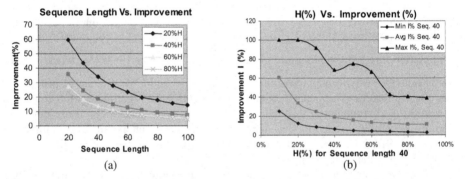

**Fig. 4.** (a) Sequence length and corresponding pruning improvement (%). (b) Improvement I(%) comparing maximum, average and minimum

## 7  Conclusions

This paper has presented a novel pruning strategy for Hydrophobic-Hydrophilic (HP) model to reduce the computation overhead during an ordered traversal of amino acid chain sequences. The new approach guarantees a minimum pruning for any sequence, thereby ensuring a speed up in the search process for protein folding prediction using GA. A series of lemma have been postulated in the development of the theoretical basis of this new strategy and simulation results for both randomly-generated and benchmark sequences confirm the improvement achieved. While the focus of the pruning algorithm has been on a 2D HP model, the strategy can be extended in a straightforward manner to a 3D HP model. Also, our algorithm can be easily extended to be embedded within current caching approaches [16-17], thereby providing a further reduction in the computational load.

**Table 1.** Pruning results for 2D benchmark sequences. Pruning improvement is shown respect to all hydrophobic residues traversal

| Sequence Length | Total # of H | Max-Prune | Improvement (%) | Traversal Direction | Sequence |
|---|---|---|---|---|---|
| 20 | 10 | 3 | **30.00** | H2L | HHHPPHPHPHPPHPHPHPPH |
| 20 | 10 | 2 | **20.00** | L2H / H2L | HPHPPHHPHPPHPHHPPHPH |
| 24 | 10 | 2 | **20.00** | L2H / H2L | HHPPHPPHPPHPHPPHPHPHPPHH |
| 25 | 9 | 2 | **22.22** | L2H | PPHPPHHPPPPHHPPPPHHPPPPHH |
| 36 | 16 | 2 | **12.50** | H2L | PPPHHPPHHPPPPPHHHHHHHPPHHPPPPHHPPHPP |
| 45 | 27 | 3 | **11.11** | H2L | PHHHPHPHHPPPHPHHPHHPPHPHHHHPHPPHHHHHHPHPHHPPHHP |
| 48 | 25 | 3 | **12.00** | L2H | PPHPPHHPPHHPPPPPHHHHHHHHHHPPPPPPHHPPHHPPHPPHHHHHH |
| 50 | 24 | 2 | **8.33** | L2H / H2L | HHPHPHPHPHHHHPHPPPHPPPHPPPHPPPHPHHHHPHPHPHP HH |
| 57 | 30 | 2 | **6.67** | L2H | HPHHHPHHHPPHHPHPHHHPHHHHPHPHPHHHPHHHPHPHPHPPPHPPH PPHHPPHPPH |
| 60 | 43 | 3 | **6.98** | L2H | PPHHHPHHHHHHHHPPHHHHHHHHHHHPHPPPHHHHHHHHHHHHHHHP PPPHHHHHHPHHPHP |
| 64 | 42 | 3 | **7.14** | L2H / H2L | HHHHHHHHHHHPHPPHPHHPPHHPPHPHPPHHPPHHPPHPPHHPPHHP PHPHPHHHHHHHHHHHH |
| 102 | 37 | 2 | **5.41** | H2L | PHHPPPPPPHHPPPHHHPPHHPPPPPPPHPPPHHPHHPPPPPPHPPHPHPPH PPPPHHHPPPHHPHHPPPPPHHPPPPHHHHPHPPPPPPPPHHHHHHPP HPP |
| 123 | 47 | 5 | **10.65** | L2H | PPHHHHPPPPPHPPPPPPHHPPPPHHPPPHHHPPPPPHPPPHPHPHHPPPH HPHPHHHPPPPHHHPPPPPPHHHPHPHPHPPHHPPPPPPPPHHPHHHHPPPP HPPPHHHHHPPPPHHHPHPHPHPH |
| 136 | 50 | 2 | **4.00** | L2H / H2L | HPPPPPHPPPPHHPHHPPPPPHPHHHHHPPPHPHPHHHHPPPPPPPPPP HPPHPPPPHPHPPHHPHHHPPHPHPHPHPPPPPPPPPHPPPHHHHHHHPPPHH PPHHHPPPHHPHHHHHHPPPPPPPPPPHPPPPHPHPPPP |

# References

1. Allen, et al.: Blue Gene: A vision for protein science using a petaflop supercomputer. IBM System Journal (2001), Vol 40, No 2
2. Gary, B.F. and David, W.C. (eds.): Evolutionary Computation in Bioinformatics. Elsevier Science (2003) USA
3. Lathrop, R.H.: Protein Structure Prediction. http://helix-web.stanford.edu/psb98/lathrop.pdf (1998)
4. Beutler, T.C. and Dill, K.A.: A fast conformational search strategy for finding low energy structures of model proteins. Protein Science (1996)
5. Unger, R. and Moult J.: Genetic Algorithm for Protein Folding Simulations, J. Mol. Biol. (1993), 231, 75-81
6. Dill, K.A., Fiebig, K.M., Chan H.S.: Cooperativity in Protein-Folding Kinetics. Proceedings of the National Academy of Sciences USA, Biophysics (1993) Vol 90, pp.1942-1946
7. Toma, L. and Toma, S.: Contact interactions method: A new algorithm for protein folding simulations. Protein Science (1996) 5: 147-153
8. Backofen, R.: Using Constraint Programming for Lattice Protein Folding. http://helix-web.stanford.edu/psb98/backofen.pdf (1998)

9. Dill, K.A.: Theory for the Folding and Stability of Globular Proteins. Biochemistry (1985) 24: 1501
10. Berger, B. and Leighton, T.: Protein Folding in the Hydrophobic-Hydrophilic (HP) model is NP-Complete. ACM, Proceedings of the second annual international conference on Computational molecular biology (1998)
11. Hoque M.T., Chetty, M. and Dooley L.S.: An Efficient Algorithm for Computing the Fitness Function of a Hydrophobic-Hydrophilic Model. 4th International Conference on Hybrid Intelligent Systems (HIS 2004), pp. 285-290, ISBN 0-7695-2291-2.
12. Hoque M.T., Chetty, M. and Dooley L.S.: Partially Computed Fitness Function Based Genetic Algorithm for Hydrophobic-Hydrophilic Model. 4th International Conference on Hybrid Intelligent Systems (HIS 2004), pp. 291-296, ISBN 0-7695-2291-2.
13. König, R. and Dandekar, T.: Refined Genetic Algorithm Simulation to Model Proteins. Journal of Molecular Modeling (1999)
14. Takahashi, O., Kita, H. and Kobayashi, S.: Protein Folding by A Hierarchical Genetic Algorithm. 4th Int. Symp. On Artificial Life and Robotics (AROB) (1999)
15. Voelz, V.: Zipping as a fast conformational search strategy for protein folding. http://laplace.compbio.ucsf.edu/~voelzv/orals/orals_proposal.pdf (2004)
16. Santos, E.E. and Santos, E.J.: Effective and Efficient Caching in Genetic Algorithms. International Journal on Artificial Intelligence Tools, © Worlds Scientific Publishing Company (2000)
17. Santos, E.E. and Santos, E.J.: Reducing the Computational Load of Energy Evaluations for Protein Folding", 4th IEEE Symp. on BIBE'04.
18. Newman, A.: A new algorithm for protein folding in the HP model. Proceedings of the thirteenth annual ACM-SIAM symposium on Discrete Algorithms (2002)
19. Hart, W. and Istrail, S.: HP Benchmarks, http://www.cs.sandia.gov/tech_reports/compbio/tortilla-hp-benchmarks.html

# Evaluation of Fuzzy Measures
# in Profile Hidden Markov Models for Protein Sequences

Niranjan P. Bidargaddi[1,2], Madhu Chetty[1,2], and Joarder Kamruzzaman[1]

[1] Gippsland School of Computing and Information Technology,
Monash University, Churchill, Victoria-3842, Australia
[2] Victorian Bioinformatics Consortium, Clayton, Victoria-3800, Australia

**Abstract.** In biological problems such as protein sequence family identification and profile building the additive hypothesis of the probability measure is not well suited for modeling HMM based profiles because of a high degree of interdependency among homologous sequences of the same family. Fuzzy measure theory which is an extension of the classical additive theory is obtained by replacing the additive requirement of classical measures with weaker properties of monotonicity, continuity and semi-continuity. The strong correlations and the sequence preference involved in the protein structures make fuzzy measure architecture based models as suitable candidates for building profiles of a given family since fuzzy measures can handle uncertainties better than classical methods. In this paper we investigate the different measures(S-decomposable, $\lambda$ and belief measures) of fuzzy measure theory for building profile models of protein sequence problems. The proposed fuzzy measure models have been tested on globin and kinase families. The results obtained from the fuzzy measure models establish the superiority of fuzzy measure theory compared to classical probability measures for biological sequence problems.

## 1   Introduction

Probability measure theory which is based on the classical measure theory is additive. It assigns 0 to the empty set and a non zero positive number to the nonempty set. By also assigning 1 to the Universal set the probability theory obeys the additivity of classical theory. In real error prone physical conditions the additive property is too restrictive [16]. It has also been verified in various domains([5], [14], [12] and [15]) that some of the measurements involving subjective judgements are intrinsically non-additive. It has also been proved that all measurements are inherently fuzzy due to unavoidable measurement errors [16]. Fuzzy measure theory which deals with all the non probability measures is an extension of the classical additive theory and is obtained by replacing the additive requirement of classical measures with weaker properties of monotonicity, continuity and semi-continuity [13]. Dempster and Shafer formulated a generalized theory called as evidence theory based on the belief and plausibility measures. Belief measures are superadditive and upper semicontinuous where as plausibility measures are subadditive and lower semicontinuous. Possibility measure theory is another set of non-additive theory based on fuzzy set theory. Possibility measure assigns the supremum of the possibility distribution function in each crisp set of concern. The $\lambda$ measures are normalized non additive measures formulated by Sugeno. Fuzzy integrals are

J.L. Oliveira et al. (Eds.): ISBMDA 2005, LNBI 3745, pp. 355–366, 2005.

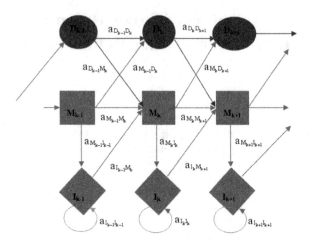

**Fig. 1.** Plan 7 architecture of Profile HMM used in HMMER 2 [7]. This architecture differs from the original (Plan 9) Krogh/Hausler architecture also used in earlier versions of HMMER by reducing the number of transitions from 9 to 7. It has no $D \rightarrow I$ and $I \rightarrow D$ transitions. The equations for fuzzy profile HMM are formulated based on the plan 9 architecture.The fuzzy model for plan 7 architecture can be easily obtained by setting the transition parameters $a_{D_k I_k}$ and $a_{I_{k-1} D_k}$ to zero in all the equations wherever they appear

the non-linear functionals which serve the purpose of aggregation of non-additive measures. The two most common aggregation operators are Sugeno integrals and Choquet integrals [11]. The integrals combine the partial support for a hypothesis from the viewpoint of each information source and the importance of various subsets of sources. This intrinsic property of fuzzy integrals to combine both the objective evidence and hypothesis serves as a motivation for their application in profile hidden Markov models. However, HMMs have several limitations. HMMs do not capture any higher-order correlations of the amino acids in the protein sequences and the identity of an amino acid at a particular position is independent of the identity of all other positions. They are also constrained by the statistical independence assumptions during the formulation of the forward and backward variables which are used to compute the matching scores of an unknown sequence to a known family by the first order Markov assumption. Due to the statistical independence assumptions, the joint measure variables (forward and backward) are decomposed as a combination of two measures defined on amino acid emission probabilities and state probabilities. To relax the statistical independence assumptions and achieve improved performance and flexibility we presented a fuzzy hidden Markov model based on fuzzy forward and backward variables in [3] and [4] based on globin and kinase families. The fuzzy model has its strength in its ability to incorporate the objective evidence (the importance of each consensus column for profiles which serve as the source). The authors have reported in [3] and [4] improvements resulting from fuzzifying the forward-backward algorithm for decoding protein sequence,viterbi search algorithm and parameter estimation using possibility measures. In this paper, a comparative evaluation of the effect of different fuzzy measures on the fuzzy profile

HMM is investigated. In the next section a brief overview on the classical and fuzzy profile HMM is presented. This is followed by a section on various fuzzy measures and their application in fuzzy profile HMM. In the last section we present the results along with conclusion and future works.

## 2  Fuzzy Profile HMM

Classical profile HMMs shown in Fig. 1 are statistical models of multiple sequence alignments based on classical probability theory. A profile HMM is capable of modelling gapped alignments including insertions and deletions, which allows modelling of a complete conserved domain (rather than just a small ungapped motif). Due to its probabilistic nature, a profile HMM can be trained from unaligned sequences, if a trusted alignment is not yet known. Profile HMM architecture shown in Fig. 1 is characterised by the following parameters:

$M_k$ Match state k, with 20 emission probabilities

$D_k$ Delete state k, non-emitter

$I_k$ Insert state k, with 20 emission probabilities

It is made up of a linear set of match (M), insert (I) and delete (D) states. There is one M state per consensus column in the multiple alignments. Each match state carries a vector of 20 probabilities, for scoring the 20 amino acids. Each match state has an I and a D state associated with it. The group of three states (M/D/I) at the same consensus position in the alignment is called a node. The states are interconnected by arrows as shown in Fig. 1 which represent state transition probabilities. The transitions are arranged so that at each node, either an M state is triggered (a residue is aligned and scored) or a D state is triggered (no residue is aligned, resulting in a deletion-gap character, '-'). Insertions occur between nodes, and an I state can have a self-transition, allowing one or more inserted residues to occur between consensus columns. The transition to an I state for the first inserted residue, followed by zero or more I→I self transitions for each subsequent inserted residue, is the probabilistic equivalent of the familiar gap-open and gap-extend affine gap penalty system. Like all HMMs, profile HMMs have emission and transition probabilities with probability distribution over the whole space of sequences which is parameterised using Baum-Welch re-estimation formulas to peak the distribution around the members of the family [6]. In fuzzy profile HMMs, the additive property of probability measures is replaced with the weaker condition of monotonicity. The fuzzy profile hidden Markov model, $\overline{\lambda} = (\widehat{A}, \widehat{B}, \widehat{\pi})$ is characterised by the following parameters [3].

  O protein sequence

  T length of the protein sequence

  N profile model length

  $\Omega$ set of protein sequences of the family

  X states at time $t$

  Y states at time $t + 1$

 $\widehat{\pi}_s(\cdot)$ initial state fuzzy measure

 $\widehat{\pi}_s(\{s_i\})$ initial state fuzzy density

 $\widehat{b}_j(O_t)$ symbol fuzzy density

 $\widehat{a}_y(\cdot|X)$ transition fuzzy measure

 $\widehat{a}_y(y_j|x_i)$ transition fuzzy density

where $\widehat{A} = [\widehat{a}_y(y_j|x_i) = \widehat{a}_{ij}]$, $\widehat{B} = [\widehat{b}_j(O_t)]$ and $\widehat{\pi} = [\widehat{\pi}_s(\{s_i\})]$. The fuzzy measure coefficients are called fuzzy densities. The homologous sequences of a given family have more than one kind of interactions among themselves. The classical profile HMM which is based on probability measures fixes these interactions based on probability densities. The fuzzy profile HMM overcomes this problem of considering only one type of interaction by not fixing the relationship between fuzzy densities. Fuzzy densities are similar to probability densities in classical profile HMM except for the fact that they are independent of each other. We formulated the fuzzy forward variables which resulted in an improved score compared to classical forward variables, for searching sequences of a family . This algorithm is briefly described in this section for easy reference. In classical profile HMM, the joint probability measure $P(O_1, O_2, \cdots, O_t, O_{t+1}, q_{t+1} = S_j)$ is written as the product $P(O_1, O_2, \cdots, O_t, O_{t+1} | q_{t+1} = S_j) \cdot (q_{t+1} = S_j)$ thus making the following two assumptions of statistical independence.

- The amino acid $O_{t+1}$ emitted by the HMM at time $t + 1$, is independent of the previous amino acid sequences $(O_1, O_2, \cdots, O_t)$ emitted.
- The state active at time $t + 1$ (M, I, D) is independent of the previous subsequence of amino acids $(O_1, O_2, \cdots, O_t)$ observed.

These assumptions are not realistic for the homologous sequences of a family, since they have a high correlation in them. Improved results for profiles can be expected through the relaxation permitted by fuzzy measures. For the fuzzy profile HMM, we introduce the fuzzy forward variable based on the joint fuzzy measure $\widehat{f}_{\Omega_y}(\{O_1, O_2, \cdots, O_t\} \times \{y_j\})$ which can be reduced to the combination of measures defined on $\{O_1, O_2, \cdots, O_t\}$ and on the states $y_j = (M_{t+1}, I_t, D_{t+1})$. This avoids the assumption of decomposition of measures as done in classical HMM. At any time, the fuzzy measure $\widehat{f}_{\Omega_y}$ on $\Omega_{1,t+1} \times Y$ can be constructed from its constituent forward variables through recursion , after integrating (using Choquet integral) with respect to any arbitrary fuzzy measure and using multiplication as as intersection operator. This is shown by the following equations.

$$f_{s_i}(t+1) = \widehat{f}_{\Omega_y}(\{O_1, O_2, \cdots, O_t\} \times \{y_j\}) \tag{1}$$

$$= \int_X \widehat{a}_y(\{y_j\}|x) \circ \widehat{f}_{\Omega_X}(\{O_1, O_2, \cdots, O_t\}) \wedge \widehat{b}_j(O_{t+1}) \tag{2}$$

where $\wedge$ is the fuzzy intersection operator and $s_i$ is the state at time $t + 1$. Accordingly, the forward variables $f_{M_k}(i), f_{I_k}(i)$ and $f_{D_k}(i)$ for $k^{th}$ Match, Insert and Delete states respectively are reformulated using possibility measure as shown in (15)-(16), respectively [3].

$$f_{M_k}(i) = e_{M_k}(i)[f_{M_{k-1}}(i-1)a_{M_{k-1}M_k}\rho_{i-1}(M_{k-1}, M_k) + f_{I_{k-1}}(i-1)a_{I_{k-1}M_k}\rho_{i-1}(I_{k-1}, M_k) + f_{D_{k-1}}(i-1)a_{D_{k-1}M_k}\rho_{i-1}(D_{k-1}, M_k)] \tag{3}$$

$$f_{I_k}(i) = e_{I_k}(i)[f_{M_k}(i-1)a_{M_kI_k}\rho_{i-1}(M_k, I_k) + f_{I_k}(i-1)a_{I_kI_k}\rho_{i-1}(I_k, I_k) + f_{D_k}(i-1)a_{D_kI_k}\rho_{i-1}(D_k, I_k)] \tag{4}$$

$$f_{D_k}(i) = f_{M_{k-1}}(i)a_{M_{k-1}D_k}\rho_{i-1}(M_{k-1}, D_k) + f_{I_{k-1}}(i)a_{I_{k-1}D_k}\rho_{i-1}(I_{k-1}, D_k) + f_{D_{k-1}}(i)a_{D_{k-1}D_k}\rho_{i-1}(D_{k-1}, D_k)] \tag{5}$$

The new term $\rho$ in the above equations represents the fuzzy measure difference which is calculated using the Choquet integral [3]-[3].

## 3    Fuzzy Measures

For protein sequences which have high degrees of interdependencies the additive hypothesis of probability measure is not well suited. The classical model based on fuzzy measures assigns the same level of importance for the source. The fuzzy integration takes into account the relative importance of the source along with the information. In this section, we give a brief definition of various other fuzzy measures [16]. Let $\Omega$ be the powerset of a set X. A set function $g : \Omega \rightarrow [0,1]$ defined on $\Omega$ which satisfies the conditions of boundary, monotonicity and continuity shown in (7)-(8) is called a fuzzy measure.

$$g(\phi) = 0, g(X) = 1 \tag{6}$$
$$\text{If A, B} \subset \Omega \text{and A} \subset \text{B, then g(A)} \leq g(\text{B}) \tag{7}$$

For any increasing sequence $A_1 \subseteq A_2 \subseteq \cdots \subseteq A_i \cdots$ of sets in $\Omega$, if $\bigcup_{i=1}^{\infty} A_i \in \Omega$ then

$$\text{limit}_{i \rightarrow \infty} g(A_i) = g(\bigcup_{i=1}^{\infty} A_i) \tag{8}$$

Based on the above definitions some of the various fuzzy measures are discussed in brief below.

### 3.1    S-Decomposable Measures

A t-conorm S is a operation on the unit interval [0,1] satisfying the following conditions [8].

$$S(a,0) = a \text{ (neutrum element)} \tag{9}$$
$$b \leq d \text{ implies } S(a,b) \leq S(a,d) \text{ (monotonicity)} \tag{10}$$
$$S(a,b) = S(b,a) \text{ (commutitivity)} \tag{11}$$
$$S(a,S(b,d)) = S(S(a,b),d) \text{ (associativity)} \tag{12}$$

We have maximum and drastic t-conorm operators as shown in (14)-(15) respectively.

$$S(a,b) = max(a,b) \text{ (Maximum)} \tag{13}$$
$$S(a,b) = a \text{ (when b = 0)}$$
$$= b \text{ (a = 0)} \tag{14}$$
$$= 1 \text{ (for other cases)}$$

An important uncertainty theory measure known as possibility measure is based on the above definitions of max t-conorm operation. If X is a universal set with $\Omega$ consisting of all the subsets of X, a possibility measure $g_P$ is a function,

$$g_P : \Omega \rightarrow [0,1] \tag{15}$$

where in $g_P(\phi) = 0$, and $g_P(X) = 1$. It satisfies the constraints shown in (16) along with the ones defined above for the fuzzy measures.

$$g_P(\bigcup_{i \in \Omega} A_i) = max(g_P(A_i)) \forall \, i \in X \tag{16}$$

The possibility measures on each element of the set X denoted by $g_P(x_i)$ are called as the possibility-density measures. Using these density measures we can calculate the possibility measures for all the sets in $\Omega$ using (17).

$$g_P(A_i) = max(g_P(x)) \forall x \in A_i \tag{17}$$

## 3.2   $\lambda$ Measures

$\lambda$ measures were introduced by Sugeno [13]. If X is a universal set with $\Omega$ consisting of all the subsets of X, then the $\lambda$-fuzzy measure $g_\lambda$ is a function given by (18),

$$g_\lambda : \Omega \to [0,1] \tag{18}$$

where in $g_\lambda(\phi) = 0$, and $g_{lambda}(X) = 1$. It satisfies the constraints shown in (6) along with the ones defined above for the fuzzy measures. They satisfy the additional additive properties shown in (19) below, along with the conditions of monotonicity, boundary conditions and continuity.

$$g_\lambda(A \cup B) = g_\lambda(A) + g_\lambda(B) + \lambda g_\lambda(A) g_\lambda(B) \tag{19}$$
$$\text{where } \lambda > -1 \text{ and } A \cap B = \phi, \forall \, A, B \, \in \Omega$$

When we have a finite set X = $\{x_1, x_2, \cdots, x_n\}$, the $i^{th}$ density of the $\lambda$-fuzzy measure $g_\lambda$, $g_\lambda^i$ is given by $g\lambda\{x_i\}$. The value of $\lambda$ in (19) can be calculated from the equation $g_\lambda(X) = 1$ as shown in (20).

$$\lambda + 1 = \prod_{i=1}^{n}(1 + \lambda g_\lambda^i) \tag{20}$$

The $\lambda$ fuzzy measures when used to represent uncertainty are also referred to as Sugeno measures.

## 3.3   Belief Measures

Belief measures are uncertainty measures which form the basis of Dampster-Shafer theory [16]. For a universal set X, if $\Omega$ consists of all the subsets, the belief measure $g_{bel}$ is a function shown in (21).

$$g_{bel} : \Omega \to [0,1] \tag{21}$$

where in $g_{bel}(\phi) = 0$, and $g_{bel}(X) = 1$. The additive constraints for belief measures are given by (22) as shown below, where $A_i$ is the $i^{th}$ subset of X.

$$g_{bel}(A_1 \cup A_2 \cup \cdots \cup A_N) \geq \sum_j g_{bel}(A_j) - \sum_{j<k} g_{bel}(A_j \cap A_k) + \cdots + (-1)^{N+1}$$
$$g_{bel}(A_1 \cap A_2 \cap \cdots \cap A_N) \tag{22}$$

The belief measures are super additive in the sense that $g_{bel}(A \cup B) \geq g_{bel}(A) + g_{bel}(B)$ for A,B $\in \Omega$. Belief measures are determined for all sets $A \in \Omega$ by the following equation.

$$g_{bel}(A) = \sum_{B|B \subseteq A} g_{bel}(B) \tag{23}$$

## 4   Fuzzy Measures Implementations

According to the definition of fuzzy measures and fuzzy integrals on a discrete set $X$, with functions $h : X \to [0,1]; g : 2^X \to [0,1]$, the term $\rho$ is estimated using Choquet-integral ($e_{choquet}$) given by (26)-(27) after satisfying the constraints in (24)-(25) . The term $\rho$ appears in fuzzy forward-backward variables, fuzzy Viterbi algorithm and fuzzy Baum-Welch algorithm described in detail in [3]-[4].

$$h(x_1) \leq h(x_2) \leq \ldots \leq h(x_i) \leq \ldots \leq h(x_N) \tag{24}$$

$$k_i = \{x_i, x_{i+1}, \ldots, x_N\} \tag{25}$$

$$e_{Chouqet} = \sum_{i=1}^{N} h(x_i)[g(k_i) - g(k_{i+1})] \tag{26}$$

$$e_{Chouqet} = \sum_{i=1}^{N} h(x_i)d_i \tag{27}$$

$d_i$ represents the difference between successive fuzzy measures and $g(k_i)$ represents the fuzzy measure. For the fuzzy profile HMMs the matrix $A$ containing all the transition parameters (Insert, Match and Delete) represents the function $h$ which is sorted at $j^{th}$ row to obtain $k_i(j)$ as

$$k_i(j) = \{s_i, s_{i+1}, \ldots, s_N\} \tag{28}$$

where $S_i$ is the state number at $i^{th}$ position according to constraints in (24) based on transition to $j^{th}$ state from all other states. After obtaining $k_i(j)$ f fuzzy measure $g(k_i(j))$ is calculated using (31)-(34) for for $\lambda$, possibility belief and probability measures respectively measures. The calculation of fuzzy integral with respect all of the above fuzzy measures only requires the knowledge of the fuzzy densities, $g(\{s_i\})$ which are calculated as shown in (29).

$$g(\{s_i\}) = \alpha_t(i) \tag{29}$$

$\lambda$-fuzzy measure:

$$g(k_i(j)) = g(\{s_i, s_{i+1}, \ldots, s_N\}) \tag{30}$$

$$g(k_i(j)) = g(\{s_i\})g(k_{i+1}(j)) + \lambda g(\{s_i\})g(k_{i+1}(j)) \tag{31}$$

Possibility measure:

$$g(k_i(j)) = \wedge(g(\{s_i\}), g(k_{i+1}(j))) \tag{32}$$

*Belief measure:*

$$g(k_i(j)) = \sum_{B|B \subseteq k_i(j)} (B) \tag{33}$$

*Probability measure:*

$$g(k_i(j)) = g(\{s_i\}) + g(k_{i+1}(j)) \tag{34}$$

To avoid data overflow and speed up the computing process we use possibility measure with $\wedge$ operator for computing the fuzzy measures. After obtaining the fuzzy measures, the difference between successive fuzzy measures $d_t(i, j)$ is calculated as

$$d_t(i, j) = g(k_i(j)) - g(k_{i+1}(j)) \tag{35}$$

$d_t(i, j)$ is normalized with respect to fuzzy densities and stored in $\rho_t(i, j)$ as shown in (36).

$$\rho_t(i, j) = \frac{d_t(i, j)}{\alpha_t(i)} \tag{36}$$

## 5   Experiment Results

We evaluated the performance of Possibility (S-decomposable), $\lambda$, belief and probability measures for profile HMMs on two different families, namely, globins and kinases. We used the fuzzy estimation for model building, and fuzzy Viterbi alignment for Z-score calculation [4]. Due to similar trends observed in kinase family we analyze only the globin results here.

### 5.1   Globin Family

The measure analysis modeling was first tested on the widely studied globins, a large family of heme containing proteins involved in the storage and transport of oxygen that have different oligomeric states and overall architecture. Globin sequences were extracted from the Pfam [1] database by searching for the keyword "globin". The globin data set sample used in the experiment consists of 625 different globin sequences. The sequences in the family vary in length from 109 to 428 amino acids. We have divided the data set into training data set of 50 sequences which were used in building the profile model, and the remaining 575 sequences along with 1953 non-globin sequences as the test data set (1969 sequences in total). We align the sequences in the training and test data set against the globin profile model using fuzzy Viterbi algorithm. The profile model parameters are estimated using the fuzzified method described in [3]. For each measure the corresponding fuzzy profile model is built and the fuzzy Viterbi algorithm is used for classification [3]-[4]. We also built the profile model using classical probability theory and use the classical Viterbi algorithm [7].

### 5.2   Z-Score Plot

Since the profile HMM model is concentrated on a subset of the protein family, we calculate Z-scores for each sequence to have a clear view of log-odd scores obtained

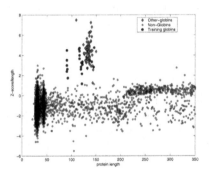

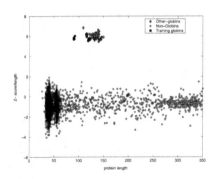

**Fig. 2.** Z-score plot for sequence classification using probability measure based profiles

**Fig. 3.** Z-score plot for sequence classification using S-dec measure based profiles

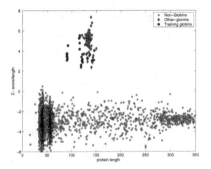

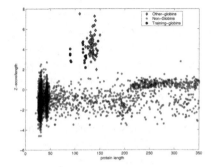

**Fig. 4.** Z-score plot for sequence classification using $\lambda$-dec measure based profiles

**Fig. 5.** Z-score plot for sequence classification using belief measure based profiles

from Viterbi algorithm. The Z-score is a measure of how far an observation is from the mean, as measured in units of standard deviation. For a given data set X with mean $\mu$ and standard deviation $\sigma$, Z-score is given by

$$Z = (X - \mu)/\sigma \qquad (37)$$

The Z-score plots for probability, possibility, $\lambda$ and belief measure based models are shown in Fig. 2-5 respectively. For possibility measures as seen in Fig. 3 it can be observed that all the member sequences(both from training and test data set) are clustered dense with the Z-score $\approx 5.5- \approx 7.0$. This is mainly because of the max operation performed by the possibility measure. For probability measures the member sequences are scattered sparsely with the Z-score varying from $\approx 1.0- \approx 8.0$. Similarly, the Z-score varies from $\approx 2.0- \approx 8.0$ and $\approx 1.5- \approx 8.0$ for $\lambda$ and belief measures respectively. The maximum Z-score value for non member sequences in $\lambda$ measures was 0.6, which was the least compared to the other measures. As a result $\lambda$ measures performed better at lower Z-score cut-offs compared to the other measures. The highest Z-score approximate mean distance between member and non-member sequences which was $\approx 3.0$ was observed in possibility measures. This indicates that possibility measure has the highest accuracy for a wide region of Z-scores. The plots of the belief measure and probability measure models had the least performance and are very similar.

## 5.3   ROC

Receiver Operating Characteristic (ROC) curve is a plot of the true positive rate (sensitivity) and false positive rate (alarm rate) against the different possible Z-score cutoff points. Sensitivity measures the proportion of true positives or the proportion of cases correctly identified by the test with respect to the total correct cases. Specificity measures the proportion of cases correctly identified by the test with respect to all cases identified as correct by the test. False alarm rate measures the proportion of cases wrongly identified by the test with respect to the total non-member cases. Accuracy measure gives the overall classification percentage. Figures 6-9 shows a plot of sensitivity, specificity, FAR and accuracy versus Z-score respectively for profiles of globin family built using classical and fuzzy model with S-decomposable, $\lambda$ and belief measures at different Z-score cut-off values. The accuracy values indicate that fuzzy model built using possibility measures performs better than other models at all Z-score cut-offs values. Among the fuzzy models again S-decomposable performs better than $\lambda$ and belief measure. Belief measure performs very similar to the probability measures(classical). The sensitivity decreases starting from Z-score = 2.0 for all measures except possibility measures. At Z-score = 5.5 the sensitivity starts decreasing for the possibility measures. The possibility measure performs best when it comes correctly identifying the test sequences. False alarm rate decreases with the increase in Z-score for all the measures. For $\lambda$ measures the false alarm rate reaches 0 at Z-score cut-off = $\approx$ 1.0, where as the probability and belief measures have the highest FAR(7.32 and 7.07) at Z-score = $\approx$ 1.0. This indicates that $\lambda$ measure performs slightly better that possibility measures in terms of false classification. Possibility measure has accuracy = 100% for Z-scores = $\approx$ 2.0− $\approx$ 5.0.Choosing a Z-score cut off = 2 distinguishes all the globins from non globins with lambda and possibility measures, whereas the classical and belief measure based models produces 8 and 10 false positive globins respectively and 2 globins protozoan/cyanobacterial globin and Hypothetical protein R13A1.8) are missed out.

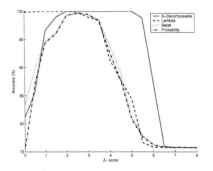

**Fig. 6.** Accuracy curve for different measures with respect to Z-score cutoffs

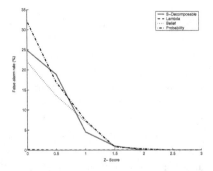

**Fig. 7.** False alarm rate curve for different measures with respect to Z-score cutoffs

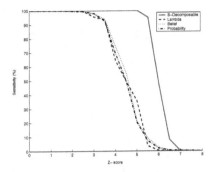

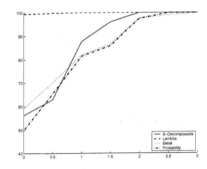

**Fig. 8.** Sensitivity curve for different measures with respect to Z-score cutoffs

**Fig. 9.** Specifity curve for different measures with respect to Z-score cutoffs

## 6    Discussions

Fuzzy measures which are a generalization of classical measures are non-additive. They are very descriptive in the sense that they can combine objective evidence along with the information hypothesis. The fuzzy integrals which serve as the aggregation operators also remove the constraints of statistical independence assumption. The fuzzy measures consider the importance of sets of consensus columns into account. A further improvement can be done by taking specific biological and physio-chemical factors of the family into account for assigning the fuzzy measures. This is an advantage compared to the classical probability models which assign uniform importance to the evidence of the source. A general classical model, $|X| = n$ has $n$ parameters, where as the fuzzy model has at least $2^{n-1}$ parameters in terms of sets. The complexity of the model increases with the number of parameters. New techniques needs to be investigated to optimize the parameters. One of the possibilities is to reduce the sets of the consensus columns which are below a threshold values and thus reducing the number of sets. Markov Chain Monte Carlo based type switching techniques from Bayesian networks can also be explored to do approximations of sets merge or divide. We have done a comparative analysis of different fuzzy measures in this paper for sequence alignment and family identification. Different measures can be suitable for different kinds of families based on the physio-chemical properties of that family due to their non-additive nature. A systematic analysis at a larger scale will help us to establish the correlation and dependencies between the family types and the various fuzzy measures. It may be necessary to focus on developing non-additive measures based graphical models based on fuzzy measures and integrals for protein phylogeny trees and 3-D structure. In this paper we investigated the performance of possibility, $\lambda$, belief and probability measures on the fuzzy model using globin family. Experimental results obtained in terms of Z-score plots and ROC analysis showed the superiority of fuzzy measures over probability measures. S-Decomposable measures had the best sensitivity performance compared to other measures and hence is the most sensitive one for classifying the member sequences correctly. $\lambda$ measures had the lowest false alarm rate for all values of Z-score cut-offs. This indicates that $\lambda$ measures misclassify non member sequences at a very low rate compared to other measures. The average accuracy was highest in S-decomposable measures even though $\lambda$ measures had a higher accuracy for Z-scores varying from $\approx 0.0- \approx 2.0$. The performance of belief measures was very similar to the probability measures. Similar trends were also observed for the kinase family (data not shown here).

## Acknowledgments

The authors wish to thank Dr. M.A. Magdi for his valuable discussions and suggestions on generalized HMM. We also wish to thank Dr. James Whisstock, Scientific Director, Victorian Bioinformatics Consortium Melbourne for his discussions and suggestions.

## References

1. Bateman, A. (2002) The pfam protein families database, *Nucleic Acids Research*, **30**, 276-280.
2. Baldi, P. Brunak, S. (2001) Bioinformatics-the machine learning approach, *MIT press.*
3. Bidargaddi, N. P., Chetty, M., Kamruzzaman, J. (2004) Fuzzy decoding in profile hidden Markov models for protein family identification, *Advances in Bioinformatics and its Applications*, Series in Mathematical Biology and Medicine, **8**.
4. Bidargaddi, N. P., Chetty, M., Kamruzzaman, J. (2005) Fuzzy Viterbi algorithm for improved sequence alignment and searching of proteins, *proc. 3$^{rd}$ European workshop. Evolutionary computation and bioinformatics.* LNCS, **3449** 11-21.
5. Cheok, A. D., (2001) Use of a novel generalized fuzzy hidden Markov model for speech recognition, *IEEE Conf. Fuzzy System*, 1207-1210.
6. Durbin, R., Eddy, S., Krogh, A., Mitchison, G., (2003) Biological sequence analysis- probabilistic models of proteins and nucleic acids, *Cambridge University Press, Cambridge.*
7. Eddy, S. R. (1998) Profile hidden Markov models, *Bioinformatics*, **14**, 755-763.
8. Grabisch, M., Murofushi, T. and Sugeno, M. (2000) Fuzzy measures and integrals - theory and applications, *Physica-Verlag, Heidelberg, New York 2000..*
9. Koski, T.(2001) Hidden Markov models in bioinformatics, *Kluwer academic publishers.*
10. Krogh, A. (1998) An introduction to hidden Markov models for biological sequences, *Computational Methods in Molecular Biology,Elsvier Science*, **99**, 45-63.
11. Magdi, M. A., Gader, P. (2000) Generalized hidden Markov models- part I: theoretical frameworks, *IEEE Trans. Fuzzy Systems*, **8**, 67-80.
12. Shi, H., Gader, P. D. (1996) Lexicon-driven handwritten word recognition using Choquet fuzzy integral, *IEEE Conf*, **99**, 412-417.
13. Sugeno, M., (1977) Fuzzy measures and fuzzy integrals- a survey, *Fuzzy Automata and Decision Processes, M.M. Gupta, G. N. Saridis, and B. R. Gaines, Eds, New York: North-Holland*, 89-102.
14. Tran, D., Wagner, M. (1999) Fuzzy hidden Markov models for speech and speaker recognition, *IEEE Conf. Speech Processing*, 426-430.
15. Valsan, Z., Gavat, I., Sabac, B. (2002) Statistical and hybrid methods for speech recognition in Romanian, *International Journal of Speech Technology*, **5**, 259-268.
16. Wang, Z., and Klir, G.J. (1992) Fuzzy measures and integrals - theory and applications, *Physica-Verlag, Heidelberg, New York.*

# Relevance, Redundancy and Differential Prioritization in Feature Selection for Multiclass Gene Expression Data

Chia Huey Ooi, Madhu Chetty, and Shyh Wei Teng

Gippsland School of Information Technology
Monash University, Churchill, VIC 3842, Australia
{chia.huey.ooi,madhu.chetty,shyh.wei.teng}
@infotech.monash.edu.au

**Abstract.** The large number of genes in microarray data makes feature selection techniques more crucial than ever. From various ranking-based filter procedures to classifier-based wrapper techniques, many studies have devised their own flavor of feature selection techniques. Only a handful of the studies delved into the effect of redundancy in the predictor set on classification accuracy, and even fewer on the effect of varying the importance between relevance and redundancy. We present a filter-based feature selection technique which incorporates the three elements of relevance, redundancy and differential prioritization. With the aid of differential prioritization, our feature selection technique is capable of achieving better accuracies than those of previous studies, while using fewer genes in the predictor set. At the same time, the pitfalls of over-optimistic estimates of accuracy are avoided through the use of a more realistic evaluation procedure than the internal leave-one-out-cross-validation.

**Keywords:** Molecular classification, microarray data analysis, feature selection

## 1 Introduction

When it comes to multiclass microarray datasets, most of the previous classification studies have taken one of the following stances:

1. Feature selection does not aid in improving classification accuracy [1, 2], at least not as much as the type of classifier used.
2. Feature selection is often rank-based, and is implemented mainly with the intention of merely reducing cost/complexity of subsequent computations (since the transformed dataset is smaller), rather than also finding the feature subset which best explains the dataset [1, 3].
3. Studies proposing feature selection techniques with sophistication above that of rank-based techniques resort to an evaluation procedure which often gives overly-optimistic estimate of accuracy, but has the advantage of costing less computationally than procedures which yield a more realistic estimate of accuracy [4, 5].

From these stances, we see the three levels with which feature selection has been, and still is regarded for multiclass microarray datasets: 1) should not be considered at all, 2) simple rank-based methods for dataset truncation, and finally, 3) more complicated methods with sound theoretical foundation, but with dubious empirical results.

An important axiom governing the principles behind most feature selection works of the third level can be summarized by the following statement: A good predictor set

J.L. Oliveira et al. (Eds.): ISBMDA 2005, LNBI 3745, pp. 367–378, 2005.

should contain features highly correlated with the target class distinction, and yet uncorrelated with each other [6]. The attribute referred to in the first part of this statement is encapsulated in the term 'relevance', and has been the backbone for simple rank-based feature selection techniques, where genes are selected into the predictor set based on the score of their correlation to the target class distinction. The measurement of the aspect alluded to in the second part, 'redundancy' however, is not as straightforward, since the pairwise relationship between each pair of genes in the predictor set needs to be taken into account.

Previous studies [4, 6] have based their filter-based feature selection techniques on the concept of relevance and redundancy having equal role in the formation of a good predictor set. On the other hand, Guyon and Elisseeff demonstrated using a 2-class problem that seemingly redundant features may improve the discriminant power of the predictor set instead [7], although it remains to be seen how this scales up to multiclass domains with thousands of features. A study was implemented on the effect of varying the importance of redundancy in predictor set evaluation in [8]. However, due to its use of a relevance score that was inapplicable to multiclass problems, the study was limited to binary classification.

From here, we can rephrase the three levels of feature selection for tumor classification as follows: 1) no selection, 2) pick based on relevance alone, and finally, 3) pick based on relevance and redundancy. Thus, currently, relevance and redundancy are the two existing components used in predictor set scoring methods to evaluate the goodness of a predictor set.

We propose going one step further, by introducing the third element, this element being the relative importance placed between relevance vs. redundancy. This third element compels the search method to prioritize the optimization of one of the elements (of relevance and redundancy) at the cost of the optimization of the other. The degree of differential prioritization is determined by this third element. That is, unlike other existing redundancy-based feature selection studies, with our proposed feature selection technique, it is not taken for granted that the optimizations of both elements of relevance and redundancy are to have equal priorities in the search for the optimal predictor set.

The effectiveness of our proposed feature selection technique on the tumor classification of a multiclass microarray dataset has been reported in [9]. However, this paper aims to do more than illustrate the efficacy of the technique on various other multiclass microarray datasets. More importantly, applying our technique to *multiple* such datasets makes it possible for us to discern the relationship between dataset characteristics and the optimal degree of differential prioritization for a particular dataset.

Having introduced the element of differential prioritization, we go on to demonstrate the importance of applying evaluation procedure which yields more realistic estimate of accuracy than the internal cross validation procedure used in recent tumor classification studies [3, 4, 5]. This is done by evaluating our feature selection techniques using two evaluation procedures: the first being the $F$-splits procedure, the second is the aforementioned internal cross validation procedure.

The contributions of this study are threefold: 1) to show that a degree of freedom in adjusting the priorities between maximizing relevance and minimizing redundancy is necessary to produce the best classification performance (i.e. equal-priorities techniques might not yield the optimal predictor set); 2) to demonstrate the relationship between dataset characteristics and the optimal degree of differential prioritization; and 3) to highlight the importance of using a realistic evaluation procedure.

## 2  Methods

The training set upon which feature selection is to be implemented, T, consists of $N$ genes and $M_t$ training samples. Sample $j$ is represented by a vector, $\mathbf{x}_j$, containing the expression of the $N$ genes $[x_{1,j},..., x_{N,j}]^T$ and a scalar, $y_j$, representing the class the sample belongs to. The target class vector $\mathbf{y}$ is defined as $[y_1, ..., y_{Mt}]$, $y_j \in [1,K]$ in a $K$-class dataset. Gene $i$, on the other hand, is represented by vector $\mathbf{g}_i$, containing expression of gene $i$ across the $M_t$ samples in the training set, $[x_{i,1},..., x_{i,Mt}]$. From the total of $N$ genes, the objective of feature selection is to form the subset of genes, called the predictor set $S$, which would give the optimal classification accuracy.

### 2.1  The Antiredundancy-Based Scoring Method

A score of goodness incorporating both the elements of maximum relevance and minimum redundancy ensures that the optimal predictor set should possess maximal power in discriminating between different classes (maximum relevance), while at the same time containing features with minimal correlation to each other (minimal redundancy).

For the purpose of defining our predictor set scoring method, without loss of generality, we define the following parameters.

- $V_S$ is the measure of relevance for the candidate predictor set $S$.
- $U_S$ is the measure of antiredundancy for the candidate predictor set $S$.

Both $V_S$ and $U_S$ are to be maximized in the search for the optimal predictor set.

$U_S$ quantifies the *lack of redundancy* in $S$. With $U_S$, we have an antiredundancy-based scoring method in which the measure of goodness for predictor set $S$ is given as follows.

$$W_{A,S} = (V_S)^\alpha \cdot (U_S)^{1-\alpha} \tag{1}$$

where the power factor $\alpha \in (0, 1]$ denotes the degree of differential prioritization between maximizing relevance and maximizing antiredundancy.

### 2.2  Significance of the Differential Prioritization Factor, $\alpha$

In the previous section it has been stated that an optimal predictor set is to be found based on two criteria: maximum relevance and maximum antiredundancy. However, the quantification of the priority to be assigned to each of these two criteria remains an unexplored area.

In the antiredundancy-based scoring method, decreasing the value of $\alpha$ forces the search method to put more priority on maximizing antiredundancy at the cost of maximizing relevance. Raising the value of $\alpha$ increases the emphasis on maximizing relevance (at the same time decreases the emphasis on maximizing antiredundancy) during the search for the optimal predictor set. A predictor set found using larger value of $\alpha$ has more features with strong relevance to the target class vector, but also more redundancy among these features. Conversely, a predictor set obtained using smaller value of $\alpha$ contains less redundancy among its member features, but at the same time also has fewer features with strong relevance to the target class vector. At $\alpha = 0.5$, we get an equal-priorities scoring method. At $\alpha = 1$, the feature selection technique becomes rank-based.

We posit that different datasets will require different degrees of prioritization between maximizing relevance and maximizing antiredundancy in order to come up with the most efficacious predictor set. Therefore the optimal range of $\alpha$ (optimal as in leading to the predictor set giving the best estimate of accuracy) is dataset-specific.

## 2.3  Definitions of Relevance and Antiredundancy

The measure of relevance for $S$ is computed by averaging up the score of relevance, $F(i)$ of all members of the predictor set, as recommended in [4]:

$$V_S = \frac{1}{|S|} \sum_{i \in S} F(i) \tag{2}$$

$F(i)$ is the score of relevance for gene $i$. It indicates the correlation of gene $i$ to the target class vector $\mathbf{y}$. For continuous-valued datasets, a popular parameter for computing $F(i)$ is the BSS/WSS ratios (the $F$-test statistics) used in [4, 10]. For gene $i$,

$$F(i) = \frac{\sum\limits_{j=1}^{M_t} \sum\limits_{k=1}^{K} I(y_j = k)(\bar{x}_{ik} - \bar{x}_{i\bullet})^2}{\sum\limits_{j=1}^{M_t} \sum\limits_{k=1}^{K} I(y_j = k)(x_{ij} - \bar{x}_{ik})^2} \tag{3}$$

where $I(.)$ is an indicator function returning 1 if the condition inside the parentheses is true, otherwise it returns 0. $\bar{x}_{i\bullet}$ is the average of the expression of gene $i$ across all training samples, while $\bar{x}_{ik}$ is the average of the expression of gene $i$ across training samples belonging to class $k$. The BSS/WSS ratio, first used in [10] for multiclass tumor classification, is a modification of the $F$-ratio statistics for one-way ANOVA (Analysis of Variance). It indicates the gene's ability in discriminating among samples belonging to the $K$ different classes.

The measure of antiredundancy for $S$ is computed by summing up one minus absolute values of the measures of correlation between all possible pairwise combinations of the members of $S$, and normalizing by division with the square of the size of $S$. Since both correlation and anti-correlation contribute to redundancy in $S$, absolute values of correlation are used.

$$U_S = \frac{1}{|S|^2} \sum_{i,j \in S} 1 - |R(i, j)| \tag{4}$$

For continuous-valued datasets, a conventional measure of correlation between pairs of genes is the absolute value of the Pearson product moment correlation coefficient, which measures similarity between two genes. Between genes $p$ and $q$, the measure of correlation $R(p,q)$ is the Pearson product moment correlation coefficient between genes $p$ and $q$. Larger $U_S$ indicates lower average pairwise correlation in $S$, and hence, smaller amount of redundancy among the members of $S$.

## 2.4  The Search Method

An exhaustive search for the optimal predictor set is computationally expensive. For instance, in searching for the best $S$ of a certain size $P$, the order of complexity is $O(N^P)$. We employed the linear incremental search method, where the first member of $S$ is chosen by selecting the gene with the highest $F(i)$ score. To find the second and the subsequent members of the predictor set, the remaining genes are screened one by one for the gene that would give the maximum $W_{A,S}$. This search method, with a lower computational complexity of $O(NP)$, has been applied in previous feature selection studies [4, 5].

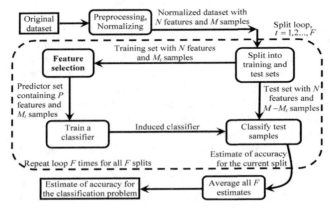

**Fig. 1.** $F$-splits procedure

## 2.5  Over-Optimistic Estimate of Accuracy

In several previous studies on feature selection for microarray datasets [3, 4, 5], feature selection techniques have been applied *once* on the *full* dataset before leave-one-out-cross-validation (LOOCV) procedure is employed to evaluate the classification performance of the resulting predictor sets. We denote this evaluation procedure the Internal LOOCV (ICV) procedure. ICV is known to produce selection bias, which leads to an overly-optimistic estimate of accuracy [11].

To avoid this pitfall, we propose the use of different splits of the dataset into training and test sets and *repeating feature selection for each of the splits*. During each split, our feature selection techniques will be applied only on the training set of that particular split. It is very important that no information from the test set is ever 'leaked' into the process of forming the predictor set (which is precisely what happens during the ICV procedure). Classifier trained on the predictor set and the training samples will then be used to predict the class of the test samples of the current split. The test set accuracies obtained from each split will be averaged to give an estimate of the classification accuracy. We call this procedure of accuracy estimation the $F$-splits procedure ($F$ being the number of splits used) (Figure 1).

In addition to accuracy, we used an approximation of the area under the Receiver Operating Characteristic (ROC) curve (AUC) as a performance evaluation parameter. The approximation used is the modified macro-average of class accuracies (MAVG-MOD) recommended in [12] for multiclass problems employing crisp classifiers.

$$\text{MAVG - MOD} = \left( \frac{1}{K} \sum_{k=1}^{K} a_k^{\tau} \right)^{\frac{1}{\tau}}$$    (5)

For best performance, the value of $\tau$ has been determined as 0.76 [12]. The class accuracy for class $k$ is represented by $a_k$.

## 3  Results

### 3.1  Benchmark Datasets and Evaluation Procedures

Several multiclass microarray datasets are used as benchmark datasets (Table 1). The GCM dataset [2] contains 14 tumor classes: breast, prostate, lung, colorectal, lymphoma, bladder, melanoma, uterus, leukemia, renal, pancreas, ovarian, mesothelioma and CNS (central nervous system). For NCI60 [13], only 8 tumor classes (breast, CNS, colon, leukemia, melanoma, ovarian, renal and non-small-cell lung cancer) are analyzed; the 2 samples of the prostate class are excluded due to the small class size. In the 5-class lung dataset [14], 4 classes (lung adenocarcinoma, squamous cell lung carcinoma, pulmonary carcinoid and small-cell lung cancer) are subtypes of lung cancer; the fifth class comprises of normal samples. The MLL dataset [15] contains 3 subtypes of leukemia: ALL, MLL and AML. The AML/ALL dataset [16] also contains 3 subtypes of leukemia: AML, B-cell and T-cell ALL. Datasets are preprocessed and normalized based on the recommended procedures in [10] for Affymetrix and cDNA microarray data.

Different degrees of importance were placed on antiredundancy measure by varying the values of $\alpha$ from 0.1 up to 1. Optimal predictor sets ranging from sizes $P=2,3,\ldots,100$ were formed in the different runs.

For each dataset we implemented two different evaluation procedures: the 10-splits procedure, and the ICV procedure employed in previous tumor classification studies [3, 4, 5]. The difference between the estimates of accuracy obtained from the two procedures offers us an insight into the effect of selection bias which occurs when feature selection is not repeated for different splits or subsets of the dataset.

**Table 1.** Descriptions of benchmark datasets. $N$ is the number of features after preprocessing

| Dataset | Type | $N$ | $K$ | Training:Test set ratio (10-splits procedure) |
|---------|------|-----|-----|-----------------------------------------------|
| GCM     | Affymetrix | 10820 | 14 | 144:54 |
| NCI60   | cDNA       | 7386  | 8  | 40:20  |
| Lung    | Affymetrix | 1741  | 5  | 135:68 |
| MLL     | Affymetrix | 8681  | 3  | 48:24  |
| AML/ALL | Affymetrix | 3571  | 3  | 48:24  |

The DAGSVM classifier is used throughout the performance evaluation for both 10-splits and ICV procedures. The DAGSVM is an SVM-based multi-classifier which uses substantially less training time compared to either the standard algorithm or Max Wins, and has been shown to produce accuracy comparable to both of these algorithms [17].

## 3.2 Best Estimates of Accuracy for the Benchmark Datasets

The best estimate of accuracy obtained from each dataset is shown in Table 2. Where a draw occurs in terms of the estimate of accuracy, the $\alpha$ value giving the smaller predictor set size is proclaimed as the optimal $\alpha$. Comparisons with previously reported results will only be made for the 2 datasets which have been known to produce low realistic estimates of accuracy (<90%), the GCM and NCI60 datasets [1].

For the GCM dataset, with a predictor set containing no more than 94 genes at most, an accuracy of 80.6–84.3% is achievable with our predictor set scoring method when the value of $\alpha$ is set within the range of 0.2–0.3. This is a significant improvement compared to the 78% accuracy obtained, using all available 16000 genes, in the original analysis of the same dataset [2]. However, strict comparison cannot be made against this 78% accuracy of [2] and the 81.5% accuracy (using 84 genes) achieved in [18] since the evaluation procedure in both studies [2, 18] is based on a single (the original) split. We can make a more appropriate comparison, however, with a comprehensive study on various rank-based feature selection techniques [1]. The study uses external 4-fold cross validation to evaluate classification performance. In [1], the best accuracy for the GCM dataset is 63.3%, when *no* feature selection is applied prior to classification!

For the NCI60 dataset, the best accuracy of 74% from the 10-splits evaluation procedure occurs at $\alpha = 0.3$, and is better than the best accuracy obtained from the two studies employing a similar evaluation procedure [1, 10]. In [10], the best averaged accuracy is around 63% (using the top 30 BSS/WSS-ranked genes), whereas the study in [1] performs slightly better with best accuracy of 66.7% (150 genes) achieved using the sum minority rank-based feature selection technique [1]. For the estimate of accuracy from the ICV procedure, the ICV estimate of 96.7% for our predictor set scoring method is significantly higher than the best ICV estimate in [4] for the continuous-valued version of their predictor set scoring method (80.6%).

**Table 2.** Best accuracy estimated from the 10-splits and ICV procedures, followed by the corresponding differential prioritization factor and predictor set size

| Dataset | 10-splits | ICV |
|---------|-----------|-----|
| GCM | 80.6%, $\alpha$=0.2, 85 genes | 84.3%, $\alpha$=0.3, 94 genes |
| NCI60 | 74.0%, $\alpha$=0.3, 61 genes | 96.7%, $\alpha$=0.2, 89 genes |
| Lung | 95.6%, $\alpha$=0.5, 31 genes | 96.1%, $\alpha$=0.4, 43 genes |
| MLL | 99.2%, $\alpha$=0.6, 12 genes | 98.6%, $\alpha$=0.6, 4 genes |
| AML/ALL | 97.9%, $\alpha$=0.8, 11 genes | 98.6%, $\alpha$=0.6, 5 genes |

From Table 2 and Figure 2, it can be seen that the optimal value of $\alpha$, (i.e. the value of $\alpha$ where the best accuracy is obtained) is not necessarily 0.5 (denoting equal priorities for maximization of relevance and maximization of antiredundancy) or 1 (in which our feature selection technique becomes rank-based selection technique). As mentioned previously in Section 2.2, for a given predictor set scoring method and a set of definitions of $F(i)$ and $R(i,j)$, the optimal range of $\alpha$ is most likely dataset-dependent, as shown by the results.

For the GCM dataset, the highest 10-splits accuracy, 80.6% is obtained when $\alpha$ is set to 0.2. If we had limited ourselves to equal priorities for the maximizations of relevance and antiredundancy (i.e. equivalent to setting $\alpha$ to 0.5), we would have achieved only a 75.9% accuracy, and if we had been content with rank-based techniques (i.e. equivalent to setting $\alpha$ to 1), we would have fared worse, with only a measly 65.6% if the maximum size of the predictor set is set to 100.

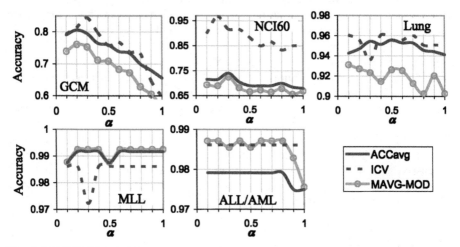

**Fig. 2.** MAVG-MOD and best accuracy estimate from the 10-splits procedure (ACC$_{avg}$) and best accuracy estimate from the ICV procedure respectively plotted against $\alpha$

The same can be said for the NCI60 dataset. Using $\alpha = 0.3$ we obtain the best accuracy of 74% based on the 10-splits evaluation procedure, whereas at $\alpha = 0.5$ and $\alpha = 1$, much lower accuracies (69% and 68% respectively) are achieved.

The lung dataset is the only dataset tested where the best 10-splits accuracy is obtained when $\alpha$ is set to 0.5. Rather than presenting a conflict against the results from the other datasets, in Section 4.2 we will prove that in case of the lung dataset 0.5 merely happens to be the optimal $\alpha$ due to the characteristics of the lung dataset itself, in the same way that the values of the optimal $\alpha$ for each of the other datasets are influenced by the characteristics of the respective datasets.

For the MLL dataset, we have an accuracy of 99.2% at $\alpha = 0.6$. When the predictor set scoring method gives equal priorities to relevance and redundancy ($\alpha = 0.5$), the accuracy achieved drops to 98.7%. However, when the selection is rank-based ($\alpha = 1$), the same accuracy of 99.2% is obtained, but using *twice* the number of genes (24 instead of 12 genes) compared to the predictor set scoring method run with $\alpha = 0.6$. This is not surprising, considering that at $\alpha = 1$, genes are selected based on relevance without regards to redundancy, hence the bigger size of predictor set due to the inclusion of redundant genes.

For the AML/ALL dataset, we get an accuracy of 97.9% at $\alpha = 0.8$ using an 11-gene predictor set. At $\alpha = 0.5$, the same accuracy is achieved but using twice the number of genes (20 genes). Accuracy drops to 97.5% when the value $\alpha = 1$ is used.

With the single exception of the lung dataset, Figure 2 shows that our alternative performance evaluation parameter, MAVG-MOD, demonstrates the same trend against $\alpha$ as accuracy does, i.e., the peak of the accuracy curve always coincides with the peak of the MAVG-MOD curve. Even for the lung dataset, the peak of the MAVG-MOD at $\alpha = 0.5$ is only slightly lower than the peak at $\alpha = 0.1$ (by 0.005). Looking at the class accuracies for this dataset, we found the underlying reason: the class accuracies of the 4 classes with the best class accuracies peak at $\alpha = 0.1$, whereas only one class, the worst-performing class (squamous cell lung carcinoma) has its best accuracy at $\alpha = 0.5$. Therefore, being capable of producing the highest class accuracy for the worst-performing class, $\alpha = 0.5$ is still the optimal value of $\alpha$ for the lung dataset.

## 4   Discussion

### 4.1   Selection Bias

Selection bias is reflected in the difference between the accuracy estimated from the 10-splits procedure and the accuracy obtained from the ICV procedure for each predictor set size ($P=2,3,\ldots,100$). By contrasting between the results from the 10-splits and the ICV procedures, it is clear that the apparently better performance reported previously [4] is a product of selection bias.

   In order to quantify selection bias susceptibility for a dataset, we use the ratio of the number of features to the median class size, $N/CS_{50}$ (Table 3). Greater number of total features ($N$) and smaller class sizes ($CS_{50}$) mean higher likelihood for a search method to find the predictor set which fits the data, thereby increasing the probability of over-fitting. In $F$-splits evaluation procedure, over-fitting can be easily detected through the resulting low accuracy estimate from the classification of test sets uninvolved in feature selection. In ICV evaluation procedure, over-fitting naturally results in higher accuracy estimate since *all* samples have been involved in feature selection.

**Table 3.** Median class size, $CS_{50}$, ratio of $N$ to $CS_{50}$ and optimal $\alpha$ values for each dataset

| Dataset | K | $CS_{50}$ | $N/CS_{50}$ | optimal $\alpha$ |
|---|---|---|---|---|
| GCM | 14 | 12 | 901.7 | 0.2 |
| NCI60 | 8 | 7.5 | 984.8 | 0.3 |
| Lung | 5 | 20 | 87.1 | 0.5 |
| MLL | 3 | 24 | 361.7 | 0.6 |
| AML/ALL | 3 | 25 | 142.8 | 0.8 |

   The $N/CS_{50}$ ratios for all datasets are correlated to the selection bias shown in Figure 3. The NCI60 dataset is the most susceptible to selection bias due to its high $N/CS_{50}$ ratio of 984.8. The GCM dataset, which has the second highest overall selection bias, also has the second highest $N/CS_{50}$ ratio (901.7). The remaining three datasets (lung, MLL and AML/ALL datasets), for each of which the $N/CS_{50}$ ratio is below 400, have relatively smaller selection bias (near 0% for most of the tested values of $\alpha$).

### 4.2   Optimal Differential Prioritization Factor and Dataset Characteristics

The relationship between the optimal value of $\alpha$ and the number of classes, $K$, for all benchmark datasets is illustrated in the left-side panel of Figure 4. The effect of the class size, represented by the median class size, $CS_{50}$, on the optimal value of $\alpha$ is shown in the right-side panel of Figure 4. It can be seen that the optimal value of $\alpha$ decreases as $K$ increases, but increases as $CS_{50}$ becomes larger.

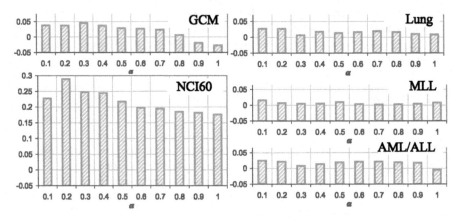

**Fig. 3.** Selection bias in GCM, NCI60, lung, MLL and AML/ALL datasets. Averaged bias among predictor sets of different sizes ($P=2,3,\ldots,100$) plotted against $\alpha$

The optimal value of $\alpha$ (Table 3) has a strong positive correlation to $CS_{50}$ (0.90) but strong negative correlation to $K$ (–0.89). This means that with **smaller** class sizes and **more** classes per dataset, the Pearson-moment-based antiredundancy plays increasingly important role in the search for the optimal predictor set than the BSS/WSS-based relevance (as reflected by the smaller optimal $\alpha$). Conversely, maximizing antiredundancy becomes less important as $K$ decreases – therefore supporting the assertion in [7] that redundancy does *not* hinder the discriminant power of the predictor set when $K$ is 2.

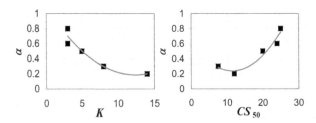

**Fig. 4.** Optimal values of $\alpha$ for all benchmark datasets plotted against $K$ (left-side panel) and $CS_{50}$ (right-side panel) of corresponding datasets

Larger number of multiclass datasets of diverse characteristics needed to be tested before a more definite rule can be determined regarding the optimal choice for the value of $\alpha$ – which we know, for now, most likely depends on at least two characteristics of the dataset, class size ($CS_{50}$) and the number of classes, $K$.

# 5 Conclusion

For majority of the datasets tested, the differential prioritization factor makes it possible to achieve an accuracy rate higher than the rates obtainable using an equal-priorities scoring method ($\alpha$ fixed at 0.5) or a rank-based selection technique ($\alpha$ fixed at 1). Therefore, instead of limiting ourselves to a fixed universal set of priorities for relevance and antiredundancy ($\alpha$ fixed to 0.5 or 1) for **all** datasets, a suitable range for $\alpha$ should be chosen based on the characteristics of the dataset of interest in order to achieve the optimal estimate of accuracy.

Estimate of accuracy from the ICV procedure, which has been popularly used for gene expression datasets due to its low computational cost, can be radically overly-optimistic, particularly when the $N/CS_{50}$ ratio is large.

# References

1. Li, T., Zhang, C., Ogihara, M.: A comparative study of feature selection and multiclass classification methods for tissue classification based on gene expression. Bioinformatics 20 (2004) 2429–2437
2. Ramaswamy, S., Tamayo, P., Rifkin, R., Mukherjee, S., Yeang, C.H., Angelo, M., Ladd, C., Reich, M., Latulippe, E., Mesirov, J.P., Poggio, T., Gerald, W., Loda, M., Lander, E.S., Golub, T.R.: Multi-class cancer diagnosis using tumor gene expression signatures. Proc. Natl. Acad. Sci. 98 (2001) 15149–15154
3. Chai, H., Domeniconi, C.: An Evaluation of Gene Selection Methods for Multi-class Microarray Data Classification. In: Proc. 2nd European Workshop on Data Mining and Text Mining in Bioinformatics (2004) 3–10
4. Ding, C., Peng, H.: Minimum Redundancy Feature Selection from Microarray Gene Expression Data. In: Proc. 2nd IEEE Computational Systems Bioinformatics Conference. IEEE Computer Society (2003) 523–529
5. Yu, L., Liu, H.: Efficiently Handling Feature Redundancy in High-Dimensional Data. In: Domingos, P., Faloutsos, C., Senator, T., Kargupta, H., Getoor, L. (eds.): Proc. 9th ACM SIGKDD International Conference on Knowledge Discovery and Data Mining. ACM Press, New York (2003) 685–690
6. Hall, M.A., Smith, L.A.: Practical feature subset selection for machine learning. In: McDonald, C. (ed.): Proc. 21st Australasian Computer Science Conference. Springer, Singapore (1998) 181–191
7. Guyon, I., Elisseeff, A.: An Introduction to Variable and Feature Selection. Journal of Machine Learning Research 3 (2003) 1157–1182
8. Knijnenburg, T.A.: Selecting relevant and non-redundant features in microarray classification applications. M.Sc. Thesis. Delft University of Technology. http://ict.ewi.tudelft.nl/pub/marcel/Knij05b.pdf (2004)
9. Ooi, C.H., Chetty, M., Gondal, I.: The role of feature redundancy in tumor classification. In: He, M., Narasimhan, G., Petoukhov, S. (eds.): Proc. International Conference on Bioinformatics and its Applications (ICBA'04). World Scientific Publishing (2004) 197–208
10. Dudoit, S., Fridlyand, J., Speed, T.: Comparison of discrimination methods for the classification of tumors using gene expression data. JASA 97 (2002) 77–87
11. Ambroise, C., McLachlan, G. J.: Selection bias in gene extraction on the basis of microarray gene-expression data. Proc. Natl. Acad. Sci. 99 (2002) 6562–6566
12. Ferri, C., Hernández-Orallo, J., Salido, M.A.: Volume under the ROC Surface for Multiclass Problems. In: Lavrac, N., Gamberger, D., Todorovski, L., Blockeel, H. (eds.): Proc. of the 14th European Conference on Machine Learning, Cavtat-Dubrovnik, Croatia. Springer (2003) 108–120

13. Ross, D.T., Scherf, U., Eisen, M.B., Perou, C.M., Spellman, P., Iyer, V., Jeffrey, S.S., Van de Rijn, M., Waltham, M., Pergamenschikov, A., Lee, J.C.F., Lashkari, D., Shalon, D., Myers, T.G., Weinstein, J.N., Botstein, D., Brown, P.O.: Systematic variation in gene expression patterns in human cancer cell lines, Nature Genetics 24(3) (2000) 227–234
14. Bhattacharjee, A., Richards, W.G., Staunton, J., Li, C., Monti, S., Vasa, P., Ladd, C., Beheshti, J., Bueno, R., Gillette, M., Loda, M., Weber, G., Mark, E.J., Lander, E.S., Wong, W., Johnson, B.E., Golub, T.R., Sugarbaker, D.J., Meyerson, M.: Classification of Human Lung Carcinomas by mRNA Expression Profiling Reveals Distinct Adenocarcinoma Subclasses. Proc. Natl. Acad. Sci. 98 (2001) 13790–13795
15. Armstrong, S.A., Staunton, J.E., Silverman, L.B., Pieters, R., den Boer, M.L., Minden, M.D., Sallan, S.E., Lander, E.S., Golub, T.R., Korsmeyer, S.J.: MLL translocations specify a distinct gene expression profile that distinguishes a unique leukemia. Nature Genetics 30 (2002) 41–47
16. Golub, T.R., Slonim, D.K., Tamayo, P., Huard, C., Gaasenbeek, M., Mesirov, J.P., Coller, H., Loh, M.L., Downing, J.R., Caligiuri, M.A., Bloomfield, C.D., Lander, E.S.: Molecular Classification of Cancer: Class Discovery and Class Prediction by Gene Expression Monitoring. Science 286 (1999) 531–537
17. Platt, J.C., Cristianini, N., Shawe-Taylor, J.: Large Margin DAGs for Multiclass Classification. Advances in Neural Information Processing Systems 12 (2000) 547–553
18. Linder, R., Dew, D., Sudhoff, H., Theegarten D., Remberger, K., Poppl, S.J., Wagner, M.: The 'subsequent artificial neural network' (SANN) approach might bring more classificatory power to ANN-based DNA microarray analyses. Bioinformatics 20 (2004) 3544–3552

# Gene Selection and Classification
# of Human Lymphoma from Microarray Data

Joarder Kamruzzaman[1], Suryani Lim[1], Iqbal Gondal[1], and Rezaul Begg[2]

[1] Faculty of Information Technology, Monash University, Australia-3842
[2] Centre For Ageing, Rehabilitation, Exercise & Sport, Victoria University,
Australia-8001

**Abstract.** Experiments in DNA microarray provide information of thousands of genes, and bioinformatics researchers have analyzed them with various machine learning techniques to diagnose diseases. Recently Support Vector Machines (SVM) have been demonstrated as an effective tool in analyzing microarray data. Previous work involving SVM used every gene in the microarray to classify normal and malignant lymphoid tissue. This paper shows that, using gene selection techniques that selected only 10% of the genes in "Lymphochip" (a DNA microarray developed at Stanford University School of Medicine), a classification accuracy of about 98% is achieved which is a comparable performance to using every gene. This paper thus demonstrates the usefulness of feature selection techniques in conjunction with SVM to improve its performance in analyzing Lymphochip microarray data. The improved performance was evident in terms of better accuracy, ROC (receiver operating characteristics) analysis and faster training. Using the subsets of Lymphochip, this paper then compared the performance of SVM against two other well-known classifiers: multi-layer perceptron (MLP) and linear discriminant analysis (LDA). Experimental results show that SVM outperforms the other two classifiers.

## 1 Introduction

The DNA microarray provides information of thousands of genes, which could be harnessed for different purposes. One common use is to separate cancerous from healthy cells using either unsupervised or supervised classifiers [1–3, 17, 21]. Alizadeh et al. [1] used unsupervised classifier to group genes having similar expression patterns in order to separate healthy from cancerous cells. In recent years, supervised methods have also been used for this classification task; for example, decision tress, linear discriminant analysis (LDA), multi-layer perceptron (MLP), support vector machines (SVM) and many others [5, 10, 12, 14]. In general, supervised methods have been shown to perform better than unsupervised methods.

Using the microarray in Alizadeh's study, Valentini [21] showed that a supervised method can achieve a significantly higher classification accuracy than that reported by Alizadeh et al. [1]. In his study, Valentini trained SVM using all

J.L. Oliveira et al. (Eds.): ISBMDA 2005, LNBI 3745, pp. 379–390, 2005.

4026 genes in the microarray to separate normal from malignant cells. However, use of such high feature dimensions reduces the efficiency of SVM.

The purpose of this paper is to test whether, by adopting gene (feature) selection techniques in conjunction with SVM, the same level of accuracy can be achieved using only a subset of the total number of genes. Fewer number of genes require less computational time for SVM as an added advantage. Identifying the contributing genes in this process also enables biologists to concentrate on few genes to explore their roles in malignancy development in greater details. We repeated Valentini's experiment by training SVM using the same microarray, and in addition, we trained more SVMs using only about 10% of the original microarray; the subset genes were derived from feature selection techniques. As MLP and LDA have been previously used for classifying microarray, we also compared the performance of SVM with these two methods on the selected subsets.

This paper is organised as follows. Section 2 discusses the methods for obtaining the subsets. Section 3 describes the three classifiers investigated. Section 4 describes the experimental set up, and Section 5 contains the results and discussion of results. Section 6 concludes the paper and provides direction for future work.

## 2    Gene Selection

Feature selection obtains a subset from a complete set of features and can increase the efficiency of the classifier by reducing redundant and irrelevant features. It can be formally defined as follows. Let $S$ be a subset of $X$ and $S = \{s_1, s_2, \cdots, s_n | s_i \in X, n << ||X||\}$, where $n$ is the number of features in the subset. The feature selection function $F$ selects $s_i$ from $X$, that is $F : X \rightarrow S$. In general, feature selection can be broadly classified into three sub-areas: embedded, filter and wrapper [15]. In this paper, we concentrated on filter-based feature selection.

The filtering method reduces redundancies within the data by selecting only relevant features. However, the definition of relevance is domain dependant, and it has been known that although irrelevant features are less useful for classifiers, not all relevant features are necessarily useful [6].

In this paper, more generic approaches were adopted for feature selection using two statistical methods: the $t$-test and Significant Analysis of Microarrays (SAM). The rationale for using statistical tests is that they are often used to validate the significance of different treatments to influence an outcome. Therefore, by performing statistical tests on the microarray features, it is possible to reduce redundancy by excluding features that are not statistically significant. The following sections briefly describe these two statistical methods.

### 2.1    Standard $t$-Test

The standard $t$-test is defined as:

$$t\text{-test} = \frac{\bar{x}_{i,1} - \bar{x}_{i,2}}{\hat{\sigma}_i \sqrt{\frac{1}{n_1} + \frac{1}{n_2}}} \tag{1}$$

where $\bar{x}_{i,1} - \bar{x}_{i,2}$ is the difference of the means between the two classes, $n_1$ and $n_2$ are the number of samples in the two classes, and $\hat{\sigma}_i$ is within-class standard deviation for gene $i$ [9].

## 2.2   Significance Analysis of Microarrays (SAM)

The standard $t$-test was proposed for testing the significance of any data, while SAM was proposed specifically for testing the significance of genes in microarrays. Tusher et al. [20] argued that the $t$-test may discover many significant genes by chance, and subsequently proposed the development of SAM. SAM assigns each gene a score calculated on the basis of change in gene expression relative to standard deviation of repeated measurements.

The statistical test in SAM is given by $d(i)$, the "relative difference" of gene $i$ and $s(i)$, the "gene-specific scatter" of gene $i$. Tusher et al. defined $d(i)$ as [20]:

$$d(i) = \frac{\bar{x}_I(i) - \bar{x}_U(i)}{s(i) + s_0} \tag{2}$$

where $\bar{x}_I(i)$ is the average level of expression for gene $(i)$ in states $I$, $\bar{x}_U$ in state $U$, $s_0$ is a data dependent constant, and

$$s(i) = \sqrt{a\left\{\sum_m [x_m(i) - \bar{x}_I(i)]^2 + \sum_n [x_n(i) - \bar{x}_U(i)]^2\right\}} \tag{3}$$

where $\sum_m$ is the summation of the expression measurements in state $I$, $\sum_n$ in state $U$, $a = \frac{1/n_I + 1/n_U}{n_I + n_U - 2}$, and $n_I$ is the number of measurements in state $I$, and $n_U$ in state $U$. The genes are then ranked in order of the magnitude of $d(i)$ and those larger than a threshold value are considered significant.

## 3   Formulation of the Three Classifiers Studied

This section describes the formulation of SVM, MLP and LDA, the three supervised classifiers used in this paper.

## 3.1   Support Vector Machines (SVM)

Support vector machine introduced by Vapnik [22] has attracted much research attention in recent years due its demonstrated improved generalization performance over other techniques in many real world applications including the analysis of microarrays [5, 21]. It has been used in classification as well as regression tasks. The main difference between this technique and many other conventional classification techniques including neural networks is that it minimizes the structural risk instead of the empirical risk. The principle is based of the fact that minimizing an upper bound on the generalization error rather than minimizing the training error is expected to perform better. The generalization error rate is bounded by the sum of training error rate and a term that depends on Vapnik-Chervonenkis (VC) dimension [13]. VC dimension is a measure of complexity of the dimension space. Support vector machines find a balance between the empirical error and the VC-confidence interval. SVMs perform by nonlinearly mapping the input data into a high dimensional feature space by means of a kernel function and then do classification in the transformed space.

Consider a data set consisting $D = (\mathbf{x}_i, y_i)_{i=1}^{L}$ of $L$, with each input $\mathbf{x}_i \in \Re^n$ and the associated output $y_i \in \{-1, +1\}$. Searching an optimal separating hyperplane (OSH) in the original input space is too restrictive in most practical cases. In SVM, each input $\mathbf{x}$ is first mapped into a higher dimension feature space $\mathcal{F}$ by $\mathbf{z} = \phi(\mathbf{x})$ via a nonlinear mapping $\phi : \Re^n \rightarrow \mathcal{F}$. Considering the case when the data are linearly separable in $\mathcal{F}$, there exists a vector $\mathbf{w} \in \mathcal{F}$ and a scalar b that define the separating hyperplane as: $\mathbf{w}.\mathbf{z} + b = 0$ such that

$$y_i(\mathbf{w}.\mathbf{z}_i + b) \geq 1, \forall i. \tag{4}$$

SVM constructs an OSH for which the margin of separation between the two classes is maximized. This margin is $2/\|\mathbf{w}\|$ according to its definition. Hence the unique hyperplane that optimally separates the data in $\mathcal{F}$ is the one that

$$min \quad \frac{1}{2}\mathbf{w}.\mathbf{w} \tag{5}$$

under the constraints of Eq. (4). When the data is linearly non-separable, the above minimization problem must be modified to allow classification error. This is done by generalizing the previous analysis with the introduction of some non-negative variables $\xi_i \geq 0$, often called *slack variables*, such that

$$y_i(\mathbf{w}.\mathbf{z}_i + b) \geq 1 - \xi_i, \forall i. \tag{6}$$

Only the misclassified data points $x_i$ yield nonzero $\xi_i$. The term $\sum_{i=1}^{L} \xi_i$ can be regarded as a measure of misclassification. Thus the OSH is determined so that the maximization of the margin and minimization of training error is achieved by adding a penalty term to Eq. (5):

$$min \quad \frac{1}{2}\mathbf{w}.\mathbf{w} + C\sum_{i=1}^{L} \xi_i \tag{7}$$

$$subject\ to \quad y_i(\mathbf{w}.\mathbf{z}_i + b) \geq 1 - \xi_i \text{ and } \xi_i \geq 0, \forall i$$

where C is a constant parameter, called regularization parameter, that determines the trade off between the maximum margin and minimum classification error. Minimizing the first term corresponds to minimizing the VC-dimension of the classifier and minimizing the second term controls the empirical risk.

Searching the optimal hyperplane in Eq. (7) is a Quadratic Programming (QP) problem that can be solved by constructing a Lagrangian and transforming in a dual. The optimal hyperplane can then be shown as the solution of

$$min \quad W(\alpha) = \Sigma_{i=1}^{L}\alpha_i - \frac{1}{2}\Sigma_{i=1}^{L}\Sigma_{j=1}^{L}\alpha_i\alpha_j y_i y_j K(\mathbf{x}_i, \mathbf{x}_j) \tag{8}$$

$$subject\ to\ \ \Sigma_{i=1}^{L}y_i\alpha_i = 0\ and\ 0 \le \alpha_i \le C, \forall i$$

where $\alpha = (\alpha_1, \alpha_2, \ldots, \alpha_L)$ is the non-negative Lagrangian multiplier. The data points $\mathbf{x}_i$ corresponding to $\alpha_i > 0$ lie along the margins of decision boundary and are support vectors (sv). The kernel function $K(.)$ describes an inner product in the $D$-dimensional space as described later and satisfies the Mercer's condition [8].

Having determined the optimum Lagrange multipliers, the optimum solution for the weight vector $\mathbf{w}$ is given by

$$\mathbf{w} = \Sigma_{i \in sv}\alpha_i y_i \mathbf{z}_i \tag{9}$$

where $sv$ are the the support vectors. For any test vector $\mathbf{x} \in \Re^n$, the output is then given by

$$y = sign(\mathbf{w}.\mathbf{z} + b) = sign(\Sigma_{i \in sv}\alpha_i y_i K(\mathbf{x}_i, \mathbf{x} + b)) \tag{10}$$

The generalization performance (i.e. classification accuracy in this study) depends on the parameters $C$ and kernel type. In this study, we used the following kernel functions which are commonly used:

Linear  $K(\mathbf{x}_i, \mathbf{x}_j) = \mathbf{x}_i.\mathbf{x}_j$
Polynomial  $K(\mathbf{x}_i, \mathbf{x}_j) = \{a(\mathbf{x}_i.\mathbf{x}_j) + b\}^d$
Radial basis (RBF)  $K(\mathbf{x}_i, \mathbf{x}_j) = \exp(-\frac{\|\mathbf{x}_i - \mathbf{x}_j\|^2}{2\sigma^2})$
Sigmoid  $K(\mathbf{x}_i, \mathbf{x}_j) = \tanh(\mathbf{x}_i.\mathbf{x}_j + c)$

## 3.2   Multi-layer Perceptron (MLP)

A three layer MLP has an input layer, a hidden layer and an output layer. Successive layers are fully connected by weights. An input vector $\mathbf{x}$ is presented to the network and multiplied by the weights. All the weighted inputs to each unit of upper layer are then summed up and produce an output governed by $\mathbf{h}$ and $\mathbf{y}$, which are defined as

$$\mathbf{h} = f(\mathbf{W}_h.\mathbf{x} + \boldsymbol{\theta}_h) and \tag{11}$$

$$\mathbf{y} = f(\mathbf{W}_o.\mathbf{h} + \boldsymbol{\theta}_o) \tag{12}$$

where $\mathbf{y}$ is the output vector produced by the network, $\mathbf{W}_h$ and $\mathbf{W}_o$ are the hidden and output layer weight matrices, respectively, $\mathbf{h}$ is the vector denoting the response of hidden layer, $\boldsymbol{\theta}_h$ and $\boldsymbol{\theta}_o$ are the output and hidden layer bias vectors, respectively and $f(.)$ is the sigmoid activation function. The standard Backpropagation training algorithm [18] uses gradient descent techniques to minimize the sum of squared error measured at the output layer. In this study, we used an Levenberg and Marquardt technique to accelerate learning speed [16].

## 3.3   Linear Discriminant Analysis (LDA)

In LDA an $n$-dimensional data is projected onto a line according to a given direction $\mathbf{w}$. The choice of the projection direction is determined by different criteria. The Fischer's linear discriminant aims at maximizing the ratio of between-class scatter to within-class scatter [4]. Let $I_y = \{i : y_i = y\}, y \in \{-1, +1\}$ be the sets of indices of training vectors belonging to the first and second class, respectively. The class separability in a direction $\mathbf{w} \in \Re^n$ is found by maximizing the function $J(\mathbf{w})$, which is defined as

$$J(\mathbf{w}) = \frac{\mathbf{w}^t \mathbf{S}_b \mathbf{w}}{\mathbf{w}^t \mathbf{S}_w \mathbf{w}} \tag{13}$$

where $\mathbf{S}_b$ and $\mathbf{S}_w$ are the between-class and within-class scatter matrices, respectively.

# 4   Experimental Setup

We used the Lympochip (a DNA microarray developed at Stanford University School of Medicine [1]); 24 samples were collected from healthy cells and 72 samples collected from malignant lymphocytes cells, and each sample consists of 4026 different genes. We then used the $t$-test and SAM implemented in BRB-ArrayTools [19] to derive two subsets of Lymphochip, so in total there are now three sets of Lymphochip (see the Appendix for the BRB-ArayTools parameters). For convenience, we refer to the complete set of genes as Lymphochip, the subset selected by $t$-test as $L_{t-test}$ and the subset by SAM as $L_{SAM}$. The total number of genes in Lymphochip, $L_{t-test}$ and $L_{SAM}$ are 4026, 387 and 418, respectively. Note that both subsets have only about 10% of the total number of genes in Lymphochip.

We trained SVM classifiers using four kernels (linear, RBF, polynomial and sigmoid) for all three data sets using libsvm [7]. As SVM is sensitive to training parameters such as the regularisation parameters $(C)$ and the parameters for each kernel type, we generated 1856 SVM by varying the values of $C$ from $2^{-31}$ to $2^{27}$ with an increment of $(2^2)$. The parameters $a$ in the polynomial kernel, $\gamma = (1/2\sigma^2)$ in RBF and $c$ in sigmoid were all varied from $2^{-31}$ to $2^9$ with an increment of $2^2$. The parameter $d$ in the polynomial kernel was varied from 2 to 11 with an increment of 1.

We also trained LDA and MLP using $L_{t-test}$ and $L_{SAM}$ and compared the best performing SVM kernel (on $L_{t-test}$ and $L_{SAM}$) with LDA and MLP. The MLP was trained using 5 to 13 hidden nodes with an increment of two nodes and the training was stopped when the mean square error reached 0.01 or smaller. As MLP settles down to different set of weights depending on initial weights and learning parameters producing different results on each run, it is a common practice to generate multiple runs of MLP and use the average results for comparison. In this experiment, MLP was trained 20 times using different initial weights and learning parameters. LDA is the simplest classifier of the three and requires no parameters setting. Like SVM, it is also a non-stochastic process, so it is adequate to run it only once.

The 6-fold cross-validation technique was used in all experiments, and each fold contained 4 healthy and 12 malignant samples. The performance of all classifiers was then analysed using the average accuracy, sensitivity and specificity of the 6-fold data measured as

$$Accuracy = \frac{TP + TN}{TP + FP + TN + FN} \times 100\% \tag{14}$$

$$Sensitivity = \frac{TP}{TP + FN} \times 100\% \tag{15}$$

$$Specificity = \frac{TN}{TN + FP} \times 100\% \tag{16}$$

where $TP$ is the number of true positives i.e. the number of malignant cells labelled as malignant; $TN$ is the number of true negatives i.e. the number of healthy cells labelled as healthy; $FP$ is the number of false positive i.e. the number of healthy cells labelled as malignant; and $FN$ is the number of false negative i.e. the number of malignant cells labelled as healthy. Accuracy measures the overall detection for both healthy and cancerous cells. Sensitivity measures the ability of a classifier in recognizing malignant cells whereas specificity measures the ability of a classifier for not failing to detect healthy cells.

To further analyse how well the classifiers are able to generalize on unseen data, we employed Receiver Operating Characteristics (ROC) curves and calculated the area under the curves. ROC curve plots sensitivity against (1-specificity) as the threshold level of the classifier is varied. ROC analysis is commonly used in medicine and healthcare to qualify the accuracy of diagnostic test and evaluate performance of intelligent system [11].

## 5   Results and Discussion

This paper first discusses the performance of SVM using four kernels in Lymphochip, $L_{t-test}$ and $L_{SAM}$ (Section 5.1) and then compared the best performing SVM kernel against MLP and LDA using $L_{t-test}$ and $L_{SAM}$ (Section 5.2)

### 5.1   Performance of SVM in Lymphochip, $L_{t-test}$ and $L_{SAM}$

The average sensitivity, specificity and accuracy for 6-fold cross validation of the four SVM kernels using the three data sets are presented in Table 1. We first compare the effectiveness of the different SVM kernels within each data sets, then across different data sets. For polynomial kernel the second degree ($d = 2$) produced the best results and is referred as Poly2 in the table.

It is clear that for Lymphochip, the accuracy of linear, RBF and sigmoid kernels are equally good and are also the best whereas the two-degree polynomial has lower specificity and accuracy. While for $L_{t-test}$, only the RBF and polynomial kernels are equally good, followed by the linear and sigmoid kernels. It is difficult to say whether the linear kernel is better than the sigmoid kernel because although its sensitivity is better than that of the sigmoid kernel,

**Table 1.** The average sensitivity, specificity and accuracy for 6-fold cross validation of four SVM kernels using the three datasets: Lymphochip, $L_{t-test}$ and $L_{SAM}$

| Kernel | Lympochip | | | $L_{t-test}$ | | | $L_{SAM}$ | | |
|---|---|---|---|---|---|---|---|---|---|
| | Sens. | Spec. | Acc. | Sens. | Spec. | Acc. | Sens. | Spec. | Acc. |
| Linear | 100.0 | 91.67 | 97.92 | 98.61 | 91.67 | 96.88 | 98.61 | 91.67 | 96.88 |
| RBF | 100.0 | 91.67 | 97.92 | 100.0 | 91.67 | 97.92 | 100.0 | 91.67 | 97.92 |
| Poly2 | 100.0 | 79.17 | 94.79 | 100.0 | 91.67 | 97.92 | 100.0 | 87.5 | 96.88 |
| Sigmoid | 100.0 | 91.67 | 97.92 | 97.22 | 95.83 | 96.88 | 97.22 | 95.83 | 96.88 |

its specificity is worse than that of the sigmoid kernel. Interestingly, the specificity of Poly2 in $L_{t-test}$ is better than in Lympochip. This means that having higher number of genes does not necessarily improves performance. For $L_{SAM}$, the RBF kernel performs the best, followed by the linear and sigmoid kernels and lastly, the Poly2 kernel. Again, it is difficult to rank the last three kernels because each performs well in different measures. As in $L_{t-test}$, the specificity of Poly2 is slightly better than in Lympochip.

After comparing the performance of kernels within each data set, we now compare the accuracy of the best performing kernel across all data sets. The results suggest that the RBF kernel is the most suitable kernel for this microarray as it performs no worse than any of the other kernels across all data sets. The observation also suggests that it is possible to achieve the same level of accuracy using reduced subsets, $L_{t-test}$ and $L_{SAM}$, especially by choosing appropriate kernel type. This is significant because both subsets have only about 10% of the number of genes in *Lymphochip* and they were derived without using any domain knowledge. The reduced number of genes reduces the training time of SVM by about 90% on average. More importantly, this added advantage comes without sacrificing the accuracy, enabling biologists to further explore the influence of the subset genes in malignancy development.

For further analysis, we studied the ROC curves and the area under the curves. The area under the ROC curve (AUC) summarizes the quality of the classification and is used as a single measure of accuracy [11]. A maximum attainable value of AUC is 1.0 and the higher value is more desirable. Figure 1 shows the ROC curves and AUC of all four kernels using the three data sets. It can be seen that apart from Poly2, the curves of the other three kernels remain reasonably close. It is interesting to note that Poly2, despite having a lower specificity in Lymphochip than in $L_{t-test}$ and $L_{SAM}$, it actually has better AUC in Lymphochip than in $L_{t-test}$ and $L_{SAM}$.

For the Lymphochip data set, it appears that the performance of all kernels except Poly2 is comparable. It is only at the subsets that we see more variation, most notably in the Poly2 kernel and, to a lesser degree, in the sigmoid kernel. This finding confirms the observation made earlier on the sensitivity, specificity and accuracy in that the performance of the kernels in Lympochip is more uniform than in the subsets. Notably, the AUC for RBF kernel in all the datasets are comparable. This result shows that RBF kernel is most suitable for

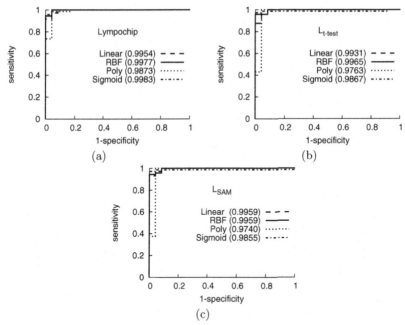

**Fig. 1.** ROC curves and areas under ROC curves (AUCs) of the four SVM kernels evaluated using (a) Lympochip and the subsets (b) $L_{t-test}$ and (c) $L_{SAM}$

lymphoma microarray data, as it performs well across three data sets, and it is also consistent with the analysis based on sensitivity, specificity and accuracy measures. Given that the performance of $L_{t-test}$ and $L_{SAM}$ are comparable to using all genes, we can reasonably conclude that it is possible to use SVM-RBF most effectively in conjunction with gene selection techniques.

The next section compares the performance of SVM with other well known classifiers (MLP and LDA) using the $L_{t-test}$ and $L_{SAM}$ subsets.

## 5.2    Comparing SVM Against MLP and LDA in $L_{t-test}$ and $L_{SAM}$

This section compares the best performing SVM kernel with MLP and LDA. To recap, the MLP were trained using 5 to 13 hidden nodes and for convenience, they are labelled from MLP-5 to MLP-13. Table 2 compares the sensitivity, specificity and accuracy of the SVM-RBF, MLP and LDA; the results for all MLP are the average from 20 runs. It can be seen from Table 2 that increasing the number of hidden nodes does not necessarily increase its effectiveness: in both $L_{t-test}$ and $L_{SAM}$, MLP-11 has higher specificity and accuracy than MLP-13.

It is clear that SVM-RBF outperforms both MLP and LDA. As LDA is only suitable for linearly separable classification, the poor results seem to suggest that the classification on the reduced dataset is non-linearly separable. This would also explain why MLP performs better than LDA; MLP is known to perform well for non-linearly separable classifications.

**Table 2.** The sensitivity, specificity and accuracy of SVM-RBF (the best performing kernel), MLP with different hidden nodes and LDA

| Classifiers | $L_{t-test}$ | | | $L_{SAM}$ | | |
|---|---|---|---|---|---|---|
| | Sens. | Spec. | Acc. | Sens. | Spec. | Acc. |
| SVM-RBF | 100.0 | 91.67 | 97.92 | 100.0 | 91.67 | 97.92 |
| MLP-5* | 95.49 | 87.71 | 93.54 | 92.85 | 91.67 | 92.55 |
| MLP-7* | 96.18 | 86.25 | 93.70 | 96.11 | 86.04 | 93.59 |
| MLP-9* | 96.11 | 90.42 | 94.69 | 95.21 | 90.42 | 94.01 |
| MLP-11* | 96.74 | 90.00 | 95.05 | 97.43 | 92.71 | 96.25 |
| MLP-13* | 96.88 | 88.96 | 94.90 | 97.22 | 88.54 | 95.05 |
| LDA | 76.39 | 70.83 | 75.0 | 72.22 | 83.33 | 75.0 |

*average of 20 runs

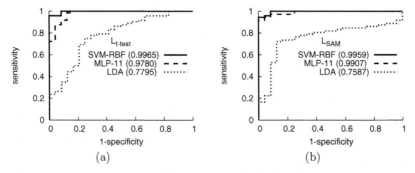

(a)     (b)

**Fig. 2.** ROC curves and AUC of the SVM-RBF, MLP-11 and LDA using the subsets (a) $L_{t-test}$ and (b) $L_{SAM}$

Figure 2 compares the ROC curves and AUC of SVM-RBF, MLP-11 and LDA in $L_{t-test}$ and $L_{SAM}$. Note that the AUC for MLP is the average of 20 runs. In both datasets, SVM-RBF perform better than the other two classifiers. As the MLP was run 20 times, we only show the curve of one run: the run having the area closest to the average (for $L_{t-test}$ the area was 0.9792, for $L_{SAM}$ was 0.9896). For both datasets, SVM-RBF yields greater AUC followed by MLP-11 and lastly LDA.

## 6   Conclusions and Future Works

This paper compared the performance of SVM in four different kernels (linear, RBF, polynomial and sigmoid) using all 4026 genes in Lymphochip and two reduced subsets extracted by employing gene selection techniques, where each set has only about 10% number of genes from the Lymphochip. The performance was measured in terms of sensitivity, specificity, accuracy and ROC analysis. This paper showed that the performance of SVM using RBF, the most suitable kernel for lymphoma microarray data, on small subsets of genes are comparable to the results of using all genes. The advantage of using only small subsets is that it requires less training time for SVM without sacrificing accuracy. Importantly, these reduced subsets were obtained using generic approaches i.e. without using any domain knowledge. The reduced sets will help biologists to concrete on fewer genes to identify their roles in malignancy development.

This paper then compared SVM with MLP and LDA using the two subsets. Experimental results showed that SVM outperforms the other two classifiers. Future experiments will involve SVM in gene selection process by determining the relative influence of selected genes for further reduction of significant gene set and further classification of malignant cells into main types of lymphoma.

## Appendix

The parameters used in generating $L_{t-test}$ using BRB-ArrayTools were as follows. P-value was 0.001, multivariate permutation: maximum number of false discover was 5, maximum portion of false discover was 0.01, confidence level was 95% and the number of permutations was 2000.

The parameters used in generating $L_{SAM}$ using BRB-ArrayTools were the following: median proportion of false discovery was 0.001 and the number of permutations was 500.

## References

1. A A Alizadeh and M B Eisen et al. Distinct types of diffuse large B-cell lyumphoma identified by gene expression profiling. *Nature*, 403:503–511, Feb 2000.
2. U Alon, N Barkai, and D A Notterman et al. Broad patterns of gene expressions revealed by clustering analysis of tumor and normal colon tissues probed by oligonucleotide arrays. In *PNAS*, volume 96, pages 6745–6750, Washington, DC, 1999. National Academy of Sciences.
3. A Ben-Dor, L Bruhn, N Friedman, I Nachman, M Schummer, and Z Yakhini. Tissue classification with gene expression profiles. In *4th Intl Conf on Comptnl Molecular Bio*, Tokyo, 2000. Universal Acad. Press.
4. C M Bishop. *Neural Networks for Pattern Recognition*. Clarendon Press, Oxford, 1995.
5. M P S Brown, W N Grundy, D Lin, N Cristianini, C Sugnet, M Agnes Jr, and D Haussler. Support vector machine classification of microarray gene expression data. Technical report, U. California (Santa Cruz), 1999.
6. R A Caruana and D Freitag. How useful is relevance? Technical report, Fall'94 AAAI Symposium on Relevance, New Orleans, 1994.
7. C C Chang and C J Lin. LIBSVM: a library for support vector machines. http://www.csie.ntu.edu.tw/~cjlin/libsvm, 2001.
8. V Chercassky and P Mullier. *Learning from Data, Concepts, Theory and Methods*. John Wiley, 1998.
9. Jay L Devore. *Probability and Statistics for Engineering and the Sciences*. Brooks/Cole, 1987.
10. S Dudoid, J fridlyand, and T Speed. Comparison of discrimination methods for the classification of tumors using gene expression data. Technical report, University of California, Berkeley, 2000.

11. L. Lukas et al. Brain tumor classification based on long echo proton mrs signals. *Artificial Intelligence in Medicine*, 31:73–89, 2004.
12. T R Golub, D K Slonim, and P Tamayo et al. Molecular classification of cancer: Class discovery and class prediction by gene expression monitoring. *Science*, 286:531–537, Oct 1998.
13. S Haykin. *Neural Network - A Comprehensive Foundation*. Prentice Hall, 1999.
14. J Khan, J S Wei, M Ringnér, L H Sall, M Ladanyi, and F Westermann. Classification and diagnostic prediction of cancers using gene expression profiling and aritifical neural networks. *Nat Med*, 7(6):673–679, 2001.
15. L C Molina, L Belanche, and A Nebot. Feature selection algorithms: A survey and experimental evaluation. In *ICDM'02*, 2002.
16. H.B. Demuth M.T. Hagan and M.H. Beale. *Neural Network Design*. PWS Publishing, Boston, MA, 1996.
17. J De Risi, V Iyer, and P Brown. Exploring the metabolic and genetic control of gene expression on a genomic scale. *Science*, 278:680–6, 1997.
18. D E Rumelhart and the PDP Research Group. *Parallel Distributed Processing*. MIT Press, New York, 1986.
19. Richard Simon and Amy Peng Lam. BRB ArrayTools v 3.2. http://linus.nci.nih.gov/BRB-ArrayTools.html, 2004.
20. Virginia Goss Tusher, Robert Tibshirani, and Gilbert Chu. Significance analysis of microarrays applied to the ionizing radiation response. In *Proc Natl Acad Sci*, volume 98, pages 5116–5121, 2001.
21. G Valentini. Gene expression data analysis of human lymphoma using support vector machines and output coding ensembles. *Artificial Intelligence in Medicine*, 26:281–304, 2002.
22. V N Vapnik. *The nature of statistical learning theory*. Springer, New York, 1995.

# Microarray Data Analysis and Management in Colorectal Cancer

Oscar García-Hernández[1], Guillermo López-Campos[1],
Juan Pedro Sánchez[1], Rosa Blanco[1], Alejandro Romera-Lopez[2],
Beatriz Perez-Villamil[2], and Fernando Martín-Sánchez[1]

[1] Medical Bioinformatics Department
Institute of Health 'Carlos III'
Ctra. Majadahonda-Pozuelo, km 2. 28220 Majadahonda, Madrid
{o.garcia,glopez,jpsanchez,rblanco,fmartin}@isciii.es
[2] Departamento de Oncología Médica
Hospital Clínico San Carlos
Martín Lagos s/n, 28040, Madrid
bperezvillamil.hcsc@salud.madrid.org

**Abstract.** The availability of microarray technologies has enabled bio-medical researchers to explore expression levels of a complete genome simultaneously. The analysis of gene expression patterns can explain the biological basis of several pathological processes. Deepening in the under-standing of the molecular processes underlying colorectal cancer might become of interest for the advance of its clinical management. This work presents the analysis of microarrays data using colon cancer samples in order to determine the differentially expressed genes underlying this dis-ease process. The comparison of gene expression levels using a complete genome approach of tumor samples versus healthy controls allows the definition of a set of genes involved in the differentiation of both tissues. The analysis of these differentially expressed genes using Gene Ontology analysis permits the location of most prevalent processes that are altered during under this disease.

## 1 Introduction

The availability of new genomic based technologies for the massive screening and analysis of data has brought new scopes for the research and analysis of both clinical and biological data. Microarray technologies [1, 2] are one of the best examples of how these new technologies have changed the research. Microar-ray technology based experiments involve several steps and processes which can generate lots of information. From the initial step of manufacture with the an-notation of probes until the final numerical data analysis several intermediate steps are done (sample processing, hybridization, scanning, etc ...). In each of those stages a great amount of data is generated and needs to be managed.

The most common application of microarray technologies since their origin has been the analysis of the gene expression under different conditions. Microar-

J.L. Oliveira et al. (Eds.): ISBMDA 2005, LNBI 3745, pp. 391–400, 2005.

rays offer the researchers the possibility of identifying and measuring simultaneously the complete set of genes expressed in a particular moment [3]. As results of this massive approach a new way of thinking the experiments has risen. In this new experimental approaches the objective is the analysis of complex systems as transcriptomes to elucidate the genes and the magnitude of the changes in their expression responsible for the adaptation of cells to the different conditions or even for complex diseases.

The evolution of the laboratory techniques for massive approaches that generates huge amounts of data has needed the evolution of bioinformatics to support them. In the case of microarray technology, bioinformatics is extremely interrelated with it. Almost all the processes in microarray experiments are supported by bioinformatics due to the huge amounts of data generated and managed. Therefore, microarray bioinformatics has been a very hot topic in recent years, specially the numerical data analysis related aspects.

The availability of public datasets as well as the increasing number of microarray publications has provided a substrate for the research in different methods to analyse gene expression data [4]. The bioinformatic analysis of gene expression studies is often related with the identification and explanation of the genes that are differentially expressed among the situations studied. There is an increasing number of techniques and algorithms and possible approaches that can be followed to achieve the final goal of understanding the underlying biology of the studied processes. These analysis usually involves several quality control steps like feature selection or outlier detection. Once this quality steps are done the data set is ready for undergoing a deeper analysis using both supervised and unsupervised techniques [5].

Cancer is a common disease nowadays in the most evolved societies. Due to the social, epidemiological and complexity cancer has become a paradigm in the application in research of new technologies for its study. Early in their development microarrays were used to study cancer processes [6]. Colon cancer is among the different types of cancer one of the most prevalent, affecting almost equally men and women. Therefore the study of the molecular basis of this type of cancer is very interesting from a clinical point of view as well as it is also an interesting problem for biology. The feasibility of analyzing with microarray technologies the gene expression profile of this disease to identify genes involved in the tumoral process is a way to deepen the available knowledge about the disease.

In this work we have used microarrays to study the gene expression profiles in colon cancer. The data generated was analysed to detect the genes differentially expressed among healthy tissue and tumour samples.

The paper is structured as follows. In section 2 we present the methodology used for microarray data preprocessing and filtering besides the information system developed for data storage and management. Section 3 presents the experimental results got from this study. Paper finishes with the conclusions got from the analysis and the future work that is going to be developed.

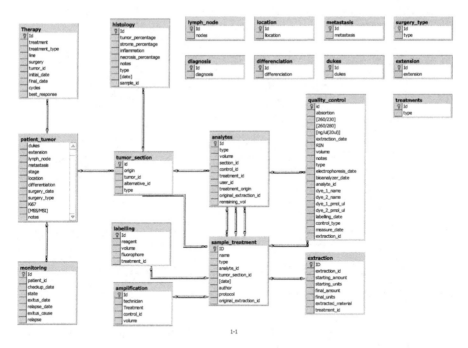

**Fig. 1.** Scheme of the patients, samples and experiments database

## 2  Microarray Data Analysis for Colorectal Cancer

Although, microarray data analysis is a wide and growing field, this study aims for detection of differentially expressed genes in colorectal cancer microarrays.

An information system is designed *ad-hoc* to correctly and easily store and manage the colorectal cancer microarray data. Furthermore, the database is extended for clinical annotation of the experiment.

### 2.1  Information System

An information system is designed and developed using SQL Server 2000 and MS Access 2003 technologies. This system is intended to manage information generated in the microarray expression analysis including patients and samples' information to be examined in the study.

The information system is composed by:

1. SQL Server databases

   (a) Managing information about the patients, samples and experiments –see Figures 1 and 2–.

   (b) Making easier the filter, transformation and normalization of microarray expression data.

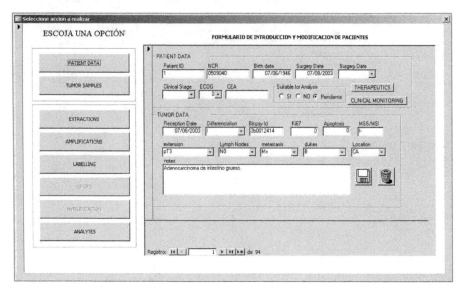

**Fig. 2.** Graphical user interface of the patients, samples and experiments database

2. Several Access clients which access to the SQL Server and allow the users to populate the databases and to browse, edit and search the data by means of the definition of different forms, views, queries and Visual Basic procedure defined in clients

Microarray expression data can be traced using the information system, which store information related to:

– Patient features. Under this topic is stored information related to the diagnosis and clinical aspects of the patients. These characteristics were selected in collaboration with clinical oncologists
– Therapies followed by the patients, including the chemotherapy drugs and surgeries suffered by the patients
– Biological samples. Biological description and annotation of the biopsies used as samples for the microarray experiments. This information contains aspects such as percentage of tumoral cells in the sample and some other parameters.
– Experimental processes applied to the samples. A series of tables are included to manage the information gathered during all the possible procedures done in the laboratory to the samples. As example of these procedures and reaction we may talk about histologies, PCR's, extractions, labellings, etc.
– Quality controls of both samples and processes done during the experiments
– Quantified expression data derived from the images generated by microarray experiments

Due to its user friendly environment and its facility to be learned and used, the MS Access 2003 technology is chosen to build the client applications. The clinical environment where this tool was going to be used, required the development of an application with an easy and friendly interface. Both physicians and biological scientists involved in gathering the information and storing it in the database were already familiar with this tool.

As MS Access 2003 has a storage limit of 2 Gigabytes, and cause of the amount of data being produced within the microarray expression experiments (several Gigabytes), it is used in combination with SQL Server. With this configuration, we define all forms, queries and procedures in the Access clients while store the data in the SQL Server database.

## 2.2   Description of Microarray Data

Microarray experiments are performed using Agilent's Whole Human Genome Oligo Microarray Kit. These oligonucleotide based arrays consist in a set of 43392 spots containing different probes for human genes as well as some internal positive and negative controls. RNA from 27 tumor samples is collected, amplified and labelled with a fluorescent label. 68 healthy colon RNA samples are collected and pooled together to act as healthy controls for the comparisons being amplified and labelled with a different fluorescent label than the one used for tumor samples. Both tumor and control samples are hybridized in pairs on the microarrays following the instructions of the microarray provider. Once the hybridization reaction is done the microarrays are washed and scanned with an Agilent scanner. The images generated by the scanner are then quantified using Agilent's Feature Extraction software using 'cookie cutter' segmentation algorithm.

## 2.3   Data Preprocessing

Microarray experiment data contain thousands of variables, several of which require a careful management due to their capacity to introduce noise into the data. Some examples of these variables are replicated probes or spot quantification processes. In some cases, these parameters are kept in mind by the quantification software. Therefore, a reliable and constant control on the setting of this software is necessary. Owing to these variability fonts, 3 preprocessing steps are followed in this study in order to polish and determine the data quality.

1. Controlling the experimental technique and reproducibility by checking labelling process for each experiment
2. Identifying required variables to perform the study and check the homogeneity in the variables and heterogeneity among the variables
3. Analyzing positive and negative controls

In this analysis, a labelling control is carried out every each 10 experiments and self-self control hybridizations are done. Subsequently, the numerical results of these hybridizations are plotted to determine whether the experimental technique is robust and reliable or not. The correlation analysis of these self-self

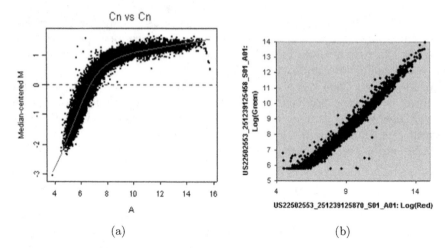

(a)                                              (b)

**Fig. 3.** Representation of self-self control graphical analysis. (a) M-A plot of a control versus control hybridization before normalization. The graph shows an intensity dependant effect of the report that is solved with normalization. (b) Scatter plot of 2 non-normalized control samples from different days and dyes. It's clearly seen the correlation of both samples showing the good behaviour and overall reproducibility of the experiments

hybridizations is high, giving an overall correlation of 0.98 for the 3 microarray experiments.

At this step of the microarray data preprocessing, we remove the labelling background to obtain sharper images following literature recommendations [7]. Next, the ratio between the signal intensity in red and green channels are calculated to better management.

**Data filtering.** Data filtering is an important task in microarray analysis to obtain a reliable subset of genes. The flags provided by Agilent Feature Extraction Software are applied to gene filtering. So that, outliers and local background are removed. Then, genes with low intensity are filtered. Therefore, intensity of negative controls is studied. Spots whose signal intensity in red channel is below 16 and in green channel is below 54 are excluded. In this way, 31134 genes are attained.

**Data transformation and normalization.** In order to obtain smoother data, the $\log_2$ transformation is applied. Due to an observed effect of the signal intensity shown in self-self control M-A plots –see Figure 4–, microarray data are normalized by the *lowess smoother* function to remove this intensity effect. Further, a perfectly normalized data are reached to perform complex analysis.

**Gene filtering.** After all the spot filtering processes, a gene filtering step is carried out. Genes under the following set of restrictions are excluded from further analyses:

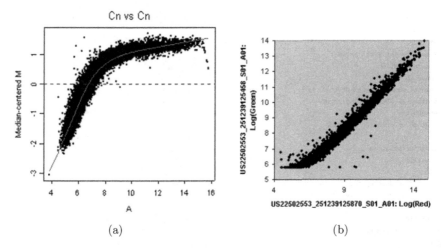

**Fig. 4.** Representation of self-self control graphical analysis. (a) M-A plot of a control versus control hybridization before normalization. The graph shows an intensity dependant effect of the report that is solved with normalization. (b) Scatter plot of 2 non-normalized control samples from different days and dyes. It's clearly seen the correlation of both samples showing the good behaviour and overall reproducibility of the experiments

- genes with less than 20% of its values differs at least $\pm 1.5$ fold from median
- genes with more than 30% missing value

Then, the number of genes that passed filtering criteria was 3676.

## 3   Experimental Results

The analysis of microarray data coming from hybridizations of colorectal cancer patients is done using BRB-Array Tools from NCI [8]. The comparison between tumor and healthy samples is done using red and green channel comparison tool, instead of the simple class comparison. In this case, resulting log ratios are analyzed by the paired $t$-test. The null hypothesis means that the log ratio distribution equals zero. The significance level of the univariate $t$-test analysis is fixed to $p = 0.001$. The confidence level of false discovery rate is fixed to 95%. Other false discovery limits used in the analysis are the number of false positives and ratio of allowed false positives. In this study, the number of total false discoveries is set to 5 and the ratio of false positives to 0.01.

From these analysis we find that there are 2412 genes which discriminate between both studied classes (tumor versus healthy tissues) with $p = 0.001$. From the false discovery assessment we conclude that 2135 genes contain no more than 5 false positive with a probability of 95%. Additionally, in 2448 genes the false positive ratio is lower than 1%. It must be noted that only the 3676 genes selected after the data preprocessing are taken to account to this class comparison.

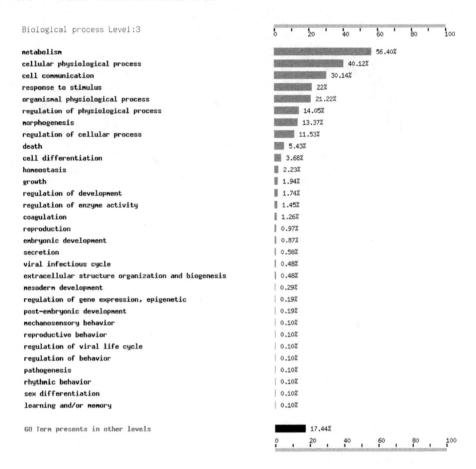

**Fig. 5.** Relation between genes and GO terms

The obtained genes are related to their biological categories and processes using the Gene Ontology (GO) annotation of genes. In order to discover the biological significance of the attained genes, FatiGO [9] is applied. The depth of the GO terms genes annotation is limited to the third level and the significance level is fixed to 0.001. A pitfall of this analysis is related to the lack of annotation for some of the probes used in the microarrays experiment. Therefore, almost half of the obtained genes are not included in the analysis.

Figure 5 shows the biological processes represented in GO. As it can be appreciated in the Figure, more than the 50% of the differentially expressed genes have a 'metabolism' annotation in GO. The 'cellular physiological process' annotation is related to more than the 40% of genes.

## 4   Conclusions and Future Work

The microarray technology allow researchers the exploration of new ways of study and analysis of diseases and therapies. Cancer has become in a common disease for the development and application of new methods and approaches.

The development of the information system described in this work allows clinicians and biologist an easy management of data related with clinical record and laboratory procedures. The use of such information system allowed the clinical annotation of the samples. This facilitates the incorporation of the clinical outcomes to the final data analysis and conclusions. Moreover, this information system facilitates the data filtering and processing. In the future the information system is going to be extended with new modules that allows the use of medical and bioinformatics standards such as MIAME [10] (Minimum Information About Microarray Experiments) and ICD (international Classification of Diseases)

In this work, a study of colorectal cancer microarrays is performed seeking for the differentially expressed set of genes. For this purpose, a group of filters are applied using Agilent's Feature Extraction and BRB-ArrayTools. In this way, the original 43392 spots in the microarrays are reduce to 2412 differentially expressed genes.

After the dimensionality reduction, the attained genes are related to GO terms in order to determinate their biological meaning. Resulting that more than the 50% of the genes have 'metabolism' GO term. This fact seems to be associated with a variation in the metabolic rate of the proliferating cells of the tumor.

As a future work line, a data filtering and gene selection based on function measure from machine learning field is developed. Moreover, a new classification of colorectal tumors type by their microarray expression profile is studied. For this purposes, techniques as unsupervised classification (clustering) and supervised classification are applied.

## Acknowledgments

Analyses were performed using BRB-ArrayTools version 3.2 developed by Dr. Richard Simon and Amy Peng Lam.

## References

1. Fodor, S., Read, J., Pirrung, M., Stryer, L., Lu, A., Solas, D.: Light-directed, spatially addressable parallel chemical synthesis. Science **251** (1991) 767–773
2. Pease, A., Solas, D., Sullivan, E., Cronin, M., Holmes, C., Fodor, S.: Light-generated oligonucleotide arrays for rapid DNA sequence analysis. Proceedings of the National Academy of Sciences **91** (1994) 5022–5026
3. Schena, M., Shalon, D., Davis, R., Brown, P.: Quantitative monitoring of gene expression patterns with a complementary DNA microarray. Science **20** (1995) 1008–1017
4. Lin, S., Johnson, K.: Methods of Microarray Data Analysis. Kluwer Academic Publishers (2000)
5. Brazma, A., Vilo, J.: Gene expression data analysis. Microbes and Infection **3** (2001) 823–829

6. DeRisi, J., Penland, L., Brown, P., Bittner, M., Meltzer, P., Ray, M., Chen, Y., Su, Y., Trent, J.: Use of a cDNA microarray to analyse gene expression patterns in human cancer. Nature Genetics **14** (1996) 457–460

7. Causton, H., Quackenbush, J., Brazma, A.: A Beginner's Guide. Microarray Gene Expression and Data Analysis. Blackwell Publishing (2003)

8. Simon, R., Lam, A.P., Ngan, M., Gibiansky, L., Shrabstein, P.: The BRB-ArrayTools development team (2005)
   http://linus.nci.nih.gov/BRB-ArrayTools.html.

9. Al-Shahrour, F., Díaz-Uriarte, R., Dopazo, J.: FatiGO: a web tool for finding significant associations of Gene Ontology terms to groups of genes. Bioinformatics **20** (2004) 578–580

10. Brazma A, Hingamp P, Quackenbush J, Sherlock G, Spellman P, Stoeckert C, Aach J, Ansorge W, Ball CA, Causton HC, Gaasterland T, Glenisson P, Holstege FC, Kim IF, Markowitz V, Matese JC, Parkinson H, Robinson A, Sarkans U, Schulze-Kremer S, Stewart J, Taylor R, Vilo J, Vingron M.: Minimum information about a microarray experiment (MIAME)-toward standards for microarray data. Nature Genetics **29** (2001) 365–371

# Author Index

Anguita, Alberto   34
Armengol, Eva   294
Arvelakis, Adam   283
Azevedo, Paulo J.   329

Begg, Rezaul   379
Benito, Mario   274
Bidargaddi, Niranjan P.   355
Blanchet, Christophe   262
Blanco, Rosa   391
Blanquer, Ignacio   22
Bonacina, Stefano   130
Bonten, Marc   161
Bremer, Eric G.   101
Brito, Rui M.M.   329
Bueno, Gloria   34

Carrilero López, Vicente   51
Chetty, Madhu   346, 355, 367
Chouvarda, Ioanna   89
Coltell, Oscar   252
Costa, Carlos   13
Crespo, José   34
Crowell, Jon   184

de la Calle, Guillermo   252, 274
Desco, Manuel   61
DeSesa, Catherine   101
Desprez, Frédéric   262
Dias, Gaspar   78
Dooley, Laurence S.   346
Dorado, Julián   34
Duart Clemente, Javier M.   51
Dubitzky, Werner   101, 305
Dusza, Jacek J.   51, 173

Ejarque, Ismael   120
Estruch, Antonio   34, 252

Feliú, Vicente   34
Flå, Tor   317

Gama Ribeiro, Vasco   13
Garamendi, Juan F.   61
García-Hernández, Oscar   391

Gerlach, Joerg C.   109
Gondal, Iqbal   379
González, José C.   211
Górriz, Juan M.   137
Guthke, Reinhard   109

Hack, Catherine J.   101
Heredia, José Antonio   34, 252
Hernández, Vicente   22
Herranz, Michel   120
Hoque, Md. Tamjidul   346
Hüntemann, Alexander   211

Jankowski, Stanisław   51, 173

Kamruzzaman, Joarder   355, 379
Kazemian, Majid   338
Kim, Eunjung   184
Koutkias, Vassilis   89
Kowalski, Mateusz   149
Krause, Antje   305
Krysztoforski, Krzysztof   149
Kurzynski, Marek   1, 242

Lang, Elmar W.   137
Lee, Hye-Jin   69
Lim, Sukhyun   69
Lim, Suryani   379
López-Campos, Guillermo   120, 391
Loureiro-Ferreira, Nuno   329
Lucas, Caro   338
Lucas, Peter   161

Maglaveras, Nicos   89
Malousi, Andigoni   89
Malpica, Norberto   61
Martín-Sánchez, Fernando   78, 120, 391
Mas, Ferran   22
Masseroli, Marco   44, 130
Molero, Eva   274
Moreno, Juan Luis   274
Moshiri, Behzad   338
Mugarra Gonzalez, C. Fernando   51
Mulay, Niranjan   101

Natarajan, Jeyakumar   101
Nikbakht, Hamid   338
Nugent, Chris   305

Oliveira, José Luís   13, 78
Ooi, Chia Huey   367
Oręziak, Artur   173

Pérez del Rey, David   34, 252
Pérez Ordóñez, Juan Luis   34
Perez-Villamil, Beatriz   391
Pfaff, Michael   109
Pinciroli, Francesco   44, 130
Pisanelli, Domenico M.   44
Plaza, Enric   294
Pless, Gesine   109
Puntonet, Carlos G.   137

Reczko, Martin   283
Ribeiro, José   13
Rodrigues, J. Rui   329
Romera-Lopez, Alejandro   391

Sánchez, Juan Pedro   252, 391
Sanz, Ferran   252
Sas, Jerzy   1
Schiavi, Emanuele   61
Schmidt, Rainer   202
Schmidt-Heck, Wolfgang   109
Schurink, Karin   161
Segrelles, Damià   22

Shin, Byeong-Seok   69
Silva, Augusto   13
Silva, Cândida G.   329
Sordo, Margarita   193
Stadlthanner, Kurt   137
Stamatakis, Alexandros   283
Symeonidis, Alkiviadis   283
Szecówka, Przemyslaw M.   149

Tapia, Elizabeth   211
Teng, Shyh Wei   367
Theis, Fabian J.   137
Thorvaldsen, Steinar   317
Tollis, Ioannis G.   283
Tomé, Ana Maria   137
Tse, Tony   184

Vernois, Antoine   262
Vicente, Francisco-Javier   78, 120
Visscher, Stefan   161

Waligora, Tina   202
Wang, Chong   305
Wierzbowski, Mariusz   173
Willassen, Nils P.   317
Wolczowski, Andrzej R.   149
Wozniak, Michal   223, 231

Zeilinger, Katrin   109
Zeng, Qing   184, 193
Zolnierek, Andrzej   242

# Lecture Notes in Bioinformatics

Vol. 3745: J.L. Oliveira, V. Maojo, F. Martin-Sanchez, A. Sousa Pereira (Eds.), Biological and Medical Data Analysis. XII, 402 pages. 2005.

Vol. 3695: M.R. Berthold, R. Glen, K. Diederichs, O. Kohlbacher, I. Fischer (Eds.), Computational Life Sciences. XI, 277 pages. 2005.

Vol. 3692: R. Casadio, G. Myers (Eds.), Algorithms in Bioinformatics. X, 436 pages. 2005.

Vol. 3680: C. Priami, A. Zelikovsky (Eds.), Transactions on Computational Systems Biology II. IX, 153 pages. 2005.

Vol. 3678: A. McLysaght, D.H. Huson (Eds.), Comparative Genomics. VIII, 167 pages. 2005.

Vol. 3615: B. Ludäscher, L. Raschid (Eds.), Data Integration in the Life Sciences. XII, 344 pages. 2005.

Vol. 3594: J.C. Setubal, S. Verjovski-Almeida (Eds.), Advances in Bioinformatics and Computational Biology. XIV, 258 pages. 2005.

Vol. 3500: S. Miyano, J. Mesirov, S. Kasif, S. Istrail, P. Pevzner, M. Waterman (Eds.), Research in Computational Molecular Biology. XVII, 632 pages. 2005.

Vol. 3388: J. Lagergren (Ed.), Comparative Genomics. VII, 133 pages. 2005.

Vol. 3380: C. Priami (Ed.), Transactions on Computational Systems Biology I. IX, 111 pages. 2005.

Vol. 3370: A. Konagaya, K. Satou (Eds.), Grid Computing in Life Science. X, 188 pages. 2005.

Vol. 3318: E. Eskin, C. Workman (Eds.), Regulatory Genomics. VII, 115 pages. 2005.

Vol. 3240: I. Jonassen, J. Kim (Eds.), Algorithms in Bioinformatics. IX, 476 pages. 2004.

Vol. 3082: V. Danos, V. Schachter (Eds.), Computational Methods in Systems Biology. IX, 280 pages. 2005.

Vol. 2994: E. Rahm (Ed.), Data Integration in the Life Sciences. X, 221 pages. 2004.

Vol. 2983: S. Istrail, M.S. Waterman, A. Clark (Eds.), Computational Methods for SNPs and Haplotype Inference. IX, 153 pages. 2004.

Vol. 2812: G. Benson, R.D. M. Page (Eds.), Algorithms in Bioinformatics. X, 528 pages. 2003.

Vol. 2666: C. Guerra, S. Istrail (Eds.), Mathematical Methods for Protein Structure Analysis and Design. XI, 157 pages. 2003.